SPACE TELESCOPE SCIENCE INSTITUTE

SYMPOSIUM SERIES: 21

Series Editor S. Michael Fall, Space Telescope Science Institute

BLACK HOLES

This volume is based on a meeting held at the Space Telescope Science Institute on April 23–26, 2007.

This collection of review papers, written by world experts in the many aspects of black hole physics and astrophysics, regarding stellar-mass, intermediate-mass, and supermassive black holes, and provides an invaluable resource, both to professional astronomers and astrophysicists, and for students. The topics covered range from black hole entropy and the fate of information to supermassive black holes at the centers of galaxies, and from the possibility to produce black holes in collider experiments to the measurements of black hole spins.

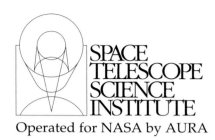

SPACE
TELESCOPE
SCIENCE
INSTITUTE

Operated for NASA by AURA

Other titles in the Space Telescope Science Institute Series.

Black holes

Proceedings of the
Space Telescope Science Institute Symposium,
held in Baltimore, Maryland
April 23–26, 2007

Edited by
MARIO LIVIO

Space Telescope Science Institute, Baltimore, MD 21218, USA

ANTON KOEKEMOER

Space Telescope Science Institute, Baltimore, MD 21218, USA

Published for the Space Telescope Science Institute

Operated for NASA by AURA

CAMBRIDGE
UNIVERSITY PRESS

CAMBRIDGE UNIVERSITY PRESS
Cambridge, New York, Melbourne, Madrid, Cape Town,
Singapore, São Paulo, Delhi, Dubai, Tokyo, Mexico City

Cambridge University Press
The Edinburgh Building, Cambridge CB2 8RU, UK

Published in the United States of America by Cambridge University Press, New York

www.cambridge.org
Information on this title: www.cambridge.org/9781107005532

© Cambridge University Press 2011

First published 2011

Printed in the United Kingdom at the University Press, Cambridge

A catalogue record for this publication is available from the British Library

Library of Congress Cataloguing in Publication data

ISBN 978-1-107-00553-2 hardback

Contents

Participants

Alexander, Tal	Weizmann Institute of Science
Balsara, Dinshaw	University of Notre Dame
Batcheldor, Dan	Rochester Institute of Technology
Bender, Pete	JILA, University of Colorado
Blandford, Roger	Stanford University
Bogdanovic, Tamara	University of Maryland
Brand, Kate	Space Telescope Science Institute
Brenneman, Laura	University of Maryland
Centrella, Joan	NASA Goddard Space Flight Center
Chiaberge, Marco	Space Telescope Science Institute
Colpi, Monica	University of Milano Bicocca
Congdon, Arthur	Rutgers University
D'Angelo, Caroline	Max Planck Institute for Astrophysics
Dai, Xinyu	The Ohio State University
Dewangan, Gulab	Carnegie Mellon University
Dinerstein, Harriet	University of Texas, Austin
Dressel, Linda	Space Telescope Science Institute
Dudik, Rachel	George Mason University
Escala, Andres	Kavli Institute for Theoretical Physics, Stanford University
Fabbiano, Giuseppina	Harvard-Smithsonian Center for Astrophysics
Fabian, Andrew	University of Cambridge
Fruchter, Andrew	Space Telescope Science Institute
Frye, Brenda	Dublin City University
Fukumura, Keigo	NASA Goddard Space Flight Center
Garcia, Michael	Harvard-Smithsonian Center for Astrophysics
Gehrels, Neil	NASA Goddard Space Flight Center
Genzel, Reinhard	Max Planck Institute for Extraterrestrial Physics
Gezari, Suvi	California Institute of Technology
Gliozzi, Mario	George Mason University
Globus, Alice	Wesleyan University
Godon, Patrick	Space Telescope Science Institute
Graber, James	
Gualandris, Alessia	Rochester Institute of Technology
Hartnett, Kevin	NASA Goddard Space Flight Center
Hasan, Hashima	NASA Headquarters
Heger, Alexander	Los Alamos National Laboratory
Horowitz, Gary	University of California, Santa Barbara
Jeletic, James	NASA Goddard Space Flight Center
Koekemoer, Anton	Space Telescope Science Institute
Krolik, Julian	The Johns Hopkins University
Landsberg, Greg	Brown University
Laor, Ari	Technion–Israel Institute of Technology
Liu, Jifeng	Harvard-Smithsonian Center for Astrophysics
Livio, Mario	Space Telescope Science Institute
Macchetto, F. Duccio	Space Telescope Science Institute
Maier, Millicent	Oxford University
Malkan, Matt	University of California, Los Angeles

McClintock, Jeffrey	Harvard-Smithsonian Center for Astrophysics
McMillan, Steve	Drexel University
McWilliams, Sean	NASA Goddard Space Flight Center and University of Maryland
Merritt, David	Rochester Institute of Technology
Mirabel, Felix	European Southern Observatory, Chile
Mountain, Matt	Space Telescope Science Institute
Müller, Andreas	Max Planck Institute for Extraterrestrial Physics
Murphy, Kendrah	The Johns Hopkins University
Narayan, Ramesh	Harvard University
Noble, Scott	The Johns Hopkins University
Noel-Storr, Jacob	Rochester Institute of Technology
Nota, Antonella	Space Telescope Science Institute
Noyola, Eva	Max Planck Institute for Extraterrestrial Physics
Onken, Christopher	Dominion Astrophysical Observatory, Herzberg Institute of Astrophysics
Paolillo, Maurizio	Università Federico II di Napoli
Pastorini, Guia	Università degli Studi di Firenze
Perkins, Kala	Stanford University
Peterson, Bradley	The Ohio State University
Pierce, Christina	University of California, Santa Cruz
Pretorius, Frans	Princeton University
Prince, Thomas	California Institute of Technology
Psaltis, Dimitrios	University of Arizona
Richstone, Douglas	University of Michigan
Russell, Kathleen	
Sambruna, Rita	NASA Goddard Space Flight Center
Schnittman, Jeremy	University of Maryland
Schreier, Ethan	Associated Universities Inc.
Shafee, Rebecca	Harvard University
Shapiro, Stuart	University of Illinois, Urbana-Champaign
Shields, Gregory	University of Texas, Austin
Siah, Javad	Villanova University
Stebbins, Robin	NASA Goddard Space Flight Center
Tanaka, Yasuo	Max Planck Institute for Extraterrestrial Physics
Tremblay, Grant	Space Telescope Science Institute
Trenti, Michele	Space Telescope Science Institute
Tundo, Elena	Università di Padova
Urry, C. Megan	Yale University
Van der Marel, Roeland	Space Telescope Science Institute
Vasudevan, Ranjan	Institute of Astronomy, University of Cambridge
Vestergaard, Marianne	University of Arizona
Volonteri, Marta	University of Michigan
Wang, Junxian	University of Science and Technology of China
White, Nicholas	NASA Goddard Space Flight Center
Yang, Yuxuan	University of Maryland

Preface

The Space Telescope Science Institute Symposium on Black Holes took place during April 23–26, 2007.

These proceedings represent a part of the invited talks that were presented at the symposium. They cover many aspects of black hole physics and astrophysics, regarding stellar-mass, intermediate-mass, and supermassive black holes. Topics range from black hole entropy and the fate of information to supermassive black holes at the centers of galaxies, and from the possibility to produce black holes in collider experiments to the measurements of black hole spins. Since these articles were written by world experts in their respective disciplines, this volume represents an extremely valuable collection for researchers and students alike.

The ST ScI Symposium on Black Holes attempted to capture all the aspects involved in the astrophysics of black holes.

We thank Sharon Toolan of ST ScI for her help in preparing this volume for publication.

Mario Livio
Anton Koekemoer
Space Telescope Science Institute
Baltimore, Maryland

Black holes, entropy, and information

By GARY T. HOROWITZ

Physics Department, University of California–Santa Barbara, Santa Barbara, CA 93106, USA

Black holes are a continuing source of mystery. Although their classical properties have been understood since the 1970's, their quantum properties raise some of the deepest questions in theoretical physics. Some of these questions have recently been answered using string theory. I will review these fundamental questions, and the aspects of string theory needed to answer them. I will then explain the recent developments and new insights into black holes that they provide. Some remaining puzzles are mentioned in the conclusion.

1. Introduction

General properties of black holes were studied extensively in the early 1970's, and the basic theory was developed. One of the key results was Hawking's proof that the area of a black hole cannot decrease (Hawking 1971). This led Bekenstein (1973) to suggest that a black hole should have an entropy proportional to its horizon area. This suggestion of a connection between black holes and thermodynamics was strengthened by the formulation of the laws of black-hole mechanics (Bardeen et al. 1973). In addition to the total mass M, angular momentum J, and horizon area A of the black holes, these laws are formulated in terms of the angular velocity of the horizon Ω, and its surface gravity κ. Recall that the surface gravity is the force at infinity required to hold a unit mass stationary near the horizon of a black hole. Of course, the force near the horizon diverges, but there is a redshifting effect so that the force at infinity remains finite. The laws of black-hole mechanics are the following:

0) For stationary black holes, the surface gravity is constant on the horizon
1) Under a small perturbation:

$$dM = \frac{\kappa}{8\pi G} dA + \Omega dJ \quad , \tag{1.1}$$

2) The area of the event horizon always increases.
The zeroth law is obvious for nonrotating black holes which are spherically symmetric, but it is also true for rotating black holes which are not. If κ is like a temperature, and A is like an entropy, then there is a striking similarity to the ordinary laws of thermodynamics:

0) The temperature of an object in thermal equilibrium is constant
1) Under a small perturbation:

$$dE = TdS - PdV \quad , \tag{1.2}$$

2) Entropy always increases.
At the time it seemed clear that the analogy between black holes and thermodynamics should not be taken too seriously, since if black holes really had a temperature, they would have to radiate, and everyone knew that nothing could come out of a black hole. Two years later, everything changed.

Hawking (1975) coupled quantum matter fields to a classical black hole, and showed that they emit black-body radiation with a temperature

$$kT = \frac{\hbar\kappa}{2\pi} \quad . \tag{1.3}$$

So adding quantum mechanics in this limited way (which was all that was known how to do) made the analogy complete. Black holes really are thermodynamic objects. For a

1

solar-mass black hole, the temperature is very low ($T \sim 10^{-7}$ K) so it is astrophysically negligible. But $T \sim 1/M$ so if a black hole starts evaporating, it gets hotter and eventually explodes. A black hole would start evaporating if it is a small primordial black hole formed in the early universe, or if we wait a very long time until the three degree background radiation redshifts to less than 10^{-7} K.

Hawking's determination of the temperature, together with the first law (Eq. 1.1), fixed the coefficient in Bekenstein's formula for the black-hole entropy:

$$S_{BH} = \frac{k}{4\hbar G} A \ . \tag{1.4}$$

This is an enormous amount of entropy. A solar-mass black hole has $S_{BH} \sim 10^{77} k$. This is much greater than the entropy of the matter that collapsed to form it: Thermal radiation has the highest entropy of ordinary matter, but a ball of thermal radiation has $M \sim T^4 R^3$, $S \sim T^3 R^3$. When it forms a black hole $R \sim M$, so $T \sim M^{-1/2}$ and hence $S \sim M^{3/2}$. On the other hand, $S_{BH} \sim M^2$. So S_{BH} grows much faster with M than the entropy of a ball of thermal radiation of the same size. Since we have suppressed all physical constants, the two entropies are equal only when M is of order the Planck mass (10^{-5} gms). We will continue to set $c = k = \hbar = 1$ in the following.

The discovery that black holes are thermodynamic objects raised the following fundamental questions:

(1) What is the origin of black-hole entropy? In all other contexts, thermodynamics is just an approximation to a more fundamental statistical description in which the entropy is the log of the number of microstates. The large entropy indicates that black holes have an enormous number of microstates. What are they?

(2) Does black-hole evaporation lose information? Does it violate quantum mechanics? Hawking argued for three decades that it did.

To understand Hawking's argument, recall that another classical property of black holes established in the 1970's was the uniqueness theorem (Robinson 1975): *The only stationary (vacuum) black-hole solution is the Kerr solution.* You can form a black hole by collapsing all kinds of different matter with different multipole moments. However, after it settles down, the black hole is completely described by only two parameters M, J. Wheeler described this by saying "black holes have no hair." The spacetime outside the horizon retains no memory of what was thrown into the black hole. Now the radiation emitted by a black hole is essentially thermal. It cannot depend on the matter inside without violating causality or locality. When the black hole evaporates, M and J are recovered, but the detailed information about what was thrown in is lost. In the language of quantum theory, pure states appear to evolve into mixed states. This would violate unitary evolution and hence one of the basic principles of quantum mechanics.

Hawking argued that the formation and evaporation of a black hole is very different from burning a book. This may seem like it is destroying information, but quantum mechanically, it can be described by unitary evolution of one quantum state into another. In principle, all the information in the book can be recovered from the ashes and emitted radiation.

2. String theory

String theory is a promising candidate for both a quantum theory of gravity and a unified theory of all the known forces and matter. One of the main successes of string theory is that it has been able to provide answers to the two fundamental questions above. To understand these answers, one needs a few basic facts about string theory.

(For more detail, see e.g., Zwiebach 2004). The first is that when one quantizes a string in flat spacetime, there are an infinite tower of massive states. For every integer N there are states with

$$M^2 \sim N/l_s^2 \ , \tag{2.1}$$

where l_s is a new length scale in the theory set by the string tension. These states are highly degenerate, and one can show that the number of string states at excitation level $N \gg 1$ is e^{S_s} where

$$S_s \sim \sqrt{N} \ , \tag{2.2}$$

i.e., the string entropy is proportional to the mass in string units. One can understand this in terms of a simple model of the string as a random walk with step size l_s. As a result of the string tension, the energy in the string after n steps is proportional to its length: $E \sim n/l_s$. If one can move in k possible directions at each step, the total number of configurations is k^n, so the entropy for large n is proportional to n, i.e., proportional to the energy.

String interactions are governed by a string coupling constant g (which is determined by a scalar field called the *dilaton*). Newton's constant G is related to g and the string length l_s by $G \sim g^2 l_s^2$ in four spacetime dimensions. It is sometimes convenient to use string units where $l_s = 1$, and sometimes to use Planck units where $G = l_p^2 = 1$. It is important to distinguish them, especially when g changes. Since g is in fact determined by a dynamical field, one can imagine that it changes as a result of a physical process, e.g., a wave of dilaton passing by. However, it will often be convenient to assume the dilaton is constant and treat g as just a parameter in the theory. In general, physical properties of a state can change when g is varied. But we will see that in some cases, certain properties remain unchanged.

The classical spacetime metric is well defined in string theory only when the curvature is less than the string scale $1/l_s^2$. This follows from the fact that fundamentally, the metric is unified with all the other modes of the string. This is easily seen in perturbation theory where the graviton is just one of the massless excitations of the string. When the curvature is small compared to $1/l_s^2$, one can integrate out the massive modes and obtain an effective low energy equation of motion which takes the form of Einstein's equation with an infinite number of correction terms consisting of higher powers of the curvature multiplied by powers of l_s. When curvatures approach the string scale, this low energy approximation breaks down.

String theory includes supersymmetry. Although this symmetry has not yet been seen in nature, there is hope that it will soon be discovered by the Large Hadron Collider being built at CERN. An important consequence of this new symmetry is the following. Supersymmetric theories have a bound on the mass of all states given by their charge, which roughly says $M \geqslant Q$. This is called the *BPS bound*. States which saturate this bound are called *BPS*. They have the special property that the mass does not receive any quantum corrections.

Quantizing a string also leads to a prediction that space has more than three dimensions. This is because a symmetry of the classical string action is preserved in the quantum theory only in ten spacetime dimensions. The idea that spacetime may have more than four dimensions was first proposed in the 1920's by Kaluza and Klein. Their motivation was to create a unified theory of the two known forces: gravity and electromagnetism. It turns out that a theory of pure gravity in five dimensions reduces to gravity plus electromagnetism (plus a scalar field) in four dimensions. The standard explanation for why we do not see the extra dimensions is that they are curled up into a small ball. However recently, it has been suggested that the extra dimensions might be large, but we do not see them because we are confined to live on a 3+1 dimensional submanifold called a *brane*.

In fact, it was realized about ten years ago that string theory is not just a theory of strings. There are other extended objects called *D-branes*; these are generalizations of membranes. They are nonperturbative objects with mass $M \sim 1/gl_s$. But the gravitational field of a D-brane is proportional to $GM \sim gl_s$ and hence goes to zero at weak coupling. This means that there is a flat spacetime description of these nonperturbative objects. Indeed, they are simply surfaces on which open strings can end. The strings we have been discussing so far have been topological circles with no endpoints. The dynamics of D-branes at weak coupling is described by open strings (topological line segments) in which the two endpoints are stuck on certain surfaces. Indeed, the "D" stands for Dirichlet boundary conditions on the ends of the string keeping it on the surface, and the surface itself is the *brane*. All the particles of the standard model (e.g., the quarks, leptons, and gauge bosons) are believed to come from these open strings and are confined to these branes. Only the graviton comes from the closed string and is free to move in the bulk spacetime.

Many types of D-branes exist, of various dimensions, and each carries a charge. If the branes are flat (or, more generally, form an extremal surface) and have no open strings attached, they are BPS states. Excited D-branes (with open strings added) lose energy when two such strings combine to form a closed string. Since the closed string has no ends, it can leave the brane.

3. Application to black holes

We now wish to apply string theory to black holes, and answer the two fundamental questions raised in the Introduction. We start with the question: What is the origin of black-hole entropy?

For two decades after Bekenstein and Hawking showed that black holes have an entropy, people tried to answer this question with limited success. The breakthrough came in a paper by Strominger & Vafa (1996). They considered a charged black hole. Charged black holes are not interesting astrophysically, but they are interesting theoretically since they satisfy a bound just like the BPS bound $M \geqslant Q$. Black holes with $M = Q$ are called *extremal* and have zero Hawking temperature. They are stable, even quantum mechanically. In string theory, extremal black holes are strong coupling analogs of BPS states. One can now do the following calculation: Start with an extremal black hole and compute its entropy S_{BH}. Imagine reducing the string coupling g. When g is very small, one obtains a weakly coupled system of strings and branes with the same charge. Strominger and Vafa count the number of BPS states in this system at weak coupling and find

$$N_{BPS} = e^{S_{\mathrm{BH}}} \ . \tag{3.1}$$

This is a microscopic explanation of black-hole entropy! Unlike previous attempts to explain S_{BH}, one counts quantum states of a system in flat spacetime where there is no horizon. One obtains a number which, remarkably, is related to the area of the black hole which forms at strong coupling.

The idea of decreasing the string coupling should be viewed as a (very useful) thought experiment in string theory. In the real world, g is fixed to some value which is difficult to change. The actual value of the string coupling depends on many details about how string theory is connected to the standard model of particle physics and is not yet known. It is likely to be of order unity.

After the initial breakthrough, the agreement between black holes and a weakly coupled system of strings and D-branes was extended in many directions (for a review, see Peet 2000). It was shown that the entropy agrees for extremal charged black holes with rotation. The entropy also agrees for near extremal black holes with nonzero Hawking temperature. Since the entropy agrees as a function of energy, it is not surprising that the radiation from the D-branes has the same temperature as the black hole. What was surprising was that the total rate of radiation from black holes agrees with D-branes. (The analog of Hawking radiation for D-branes is just the process of two open strings combining to form a closed string, which leaves the branes.) What was truly remarkable was that the deviations from black-body spectrum also agree! Neither side is exactly thermal. On the black-hole side, these deviations arise since the radiation has to propagate through the curved spacetime outside the black hole. This produces potential barriers which give rise to frequency-dependent greybody factors. On the D-brane side, there are deviations since the modes come from separate left and right moving sectors on the D-branes. The calculations of these deviations could not look more different. On the black-hole side, one solves a wave equation in a black-hole background. The solutions involve hypergeometric functions. On the D-brane side, one does a calculation in D-brane perturbation theory. Remarkably, the answers agree.

More recently, there has been further progress in counting the microstates of charged black holes. A small black hole in string theory has an entropy which is not exactly given by the Bekenstein-Hawking formula (Eq. 1.4). There are subleading corrections coming from higher curvature terms in the action. Wald (1993) derived the form of these corrections to black-hole entropy in any theory of gravity. Recently, it has been shown that for certain extremal black holes the counting of microstates in string theory reproduces the black-hole entropy *including these subleading corrections* (Dabholkar 2006). The corrections are of order the string scale divided by the Schwarzschild radius to some power.

What about neutral black holes? Susskind (1998) suggested that there should be a one-to-one correspondence between ordinary excited string states and black holes. Start with a highly excited string with mass (Eq. 2.1) and imagine increasing the string coupling g. Since $G \sim g^2 l_s^2$, two effects take place. First, the gravitational attraction of one part of the string on the other causes the string size to decrease. Second, since G increases, the gravitational field produced by the string becomes stronger and the effective Schwarzschild radius GM increases in string units. Clearly, for a sufficiently large value of the coupling, the string forms a black hole.

Conversely, suppose one starts with a black hole and decreases the string coupling. Then the Schwarzschild radius shrinks in string units and eventually becomes of order the string scale. At this point the metric is no longer well defined near the horizon. Susskind suggested that the black hole becomes an excited string state.

When I first heard this, I didn't believe it. The first half of the argument sounded plausible enough, but the second half seemed to contradict the well known fact that the string entropy is proportional to the mass while the black-hole entropy is proportional to the mass squared. It turns out that there is a simple resolution of this apparent contradiction (Horowitz & Polchinski 1997). If one changes the string coupling g, the string mass is constant in string units, while the black-hole mass is constant in Planck units. Thus M_s/M_{BH} depends on g. We expect the transition to occur when the curvature at the horizon of the black hole reaches the string scale. This implies that the Schwarzschild radius r_0 is of order the string scale. Setting $M_s \sim M_{\text{BH}}$ when $r_0 \sim l_s$ we find:

$$S_{BH} \sim r_0 M_{BH} \sim l_s M_s \sim S_s \quad . \tag{3.2}$$

So the entropies agree at this *correspondence point*. This agreement between the string entropy and black-hole entropy applies to essentially all black holes, including higher dimensional Schwarzschild black holes, and charged black holes that are far from extremality.

This leads to a simple picture for the endpoint of black-hole evaporation. In Hawking's picture, the black hole evaporated down to the Planck scale where the semiclassical approximations being used broke down. In string theory, the black hole evaporates until it reaches the string scale, at which point it turns into a highly excited string. The excited string continues to radiate until it becomes an unexcited string, i.e., just another elementary particle. The timescale for black-hole evaporation is modified slightly. In the black-hole phase, $dM/dt \sim T^2$ and $T \sim 1/M$, so the time to evaporate most of the mass is of order M^3 (in Planck units). When the temperature reaches the string scale, the black hole turns into a highly excited string. After this transition, the temperature stays at the string scale as the string radiates.

The above argument shows that strings have enough states to reproduce the entropy of all black holes, but the argument is not precise enough to reproduce the entropy exactly, including the factor of 1/4. More recently, Emparan & Horowitz (2006) showed that one can exactly reproduce the entropy of a class of neutral black holes. These are rotating black holes in five dimensions which have a translational symmetry around one compact direction (as in Kaluza-Klein theory). If one rewrites the solution as a four-dimensional black hole, there are charges associated with the Maxwell field coming from the higher dimensional metric. Using various symmetries of string theory, one can map these charges into D-brane charges and count the microstates in the same way that was done for BPS black holes.

In fact, a slight extension of this argument yields a precise calculation of the entropy of an extremal Kerr black hole (Horowitz & Roberts 2007). This black hole has an entropy which is just given in terms of its angular momentum

$$S = 2\pi |J| \quad . \tag{3.3}$$

Since J is naturally quantized, this is like the entropy of the extremal charged black holes in which the entropy is again just a function of the quantized charges. It turns out that one can lift an extremal Kerr black hole to five dimensions and map it into the class of neutral black holes who entropy was counted precisely.

We now return to the second fundamental question raised earlier: Do black holes lose information? For charged, near extremal black holes, the weak coupling limit provides a quantum mechanical system with the same entropy and radiation. This was a good indication that black-hole evaporation would not violate quantum mechanics. However, the case soon became much stronger.

By studying the black-hole entropy calculations, Maldacena (1998) was led to a remarkable conjecture now called the *gauge/gravity correspondence*: *Under certain boundary conditions, string theory (which includes gravity) is completely equivalent to a (nongravitational) gauge theory living at infinity.* At first sight this conjecture seems unbelievable. How could an ordinary field theory describe all of string theory? I don't have time to describe the impressive body of evidence in favor of this correspondence which has accumulated over the past few years. The obvious differences between string theory and gauge theory are explained by the fact that our intuition about both theories is largely based on weak coupling analyses. Under the gauge/gravity correspondence, when string theory is weakly coupled, gauge theory is strongly coupled, and vice versa.

This conjecture provides a "holographic" description of quantum gravity in that the fundamental degrees of freedom live on a lower dimensional space. The idea that quantum gravity might be holographic was first suggested by 't Hooft and Susskind, motivated by the fact that black-hole entropy is proportional to its horizon area.

The gauge/gravity correspondence has an immediate consequence: The formation and evaporation of small black holes can be described by ordinary Hamiltonian evolution in the gauge theory. It does not violate quantum mechanics. After 30 years, Hawking (2005) finally conceded this point (although his reasons were not directly related to string theory).

Let me conclude with a few open questions:

(1) Can we count the entropy of Schwarzschild black holes precisely? The recent calculation of the extremal Kerr entropy in terms of microstates gives one hope that this may soon be possible.

(2) How does the information get out of the black hole? What is wrong with Hawking's original argument? It appears that we will need some violation of locality. In other words, when one reconstructs the string theory from the gauge theory, physics may not be local on all length scales.

However, perhaps the most important open question is

(3) What is the origin of spacetime? How is it reconstructed from the gauge theory? How does a black-hole horizon know to adjust itself to have area $A = 4GS_{\mathrm{BH}}$?

REFERENCES

BEKENSTEIN, J. 1973 *Phys. Rev. D* **7**, 2333.

BARDEEN, J. M, CARTER, B., & HAWKING, S. W. 1973 *Commun. Math. Phys.* **31**, 161.

DABHOLKAR, A. 2006 *Class. Quant. Grav.* **23**, S957.

EMPARAN, R. & HOROWITZ, G. T. 2006 *Phys. Rev. Lett.* **97**, 141601.

HAWKING, S. W. 1971 *Phys. Rev. Lett.* **26**, 1344.

HAWKING, S. W. 1975 *Commun. Math. Phys.* **43**, 199. [Erratum-ibid. **46**, 206 (1976)].

HAWKING, S. W. 2005 *Phys. Rev. D* **72**, 084013.

HOROWITZ, G. T. & POLCHINSKI, J. 1997 *Phys. Rev. D* **55**, 6189.

HOROWITZ, G. T. & ROBERTS, M. M. 2007 *Phys. Rev. Lett.* **99**, 221601.

MALDACENA, J. M. 1998 *Adv. Theor. Math. Phys.* **2**, 231.

PEET, A. W. 2000 arXiv:hep-th/0008241v2.

ROBINSON, D. C. 1975 *Phys. Rev. Lett.* **34**, 905.

STROMINGER, A. & VAFA, C. 1996 *Phys. Lett. B* **379**, 99.

SUSSKIND, L. 1998. In *The Black Hole: 25 years after* (eds. C. Teitelboim & J. Zanelli). World Scientific. p. 118.

WALD, R. M. 1993 *Phys. Rev. D* **48**, 3427.

ZWIEBACH, B. 2004 *A First Course in String Theory*. Cambridge University Press, 558p.

Gravitational waves from black-hole mergers

By JOHN G. BAKER,[1] WILLIAM D. BOGGS,[2]
JOAN M. CENTRELLA,[1] BERNARD J. KELLY,[1]
SEAN T. McWILLIAMS,[2] AND JAMES R. van METER[3]

[1]Gravitational Astrophysics Laboratory, NASA Goddard Space Flight Center,
8800 Greenbelt Rd., Greenbelt, MD 20771, USA

[2]University of Maryland, Department of Physics, College Park, MD 20742, USA

[3]Center for Space Science & Technology, University of Maryland–Baltimore County,
Physics Department, 1000 Hilltop Circle, Baltimore, MD 21250, USA

Coalescing black-hole binaries are expected to be the strongest sources of gravitational waves for ground-based interferometers, as well as the space-based interferometer *LISA*. Recent progress in numerical relativity now makes it possible to calculate the waveforms from the strong-field dynamical merger, and is revolutionizing our understanding of these systems. We review these dramatic developments, emphasizing applications to issues in gravitational wave observations. These new capabilities also make possible accurate calculations of the recoil or kick imparted to the final remnant black hole when the merging components have unequal masses, or unequal or unaligned spins. We highlight recent work in this area, focusing on results of interest to astrophysics.

1. Introduction

Gravitational wave astronomy will open a new observational window on the universe. Since large masses concentrated in small volumes and moving at high velocities generate the strongest, and therefore most readily detectable waves, the final coalescence of black-hole binaries is expected to be one of the strongest sources. During the last century, the opening of the full electromagnetic spectrum to astronomical observation greatly expanded our understanding of the cosmos. In this new century, observations across the gravitational wave spectrum will provide a wealth of new knowledge, including accurate measurements of binary black-hole masses and spins.

The high frequency part of the gravitational wave spectrum, ~ 10 Hz $\lesssim f \lesssim 10^3$ Hz, is being opened today through the pioneering efforts of first-generation ground-based interferometers such as the Laser Interferometer Gravitational-Wave Observatory (LIGO), currently operating at design sensitivity. Such instruments can detect gravitational waves from coalescing stellar-mass ($M \lesssim 10^2 \ M_\odot$) and intermediate-mass ($10^2 \ M_\odot \lesssim M \lesssim 10^3 \ M_\odot$) black-hole binaries. While detections from this first generation of detectors are likely to be rare, the advanced LIGO (adLIGO) upgrade may detect the coalescence of several stellar-mass and tens of intermediate-mass black-hole binaries per year. Other high-frequency sources include binary neutron-star coalescences, supernovae, and rotating neutron stars.

The low-frequency gravitational-wave window, 3×10^{-5} Hz $\lesssim f \lesssim 1$ Hz, is especially rich in astrophysical sources and will be opened by the space-based *Laser Interferometer Space Antenna* (*LISA*) detector, currently in the formulation stage. *LISA* will be sensitive to the coalescence of massive black-hole binaries with total masses in the range $10^4 \ M_\odot \lesssim M \lesssim 10^7 \ M_\odot$ to large redshifts $z \gtrsim 10$ at relatively high signal-to-noise ratios (SNRs), and may detect 10 or more such events per year. Using such observations, the black-hole masses, spins and luminosity distances can be determined to very good precision, with errors <1% in some cases (Lang & Hughes 2006). In addition, *LISA* will detect gravitational waves from the inspiral of compact stars into central massive black holes out to $z \sim 1$, as well as tens of thousands of compact binaries in the Galaxy.

The actual merger of two comparable-mass black holes that plunge together and form a common event horizon takes place in the strong-field dynamical regime of general relativity. For many years, we were unable to calculate the expected waveforms from these very energetic events due to severe problems with the large-scale computer codes needed to simulate the mergers. Recently, however, a series of stunning breakthroughs has occurred in numerical relativity, resulting in stable, robust, and accurate simulations of black-hole mergers, as well as applications to astrophysics. In Section 2 we review these developments and present examples of the resulting gravitational waveforms. Applications of these signals to issues in gravitational-wave observations are discussed in Section 3. When the merging black holes have unequal masses, or unequal or unaligned spins, the final remnant black hole suffers a recoil; recent progress in calculating these "kicks" and their applications to astrophysics are presented in Section 4. We conclude with a summary in Section 5.

2. Calculating black-hole binary coalescence

The final coalescence of a black-hole binary is driven by gravitational wave emission, and proceeds in three stages: an adiabatic inspiral, a dynamical merger, and a final ringdown (Flanagan & Hughes 1998). During the inspiral, the black holes are well separated and can be approximated as point particles. The black holes spiral together on quasi-circular trajectories, and the resulting gravitational waveforms are *chirps*, i.e., sinusoids that increase in both frequency and amplitude as the black holes get closer together. The inspiral can be treated analytically using the post-Newtonian (PN) approach, which is an expansion in v/c, where v is the characteristic orbital velocity (see Blanchet 2006 for a review of PN results). The inspiral is followed by a dynamical merger in which the black holes plunge together to form a highly distorted single black hole, producing a powerful burst of gravitational radiation. Since the merger stage occurs in the regime of very strong gravity, a full understanding of this process requires numerical-relativity simulations of the Einstein equations. After merger, the remnant black hole then settles down, evolving towards a quiescent Kerr state by shedding its non-axisymmetric modes as gravitational waves. The late part of this ringdown stage can be treated analytically using black-hole perturbation theory, and the resulting gravitational waveforms are superpositions of exponentially damped sinusoids of constant frequency (Leaver 1986; Echeverría 1989).

In numerical relativity, the full set of Einstein's equations are solved on a computer in the dynamical, nonlinear regime. This is typically accomplished by slicing 4-D spacetime into a stack of 3-D space-like hypersurfaces, each labeled by time t (Arnowitt et al. 1962; Misner et al. 1973). The Einstein equations split into two sets. The constraints give a set of relationships that must hold on each slice, and in particular constrain the initial data for a black-hole binary simulation. This data is then propagated forward in time using the evolution equations. Four freely specifiable coordinate, or gauge, conditions give the development of the spatial and temporal coordinates during the evolution.

Simulating the merger of a black-hole binary using numerical relativity has proved to be very challenging. The first attempt to evolve a head-on collision in 2-D axisymmetry dates back to 1964 (Hahn & Lindquist 1964). In the mid-1970s, the head-on collision of two equal mass, nonspinning black holes was first successfully simulated, along with the extraction of some information about the gravitational radiation (Smarr et al. 1976; Smarr 1977, 1979). In the 1990s, fully 3-D numerical relativity codes were developed and used to evolve grazing collisions of black holes (Brügmann 1999; Brandt et al. 2000; Alcubierre et al. 2001). However, the codes were plagued by a host of instabilities that caused them to crash before any significant portion of a black-hole binary orbit could be evolved. For many years, progress was slow and incremental.

Recently, a series of dramatic developments has led to major progress in black-hole binary simulations across a broad front. The first complete orbit of a black-hole binary was achieved in 2004 (Brügmann et al. 2004). This was followed by the first full simulation of a black-hole binary through an orbit, plunge, merger and ringdown in 2005 (Pretorius 2005). In late 2005, the development of new coordinate conditions produced a breakthrough in the ability to carry out accurate and stable long-term evolutions of black-hole binaries (Campanelli et al. 2006a; Baker et al. 2006c; van Meter et al. 2006). These novel but simple "moving puncture" techniques proved highly effective. They were quickly adopted by a broad segment of the numerical relativity community, leading to stunning advances in black-hole binary modeling, starting with evolutions of equal mass, nonspinning black holes and moving quickly to include unequal masses and spins; (see, e.g., Campanelli et al. 2006b; Baker et al. 2006b; Campanelli et al. 2006d; Gonzalez et al. 2007b; Baker et al. 2007b; Herrmann et al. 2007a,b; Campanelli et al. 2007a,b; Koppitz et al. 2007; Gonzalez et al. 2007a; Baker et al. 2007a; Tichy & Marronetti 2007).

The most rapid advances in modeling black-hole binary coalescences cover the previously least understood part of the gravitational waveform, i.e., the final few cycles of radiation generated from near the "innermost stable circular orbit" (ISCO) and afterward, which we call the "merger ringdown." There is already considerable progress toward a full understanding of this important "burst" portion of the waveform, through which the frequency sweeps by a factor of ∼3 up to ringdown, and during which the gravitational wave luminosity is $\sim 10^{23}\ L_\odot$, more than the luminosity of the combined starlight in the visible universe.

A particularly significant development was the demonstration of initial data-independence of merger-ringdown waveforms for equal-mass, nonspinning black holes (Baker et al. 2006b), as summarized in Figure 1. Results from four runs with successively larger initial separations are shown; the waveforms have been aligned so that the moment of peak radiation amplitude in each simulation occurs at time $t = 0$. Here we show the gravitational wave strain from the dominant ($l = 2$, $m = 2$) mode; this represents an observation made on the equatorial plane of the system, where only a single polarization component contributes to the measured strain. The upper panel of Figure 1 shows the full simulation waveforms, while the lower panel focuses on the final merger-ringdown burst. Note that here and elsewhere in this paper, we use *geometrical units*, $G = c = 1$, to measure time, distance and mass in the same units. In particular, one solar mass $M_\odot$ is equivalent to $\sim 5 \times 10^{-6}$ sec, or ∼1.5 km.† In Figure 1, our timescale is the final mass m_f, of the post-merger hole; this will be less than the initial total binary mass M because of gravitational radiation. In the shortest run (solid line), the black holes are placed on initial orbits close to the ISCO and undergo a brief plunge followed by a merger and ringdown (Baker et al. 2006c). At the next-largest initial separation (dashed line), the black holes complete ∼1.8 orbits before merging (Baker et al. 2006b). The waveforms from the two runs with successively larger initial separations (dot-dashed and dotted lines, respectively) then lock on to the merger-ringdown part of this shorter (dashed) run. In fact, the waveforms from the three longer simulations show very strong agreement for $t \gtrsim -50\, m_f$, with differences among these waveforms $\lesssim 1\%$ in this regime.

---

† Since the simulation results scale with the masses of the black holes, they are equally applicable to *LISA* and ground-based detectors.

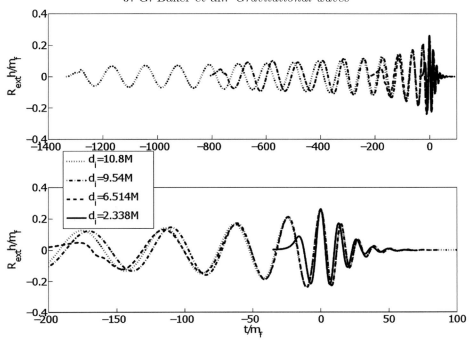

FIGURE 1. Gravitational waveforms are shown for four simulations of equal mass nonspinning black-hole binaries with successively wider initial separations. Here, d_i is the initial coordinate separation of the black-hole centers ("punctures"). All waveforms have been aligned so that the moment of maximum radiation amplitude occurs at $t = 0$ (phase is also set to zero at $t = 0$). The upper panel shows the full extent of all the runs, while the lower panel shows the final merger-ringdown burst.

Today, there is consensus that the merger of equal-mass, nonspinning black holes produces a final Kerr black hole with spin $a/m_f \sim 0.7$, and that the amount of energy radiated in the form of gravitational waves, starting with the final few orbits and proceeding through the plunge, merger and ringdown, is $E_{rad} = M - m_f \sim 0.04M$ (Pretorius 2005; Campanelli et al. 2006a; Baker et al. 2006c; Campanelli et al. 2006b; Baker et al. 2006b); see Centrella (2006) for a review. There is also agreement on the overall simple shape of the waveforms shown in Figure 1, and detailed comparison of results among numerical relativity groups has already begun (Baker et al. 2006a). Parameter-space exploration of the merger-ringdown burst, using simulations with various mass ratios (Herrmann et al. 2007a; Baker et al. 2006d; Gonzalez et al. 2007b) and spins (Campanelli et al. 2006c,d), is now underway. The next step is to push this frontier to increasingly complex mass-ratio and spin-orientation combinations, and to establish initial data independence across these parameters.

Simulations starting in the late-inspiral regime and covering more than a few orbits prior to merger are more challenging than shorter merger-ringdown evolutions (Baker et al. 2007b). Such long simulations have stronger requirements for numerical stability and place greater demands on computational resources. In addition, better accuracy is needed to control the accumulation of phase error; this in turn constrains the numerical error that can be tolerated in the rate of energy loss through gravitational-radiation reaction, which governs the inspiral.

So far, the longest simulations have been carried out for equal-mass, nonspinning black-hole binaries, starting at relatively wide separations. The two longest runs shown in Figure 1 undergo ∼4.2 (Baker et al. 2006b) and ∼7 orbits (Baker et al. 2007b), respectively, prior to merger† and demonstrate progress in long simulations (see also Husa et al. 2008). We expect that astrophysical black-hole binaries in the late inspiral regime will follow nearly circular orbits, since any initial eccentricity would have been radiated away early in their evolution. However, in the first long black-hole binary simulations, the eccentricity of the initial orbits was not very well controlled; in particular, the early part of the waveform shown by the dot-dashed line shows evidence of eccentricity which resulted from the initial data specification. The dotted curve shows more recent results, which start with very small eccentricity $\epsilon < 0.01$ (see also Pfeiffer et al. 2007; Husa et al. 2008). Such long waveforms make it possible to compare the results of numerical relativity simulations of black-hole binaries with post-Newtonian calculations in the late-inspiral regime (Buonanno et al. 2007a; Baker et al. 2006e; Hannam et al. 2008), demonstrating remarkable agreement.

3. Observing black-hole binary mergers

The final moments of a black-hole binary merger produce the most intense radiation generated through the strongest gravitational dynamics. The resulting waveforms are obviously an important part of the observable gravitational wave signature for black-hole binary events. Before these recent advances in numerical relativity, however, little was confidently known about these signals, so that observational analyses of gravitational wave data could only rely on generic "unmodeled burst" techniques (Abbott et al. 2007) rather than more effective matched-filtering techniques, which require detailed knowledge of the expected signals. Consequently the quality of scientific information which could be expected from merger observations has, to date, necessarily been significantly discounted in planning observational work.

Just how much more we can learn from observing the final stages of black-hole binary mergers with full knowledge of the waveforms, and how best to apply this knowledge in gravitational-wave data analysis, are still largely unanswered questions which the gravitational-wave community is only beginning to address. We can get some rough impressions of the observational significance of black-hole binary mergers from this new perspective by considering the SNR of the full waveforms based on matched-filtering analysis of gravitational wave data (Flanagan & Hughes 1998).

Estimating the full waveforms by stitching together the most accurate PN model with a numerical simulation of the last several cycles, we have calculated the SNR for equal-mass nonspinning binaries (Baker et al. 2007b). Figure 2 shows the contours of SNR as functions of redshift z and total binary mass M for the ground-based adLIGO detector, which will be sensitive in the frequency range 14 Hz $\lesssim f \lesssim 10^3$ Hz. AdLIGO is a planned upgrade of the initial LIGO detectors that will increase the sensitivity by roughly an order of magnitude across the frequency band. In addition, adLIGO can be tuned to optimize its sensitivity for different sources. To produce Figure 2, we used the wide-band tuning typically associated with burst sources, due to its greater sensitivity at higher frequencies, where the merger portion from many sources is predicted to occur (Shoemaker 2006). This gives an improved SNR for most black-hole masses compared to tunings that were optimized for only the early inspiral portion of the coalescence.

---

† For these quadrupolar waves, the gravitational wave and orbital frequencies are related by $f = 2f_{\mathrm{orb}}$.

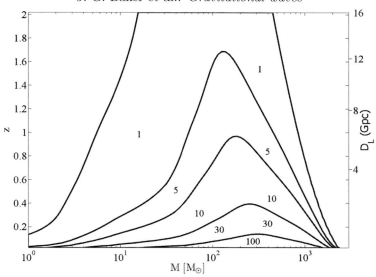

FIGURE 2. Contours of SNR for observations of black-hole binaries with adLIGO, showing mass, redshift, and luminosity-distance dependence.

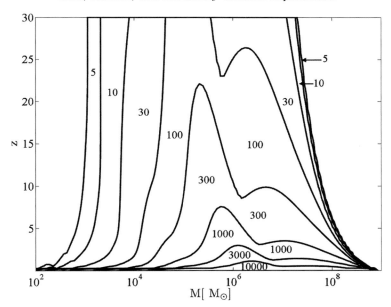

FIGURE 3. Contours of *LISA* SNR with mass, redshift, and luminosity-distance dependence. Note that black-hole binaries with masses $M > 10^7 \ M_\odot$ may not coalesce within a Hubble time (Milosavljević & Merritt 2003).

In Figure 3 we plot contours of SNR for *LISA* observations, showing that *LISA* can observe massive black-hole binaries at high SNR throughout the observable universe. *LISA* is most sensitive to systems with masses in the range $10^5 \ M_\odot \leqslant M \leqslant 10^7 \ M_\odot$, coinciding with the mass range in which models of massive black-hole binaries predict coalescence within a Hubble time (Milosavljević & Merritt 2003) and event rates for *LISA* of at least several per year (Sesana et al. 2005).

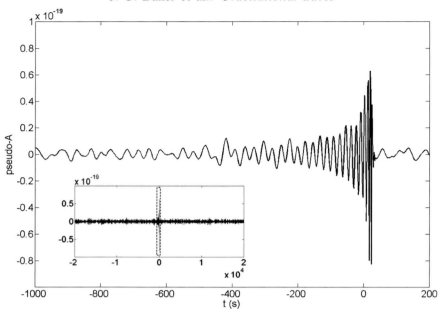

FIGURE 4. Simulated *LISA* data stream showing *LISA*'s response to a system of two equal-mass black holes. Specifically, the pseudo-A observable is shown for the case of two 4×10^5 $M_\odot$ black holes at redshift $z = 10$, inclination $\iota = \pi/2$, polarization angle $\psi = 0$, ecliptic latitude $\beta = 0$, and ecliptic longitude $\lambda = \pi$. The pseudo-A, E, and T observables are a set of linearly independent combinations of the Michelson unequal arm X, Y, and Z observables at TDI 1.5 (Shaddock et al. 2003), and are therefore useful in data analysis. The inset shows a time span of 4×10^4 seconds encompassing the time span of the main figure, the main figure being an enlargement of the dash-boxed region of the inset.

In Figure 4 we plot a simulated *LISA* data stream containing a black-hole binary signal as well as noise from the interferometer and the white dwarf confusion noise caused from the unresolvable number of background white-dwarf binaries. This is an example of a typical, non-optimal case in order to demonstrate the relative strength that these signals will generally have. Here we show two 4×10^5 $M_\odot$ black holes at redshift $z = 10$, inclination $\iota = \pi/2$, polarization angle $\psi = 0$, ecliptic latitude $\beta = 0$, and ecliptic longitude $\lambda = \pi$.

So far we have restricted our consideration to the example of equal-mass nonspinning black-hole binary mergers. More detailed considerations of black-hole binary observations must take into account the full parameter space of these systems, which, from the point of view of general relativity, are described by seven parameters, corresponding to the black-hole mass-ratio $q \equiv m_1/m_2 \leqslant 1$ (recall that the total mass M only scales the waveforms), and the spin vector of each hole (plus two more parameters if eccentric systems are considered). Much progress has already been made in conducting numerical simulations to flesh out the basic features of this parameter space (see, e.g., Campanelli et al. 2006c,d; Herrmann et al. 2007a). First impressions indicate that the waveforms in this parameter space are qualitatively simple, with smooth parameter dependencies. This suggests that our knowledge of the mergers can ultimately be applied in observations with the aid of smooth analytic waveform models, joining PN treatments of the inspiral signals with analytic approximations to the mergers. Already such a model has been proposed, which provides high overlaps with the numerical simulations for mass-ratios as low as $q = 1/4$ (Buonanno et al. 2007b).

Applying the mass scalings from Flanagan & Hughes (1998), we can estimate the effect of the mass ratio q on the computed SNRs; specifically, SNR $\sim \eta^{1/2}$ for the inspiral, and SNR $\sim \eta$ for the merger and ringdown, where $\eta \equiv m_1 m_2 / M^2 = q/(1+q)^2$ is the *symmetric mass ratio*. For stellar-mass black-hole binaries, the mass ratios are rather broadly distributed (Belczynski et al. 2007); the rates for such mergers may be low, ~ 2 yr^{-1} for adLIGO, depending on the evolution of the original binary through the common envelope phase. For intermediate-mass black-hole binaries, mass ratios in the range $0.1 \lesssim q \lesssim 1$ are expected to be the most relevant, with potential rates of ~ 10 per year (Fregeau et al. 2006); we note, however, that these rates are far more uncertain than those for stellar-mass black-hole binaries. Investigations of gravitational-wave data analysis based on unequal-mass numerical simulations are now underway: Pan et al. (2008) have investigated how numerical relativity waveforms may be used to select the best post-Newtonian template banks for searches in the LIGO, adLIGO and VIRGO data stream.

Astrophysical black-hole binaries are also expected to be spinning, which can potentially raise the SNR, for example if there is a spin-induced hangup that generates more gravitational wave cycles in the merger (Campanelli et al. 2006d). Vaishnav et al. (2007) have investigated how the addition of spins to equal-mass waveforms affects search templates, necessitating the use of higher-order modes for high matches in the LIGO band.

We also note that, with more accurate computational modeling, simulations of the merger can be used together with gravitational wave observations to probe gravity in the regime of strong fields. In particular, if the binary masses and spins can be obtained with good accuracy from observations of the inspiral with *LISA* (Lang & Hughes 2006), the merger waveform can be calculated using numerical relativity. With this, a comparison can be made between the predictions of general relativity—or any other theory of gravity used in a numerical simulation—with observations in the regime of very strong gravity.

4. Kicks from black-hole binary mergers

Black-hole binaries which possess an asymmetry, such as having unequal masses or possessing unequal or unaligned spins, will emit gravitational radiation asymmetrically. This asymmetric gravitational radiation will impart a linear momentum recoil on the binary's center of mass. Previous attempts to calculate this recoil, or "kick," analytically in the nonspinning case (Fitchett 1983), and for cases with spin (Kidder 1995) included very large uncertainties, because the dominant part of the kick occurred in the last orbit through the merger, which was very poorly modeled. With the advent of stable, accurate numerical-relativity codes, the problem of accurately calculating the kick became possible. Several groups have tackled this problem. The problem of kicks from an unequal mass binary was first investigated in Baker et al. (2006d), where recoils between 86 and 115 km s^{-1} for a $q = 2/3$ mass ratio were simulated. In Gonzalez et al. (2007b), several simulations were performed and a peak recoil velocity for nonspinning binaries of 175 km sec^{-1} was found for a mass ratio of $q \approx 1/2.78$.

A flurry of activity began when several groups began simulating kicks from spin asymmetry. Herrmann et al. (2007b) demonstrated that kicks from spin asymmetry can exceed unequal-mass kicks, and that for an equal-mass configuration with one black hole spinning prograde and the other spinning retrograde with equal magnitude $\hat{a}$ (a dimensionless spin parameter, varying from 0 for a Schwarzschild hole to 1 for a maximally spinning Kerr hole), the spin kick scales as $475\,\hat{a}$ km s^{-1}. However, it was soon found that the largest kicks are obtained when the spins are antiparallel with each other but perpendicular to the orbital angular momentum. Gonzalez et al. (2007a) were the first to perform this simulation, obtaining a kick of 2650 km s^{-1} for $\hat{a} = 0.8$. Campanelli et al. (2007b) showed that the resulting kick depends sensitively on the angle between the black-hole spin vectors and their velocities. They found a simple cosine dependence, indicating that Gonzalez et al. (2007a) somewhat fortuitously happened upon what is nearly the maximum angle. Furthermore, Campanelli et al. (2007b) projected that for maximal spin, kicks as large as 4000 km s^{-1} are possible.

Kicks of 2000 km s^{-1} are sufficiently large to kick remnant black holes out of even the largest giant elliptical galaxies. Therefore, simulation results this large had to be reconciled with the observational reality that we see massive black holes at the centers of all the galaxies that we have observed above a certain spheroidal mass threshold. Schnittman & Buonanno (2007) used the EOB formalism to perform a Monte Carlo simulation in order to try and predict how typical these large kicks are. For mass ratios between $q = 1$ and $q = 1/10$ and spins of $\hat{a} = 0.9$ on both holes, they allowed random orientations among the spin vectors and the orbital angular momentum. They found that 12% of the remnants received a kick exceeding 500 km s^{-1}, and 2.7% received a kick exceeding 1000 km s^{-1}.

Schnittman & Buonanno (2007) assume no external influence which might tend to align the spins with each other and with the orbital angular momentum. However, Bogdanovic, Reynolds, & Miller (2007) investigated the role of the Bardeen-Petterson effect (Bardeen & Petterson 1975) in mitigating the gravitational recoil from spin asymmetry. In the absence of gas accretion, the assumption of random orientation in Schnittman & Buonanno (2007) applies. However, if a black-hole binary accretes (1–10)% of its initial mass from an accretion disk, the spin of each hole will align with the angular momentum of the disk, which in turn is aligned with the orbital angular momentum of the binary. Therefore, if the majority of mergers occur in gas-rich environments, then the giant kick results, although theoretically interesting, are not astrophysically relevant. There is good reason to expect supermassive black holes to accrete substantial mass, particularly at larger redshift. In this scenario, the configuration of greatest interest is spin orientations aligned with the orbital angular momentum.

Figure 5 and Table 1 present the results of our investigation of this class of configurations, which were first presented in Baker et al. (2007a). "NE" refers to unequal mass, with the corresponding simulations having a mass ratio $q = 2/3$. "+" means prograde with respect to the orbital angular momentum, and "-" means retrograde ("0" means no spin). Figure 5 shows the accumulated kick as a function of time for all of our runs. Of particular note is the absence of an "unkick" for the equal-mass case, and the variable size of the unkick in the other cases. Also, we observe the expected accelerated merger in the NE$_{--}$ and NE$_{0-}$ cases due to spin-orbit attraction, and correspondingly the delayed merger in the NE$_{++}$ and NE$_{0+}$ cases due to spin-orbit repulsion. Using our data along with the data from Herrmann et al. (2007b) and Koppitz et al. (2007), which are also included in Table 1, we are able to construct an empirical kick formula, given by:

$$v = V_0[32\,q^2/(1+q)^5]\sqrt{(1-q)^2 + 2\,(1-q)\,K\,\cos\theta + K^2}\,, \qquad (4.1)$$

where $K = k\,(q\hat{a}_1 - \hat{a}_2)$. The parameter V_0 gives the overall scaling of the kick (note that the factor in brackets becomes unity for $q = 1$), while k gives the relative scaling of the kick contributions from spin and mass asymmetries.

Performing a least-squares fit to the data yields $V_0 = 276$ km s^{-1}, $q = 0.58$, and $k = 0.85$. Our formula yields a maximum error of 10.8% for the cases investigated. If, in fact, the majority of supermassive black-hole mergers are gas-rich, then this formula may predict a significant component of the kick for the majority of cases of astrophysical interest. Campanelli et al. (2007b) subsequently suggested an extension of the formula

Run	q	$\hat{a}_1$	$\hat{a}_2$	v_{num}	v_{pred}	$\frac{|\Delta v|}{v_{\text{num}}}(\%)$
NE$_{--}$	0.654	-0.201	-0.194	116.3	119.5	2.7
NE$_{-+}$	0.653	-0.201	0.193	58.5	58.2	0.5
NE$_{0-}$	0.645	0.000	-0.195	167.7	153.1	8.7
NE$_{00}$	0.677	0.000	0.000	95.8	98.6	2.9
NE$_{0+}$	0.645	0.000	0.194	76.9	71.7	6.8
NE$_{+-}$	0.655	0.201	-0.194	188.6	181.9	3.6
NE$_{++}$	0.654	0.201	0.194	83.4	92.4	10.8
EQ$_{+-}$	1.001	0.198	-0.198	89.8	92.6	3.2
S0.05	1.000	0.200	-0.200	96.0	93.8	2.3
S0.10	1.000	0.400	-0.400	190.0	187.6	1.2
S0.15	1.000	0.600	-0.600	285.0	281.5	1.2
S0.20	1.000	0.800	-0.800	392.0	375.3	4.3
r0	1.000	-0.584	0.584	260.0	274.0	5.4
r1	0.917	-0.438	0.584	220.0	220.8	0.3
r2	0.872	-0.292	0.584	190.0	178.1	6.3
r3	0.848	-0.146	0.584	140.0	141.9	1.4
r4	0.841	0.000	0.584	105.0	110.4	5.1

TABLE 1. Predicted versus computed kick speed. Runs labeled "S0.##" are taken from Herrmann et al. (2007b), while runs labeled "r#" are taken from Koppitz et al. (2007; from Baker et al. 2007a and reproduced with permission of the AAS).

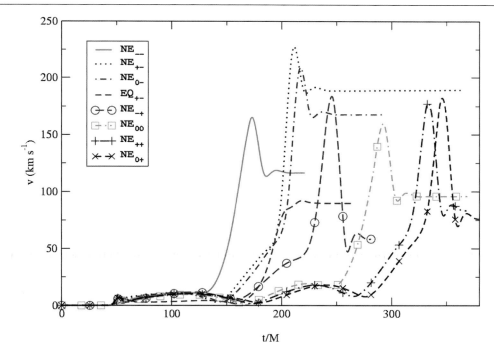

FIGURE 5. Aggregated kicks from all of our runs. The merger time for each binary matches the peak in its kick profile; the relative delay in merger times between data sets differing in initial spins is consistent with the results of Campanelli et al. (2006d). All configurations show a marked "unkick" after the peak, with the exception of the equal-mass case, EQ$_{+-}$ (from Baker et al. 2007a; reproduced with permission of the AAS).

presented in Baker et al. (2007a) for kicks out of the orbital plane, which would be more applicable in gas-poor cases where the Bardeen-Petterson effect doesn't occur, such as gas-poor galaxies and the merger of stellar-mass black holes. The achievement of simulating the merger is therefore leading us to a fairly comprehensive understanding of the recoil velocity parameter space in a strikingly brief period of time.

5. Summary

Mergers of comparable-mass black-hole binaries produce intense bursts of gravitational radiation and are very strong sources for both ground-based interferometers and the space-based *LISA*. Recent progress in numerical relativity now makes it possible to calculate the merger waveforms accurately and robustly. Today there is consensus that the merger of equal-mass, nonspinning black holes produces a final Kerr black hole with spin $a/m_f \sim 0.7$, and that the amount of energy radiated in the form of gravitational waves, starting with the final few orbits and proceeding through the plunge, merger and ringdown, is $E_{\rm rad} = M - m_f \sim 0.04M$. Simulations are now being carried out for an increasing range of black-hole component masses and spins. Computational methods continue to improve, resulting in longer simulations that start in the late inspiral regime and allow quantitative comparisons with analytic post-Newtonian methods. Applications of the resulting waveforms to gravitational wave data analysis have begun. The recoil or kick velocity that results when the black holes have unequal masses, or unequal or unaligned spins, has been calculated for a variety of interesting cases, with key applications to astrophysics. Overall, the field of black-hole binary mergers is experiencing a true "golden age," with many new results coming out across a broad front. Stay tuned!

It is a pleasure to thank our colleagues and collaborators Alessandra Buonanno, Scott Hughes, Cole Miller, Jeremy Schnittman and Tuck Stebbins for stimulating discussions. This work was supported in part by NASA grant 06-BEFS06-19. The simulations were carried out using Project Columbia at the NASA Advanced Supercomputing Division (Ames Research Center), and at the NASA Center for Computational Sciences (Goddard Space Flight Center). B.J.K. was supported by the NASA Postdoctoral Program at the Oak Ridge Associated Universities. S.T.M. was supported in part by the Leon A. Herreid Graduate Fellowship.

REFERENCES

ABBOTT, B. ET AL. 2007 *Class. Quantum Grav.* **24**, 5343.

ALCUBIERRE, M., BENGER, W., BRÜGMANN, B., LANFERMANN, G., NERGER, L., SEIDEL, E., & TAKAHASHI, R. 2001 *Phys. Rev. Lett.* **87**, 271103.

ARNOWITT, R., DESER, S. & MISNER, C. W. 1962. In *Gravitation: An Introduction to Current Research* (ed. L. Witten). p. 227. John Wiley.

BAKER, J., CAMPANELLI, M., PRETORIUS, F., & ZLOCHOWER, Y. 2006a *Class. Quantum Grav.* **24**, S25.

BAKER, J. G., BOGGS, W. D., CENTRELLA, J., KELLY, B. J., MCWILLIAMS, S. T., MILLER, M. C., & VAN METER, J. R. 2007a *ApJ* **668**, 1140.

BAKER, J. G., CENTRELLA, J., CHOI, D.-I., KOPPITZ, M., & VAN METER, J. 2006b *Phys. Rev. D* **73**, 104002.

BAKER, J. G., CENTRELLA, J., CHOI, D.-I., KOPPITZ, M. & VAN METER, J. 2006c *Phys. Rev. Lett.* **96**, 111102.

BAKER, J. G., CENTRELLA, J., CHOI, D.-I., KOPPITZ, M., VAN METER, J., & MILLER, M. C. 2006d *ApJ* **653**, L93.

BAKER, J. G., MCWILLIAMS, S. T., VAN METER, J. R., CENTRELLA, J., CHOI, D.-I., KELLY, B. J., & KOPPITZ, M. 2007b *Phys. Rev. D* **75**, 124024.

BAKER, J. G., VAN METER, J. R., MCWILLIAMS, S. T., CENTRELLA, J., & KELLY, B. J. 2006e *Phys. Rev. Lett.* **99**, 181101.

BARDEEN, J. M. & PETTERSON, J. A. 1975 *ApJ* **195**, L65.

BELCZYNSKI, K., KALOGERA, V., RASIO, F. A., TAAM, R. E., & BULIK, T. 2007 *ApJ* **662**, 504.

BLANCHET, L. 2006 *Living Rev. Rel.* **9**, 4.

BOGDANOVIC, T., REYNOLDS, C. S., & MILLER, M. C. 2007 *ApJ* **661**, L147.

BRANDT, S., CORRELL, R., GÓMEZ, R., HUQ, M. F., LAGUNA, P., LEHNER, L., MARRONETTI, P., MATZNER, R. A., NEILSEN, D., PULLIN, J., SCHNETTER, E., SHOEMAKER, D., & WINICOUR, J. 2000 *Phys. Rev. Lett.* **85**, 5496.

BRÜGMANN, B. 1999 *Int. J. Mod. Phys. D* **8**, 85.

BRÜGMANN, B., TICHY, W., & JANSEN, N. 2004 *Phys. Rev. Lett.* **92**, 211101.

BUONANNO, A., COOK, G. B., & PRETORIUS, F. 2007a *Phys. Rev. D* **75**, 124018.

BUONANNO, A., PAN, Y., BAKER, J. G., CENTRELLA, J., KELLY, B. J., MCWILLIAMS, S. T. & VAN METER, J. R. 2007b *Phys. Rev. D* **76**, 104049.

CAMPANELLI, M., LOUSTO, C. O., MARRONETTI, P., & ZLOCHOWER, Y. 2006a *Phys. Rev. Lett.* **96**, 111101.

CAMPANELLI, M., LOUSTO, C. O., & ZLOCHOWER, Y. 2006b *Phys. Rev. D* **73**, 061501(R).

CAMPANELLI, M., LOUSTO, C. O., & ZLOCHOWER, Y. 2006c *Phys. Rev. D* **74**, 084023.

CAMPANELLI, M., LOUSTO, C. O., & ZLOCHOWER, Y. 2006d *Phys. Rev. D* **74**, 041501(R).

CAMPANELLI, M., LOUSTO, C. O., ZLOCHOWER, Y., & MERRITT, D. 2007a *ApJ* **659**, L5.

CAMPANELLI, M., LOUSTO, C. O., ZLOCHOWER, Y., & MERRITT, D. 2007b *Phys. Rev. Lett.* **98**, 231102.

CENTRELLA, J. M. 2006. In *Laser Interferometer Space Antenna: 6th International LISA Symposium.* AIP Conf. Proc. 873, p. 70. AIP.

ECHEVERRÍA, F. 1989 *Phys. Rev. D* **40**, 3194.

FITCHETT, M. J. 1983 *MNRAS* **203**, 1049.

FLANAGAN, E. E., & HUGHES, S. A. 1998 *Phys. Rev. D* **57**, 4535.

FREGEAU, J. M., LARSON, S. L., MILLER, M. C., O'SHAUGHNESSY, R., & RASIO, F. A. 2006 *ApJ* **646**, L135.

GONZALEZ, J. A., HANNAM, M. D., SPERHAKE, U., BRÜGMANN, B. & HUSA, S. 2007a *Phys. Rev. Lett.* **98**, 231101.

GONZALEZ, J. A., SPERHAKE, U., BRÜGMANN, B., HANNAM, M., & HUSA, S. 2007b *Phys. Rev. Lett.* **98**, 091101.

HAHN, S. G. & LINDQUIST, R. W. 1964 *Ann. Phys.* **29**, 304.

HANNAM, M., HUSA, S., SPERHAKE, U., BRÜGMANN, B., & GONZALEZ, J. A. 2008 *Phys. Rev. D* **77**, 044020.

HERRMANN, F., HINDER, I., SHOEMAKER, D., & LAGUNA, P. 2007a *Class. Quantum Grav.* **24**, S33.

HERRMANN, F., HINDER, I., SHOEMAKER, D., LAGUNA, P., & MATZNER, R. A. 2007b *ApJ* **661**, 430.

HUSA, S., GONZALEZ, J. A., HANNAM, M., BRÜGMANN, B., & SPERHAKE, U. 2008 *Class. Quantum Grav.* **25**, 105006.

HUSA, S., HANNAM, M., GONZALEZ, J. A., SPERHAKE, U., & BRÜGMANN, B. 2008 *Phys. Rev. D* **77**, 044037.

KIDDER, L. E. 1995 *Phys. Rev. D* **52**, 821.

KOPPITZ, M., POLLNEY, D., REISSWIG, C., REZZOLLA, L., THORNBURG, J., DIENER, P., & SCHNETTER, E. 2007 *Phys. Rev. Lett.* **99**, 041102.

LANG, R. N. & HUGHES, S. A. 2006 *Phys. Rev. D* **74**, 122001.

LEAVER, E. W. 1986 *Proc. R. Soc. London Ser. A* **402** (1823).

MILOSAVLJEVIĆ, M. & MERRITT, D. 2003 *ApJ* **596**, 860.

MISNER, C. W., THORNE, K. S., & WHEELER, J. A. 1973 *Gravitation.* San Francisco: W. H. Freeman.

PAN, Y., BUONANNO, A., BAKER, J. G., CENTRELLA, J., KELLY, B. J., McWILLIAMS, S. T., PRETORIUS, F., & VAN METER, J. R. 2008 *Phys. Rev. D.* **77**, 024014.

PFEIFFER, H. P., ET AL. 2007 *Class. Quantum Grav.* **24**, S59.

PRETORIUS, F. 2005 *Phys. Rev. Lett.* **95**, 121101.

SCHNITTMAN, J. D. & BUONANNO, A. 2007 *ApJ* **662**, L63.

SESANA, A., HAARDT, F., MADAU, P., & VOLONTERI, M. 2005 *ApJ* **623**, 23.

SHADDOCK, D. A., TINTO, M., ESTABROOK, F. B., & ARMSTRONG, J. W. 2003 *Phys. Rev. D* **68**, 061303.

SHOEMAKER, D. 2006. Private communication.

SMARR, L. 1977 *Ann. N. Y. Acad. Sci.* **302**, 569.

SMARR, L. 1979. In *Sources of Gravitational Radiation* (ed. L. Smarr), p. 245. Cambridge University Press.

SMARR, L., ČADEŽ, A., DeWITT, B., & EPPLEY, K. 1976 *Phys. Rev. D* **14** (10), 2443.

TICHY, W. & MARRONETTI, P. 2007 *Phys. Rev. D* **76**, 061502.

VAISHNAV, B., HINDER, I., HERRMANN, F., & SHOEMAKER, D. 2007 *Phys. Rev. D* **76**, 084020.

VAN METER, J. R., BAKER, J. G., KOPPITZ, M., & CHOI, D.-I. 2006 *Phys. Rev. D* **73**, 124011.

Out-of-this-world physics: Black holes at future colliders

By GREG LANDSBERG

Brown University, Department of Physics, 182 Hope Street, Providence, RI 02912, USA;
landsberg@hep.brown.edu

One of the most dramatic consequences of low-scale (∼1 TeV) quantum gravity in models with large or warped extra dimension(s) is possibly copious production of mini black holes at future colliders. Hawking radiation of these black holes is expected to be constrained mainly to our three-dimensional world and results in rich phenomenology. In this talk we discuss selected aspects of mini black hole phenomenology, such as production at colliders, black-hole decay properties, and Hawking radiation as a sensitive probe of the dimensionality of extra space.

1. Introduction

Particle physics and astrophysics have much in common. Both fields build beautiful instruments to unveil hidden mysteries of space. Not only do they utilize the cutting edge (and often similar) technology to achieve best possible performance, but they also look gorgeous—shiny metal shells protecting the most precise detectors human kind ever built. We both launch big things—the astrophysicists launch things up, in the outer space; particle physicists launch things down—into enormous underground caverns where the most powerful particle accelerators collide particles to converge energy into mass and perhaps recreate the early moments of the universe. Figure 1 shows two of these spectacular launches: that of the *Hubble Space Telescope* and the largest part of the Compact Muon Solenoid (CMS) detector at the Large Hadron Collider (LHC).

The more we learn about the puzzles of the world around us, the more we find an astonishing connection between phenomena happening at the largest distance scales and physics revealed at the tiniest distances we have been able to probe. It is likely that such an intricate connection is a reflection of yet mysterious physics principles that explain the entire set of phenomena we observe—both in the microcosm and the macrocosm. This connection spans many areas in astro-particle physics: from a possible profound connection between gravity and QCD via the AdS/CFT correspondence conjecture (Maldacena 1998), to particle physics candidates for dark matter, and maybe even dark energy. The recent success of sting theory in explanation of thermodynamics or traditionally astrophysical objects—black holes—is another example of this connection. Yet another recent finding in string theory about possible existence of extremely large number of vacua, perhaps as large as 10^{1500}, may explain the extremely small, yet non-zero, value of the cosmological constant (as measured via a host of astrophysical observations) via an anthropic principle (for a related discussion, see, e.g., Susskind 2005).

Incidentally, one of the most mysterious objects conventionally associated with the vast cosmic landscape—the extremely dense end-products of stellar evolution known as black holes—may actually be accessible at the next generation of particle colliders. Of course, unlike their astronomical counterparts, these miniature black holes will be extremely small and live only for a split fraction of a second. Nevertheless, they will possess all the properties of their much larger classical cousins, thus potentially allowing us to study their features in great detail. In this talk we focus on phenomenology of mini black-hole production and decay in high-energy collisions in models with low-scale gravity. We point out exciting ways of studying quantum gravity and searching for new physics using

FIGURE 1. Spectacular launches of modern detectors. Left: the launch of the *Hubble Space Telescope* by the shuttle Discovery on April 24, 1990. Right: the "launch" of the largest (2000 ton) part of the CMS detector at the LHC 100 m down into its operational position in the underground cavern on February 28, 2007. Photos courtesy ST ScI and CERN.

large samples of black holes that may be accessible at future colliders, most notably at the LHC, which is scheduled to start operation next year at CERN, near Geneva, Switzerland.

Much of these proceedings are based on a recent review (Landsberg 2006) I have written on the subject.

2. The hierarchy problem

One of the most pressing problems in modern particle physics is the hierarchy problem, which can be expressed by a single, deceptively simple question: why is gravity (at least as we observe it) some 10^{38} orders of magnitude weaker than other forces of nature? This mysterious fact is responsible for a humongous hierarchy of energy scales: from the scale of electroweak symmetry breaking ($M_{\mathrm{EW}} \sim 1$ TeV)† to the energies at which gravity is expected to become as strong as other three forces (strong, electromagnetic, and weak), known as the Planck scale, or $M_{\mathrm{Pl}} = 1/\sqrt{G_N} \approx 1.22 \times 10^{16}$ TeV, where G_N is Newton's coupling constant.

† In what follows, we will use "natural" units: $\hbar = c = k_B = 1$, where $\hbar$ is the reduced Planck's constant, c is the speed of light, and k_B is Boltzmann's constant. This choice allows us to measure both energy and temperature in the same units, TeV ($= 10^{12}$ eV), while distance and time are measured in TeV^{-1}. The relationship between natural and SI units is as follows: 1 TeV$= 1.16 \times 10^{16}$ K, 1 TeV$^{-1} = 6.58 \times 10^{-28}$ s $= 1.97 \times 10^{-19}$ m.

If it was not for the hierarchy problem, our current understanding of fundamental particles and forces, consolidated in the standard model (SM) of particle physics, would have been nearly perfect. After all, the standard model has an impressive calculational power. Its predictions have been tested with a per mil or better accuracy, with hardly any significant deviations found so far. While certain phenomena, such as neutrino masses or existence of dark matter, are not explained within the standard model, they still can be accommodated via its minimal expansion, just like the description of CP-violation has been earlier added to the SM framework via the quark mixing matrix. While it is true that several phenomena, including the hierarchy of fermion masses, can not be explained within the standard model, nevertheless they are readily accommodated within it by introducing additional parameters, e.g., fermion couplings to the Higgs field. Even the only remaining unobserved particle predicted in the standard model—the fundamental Higgs boson responsible for the weak boson masses, while leaving photon massless—may soon be found at the LHC or even at the Tevatron (Fermilab, near Chicago, USA). Consequently, there is no convincing experimental evidence that the standard model cannot work to the highest energies, such as Planck scale. Nevertheless, there is a pretty good theoretical reason to believe that this is not the case: the hierarchy problem.

The reason that the hierarchy problem is so annoying stems from the fact that in our everyday experience, large hierarchies tend to collapse unless they are supported by some intermediate layers or by precisely tuned initial conditions (the so-called "fine tuning"). For example, while it is mechanically possible to balance a pen on its tip, the amount of fine tuning required for such a delicate balance is simply too demanding to be of practical use. Indeed, walking in a room and seeing a pen standing vertically on its point on a table would be too odd not to suspect some kind of a hidden support. Similarly, making the standard model work for all energies up to the Planck scale would require a tremendous amount of fine tuning of its parameters—to a precision of $\sim (M_{\mathrm{Pl}}/M_{\mathrm{EW}})^2 \sim 10^{-32}$! A natural reaction to this observation is to conclude that the standard model needs some "support," which should come either in a form of new physics at intermediate energy scales that adds extra layers to the hierarchy, or from certain new symmetries that would guarantee the necessary amount of fine tuning.

However, it is worth pointing out that a large amount of fine tuning, while unnatural, is not prohibited by any fundamental principle. The fact that large hierarchies tend to collapse is merely empirical. While many large hierarchies we know have indeed collapsed, some others are surprisingly stable—examples of both kinds can be found in mechanical, social, and political systems alike. Certain examples of fine tuning have been observed in nature: from the fact that the apparent angular size of the moon is the same as the angular size of the Sun within 2.5%—a mere coincidence to which we owe such a spectacular sight as a solar eclipse—to a somewhat less-abused example of the infamous Florida 2000 presidential election recount with the ratio of Republican to Democratic votes equal to 1.000061, i.e., fine tuned to unity with the precision of 0.006%!

Despite these observations, since its formulation in the late sixties, the majority of particle physicists have been working very hard to find a more rational explanation of the hierarchy of forces in the standard model. A number of viable solutions have been proposed as a result of this work: from supersymmetry (which protects the hierarchy due to nearly exact intrinsic cancelations of the effects caused by the standard model particles and by their superpartners obeying different spin statistics) to models with strong dynamics (which introduce an intermediate energy scale via new, quantum chromodynamics [QCD]-like force, and often result in a composite Higgs boson).

While each of the aforementioned solutions to the hierarchy problem easily deserves a separate talk, here we will focus on yet another, more recently proposed remedy that does not involve new particles or symmetries, but instead uses the geometry of space itself.

3. Brief history of space

Since ancient times, the idea that space may contain more than the three familiar dimensions has been one of the most popular recurring themes in the work of numerous philosophers, artists, and writers. From the very concept of Heaven and Hell to shadow or mirror worlds and parallel universes, the belief that the space around us may be a bigger place than it is commonly thought became one of the most mesmerizing and puzzling ideas which inspired many great minds of the past centuries. However, it took quite a while before this concept was truly embraced by scientists.

In fact, in the beginning of the nineteenth century, following the seminal work of Gauss, there were numerous publications claiming that additional dimensions in space would contradict the inverse square law obeyed by gravitational and electromagnetic forces. Consequently, the very concept of extra dimensions was quickly abandoned by mathematicians and physicists and remained the realm of art and philosophy.

Perhaps the first person to rigorously consider the possibility of extra dimensions in space was Riemann (1868). What started as an abstract mathematical idea of a curved Riemannian space soon became the foundation of the most profound physics theory of the last century, if not of the entire history of physics: Albert Einstein's general relativity (Einstein 1907, 1911, 1916; Einstein & Grossman 1913). While Einstein's theory was formulated in three-plus-one space-time dimensions, it soon became apparent that the theory could not be self consistent up to the highest energies in its original form.

Kaluza (1921) and Klein (1926a,b) suggested that a unification of electromagnetism and general relativity is possible if the fifth, spatial dimension of a finite size is added to the four-dimensional space-time. While this attempt has not led to a satisfactory and self-consistent unification of gravity and electromagnetism, the idea of finite (or "compactified") extra-spatial dimensions was firmly established by Kaluza and Klein, and eventually led to their broad use in string theory. Rapid progress in string theory in the 1970s helped the original idea of half-a-century earlier to regain its appeal. It was realized that an extra six or seven spatial dimensions are required for the most economical and symmetric formulation of string theory. In particular, string theory requires extra dimensions to establish its deep connection with the supersymmetry, which also leads to the unification of gauge forces. While the size of these compact dimensions is not fixed in string theory, it is natural to expect them to have radii similar to the inverse of the grand unification energy scale, or of the order of 10^{-32} m. Unfortunately, no experimental means exist to probe such short distances, so extra dimensions of string theory will likely remain untested, even if this theory turns out to be the correct description of quantum gravity.

4. Large extra-spatial dimensions

The situation with seemingly untestable extra-spatial dimensions changed dramatically in the late 1990s, when Arkani-Hamed, Dimopoulos, and Dvali (Arkani-Hamed et al. 1998, 1999; Antoniadis et al. 1998, hereafter ADD) proposed a new paradigm in which several (n) compactified extra dimensions could be as large as $\sim$1 mm! These *large extra dimensions* have been introduced to solve the hierarchy problem of the standard model by dramatically lowering the Planck scale from its apparent value of $M_{\mathrm{Pl}} \sim 10^{16}$ TeV to $\sim$1 TeV. (We further refer to this low *fundamental* Planck scale as M_D.) In the ADD model, the *apparent* Planck scale M_{Pl} only reflects the strength of gravity from the point of view of a three-dimensional (3D) observer. It is, in essence, just a virtual "image" of the fundamental, $(3 + n)$-dimensional Planck scale M_D, caused by an incorrect interpolation of the gravitational coupling (measured only at low energies and large distances) to a completely different regime of high energies and short distances.

In order for such "large" extra dimensions not to violate any constraints from atomic physics and other experimental data, all other forces except for gravity must not be allowed to propagate in extra dimensions. Conveniently, modern quantum field theory allows to confine spin 0, 1/2, and 1 particles to a subset of a multidimensional space, or a "brane." At the same time, the graviton, being a spin-2 particle, is not confined to the brane and thus permeates the entire ("bulk") space. The solution to the hierarchy problem then becomes straightforward: if the Planck scale in the multidimensional space is of the order of the electroweak scale, there is no large hierarchy and no fine-tuning problem to deal with, as non-trivial physics really "ends" at the energies ~1 TeV. Gravity appears weak to a 3D-observer merely because it spans the entire bulk space, so the gravitons spend very little time in the vicinity of our brane. Consequently, the strength of gravity in 3D is literally diluted by an enormous volume of extra space.

The compactification radius of large extra dimensions (R_c) is fixed by their number and the value of the fundamental Planck scale in the $(4 + n)$-dimensional space-time. By applying Gauss's law, one finds (Arkani-Hamed et al. 1998, 1999; Antoniadis et al. 1998)

$$M_{\text{Pl}}^2 = 8\pi M_D^{n+2} R_c^n \quad . \tag{4.1}$$

Here for simplicity we assumed that all n extra dimensions have the same size R_c and are compactified on a torus.

If one requires $M_D \sim 1$ TeV and a single extra dimension, its size has to be as large as the radius of the solar system; however, already for two extra dimensions their size is just ~1 mm; for $n = 3$ it is ~1 nm, i.e., similar to the size of an atom; and for larger n the radius further decreases to a subatomic size and reaches ~1 fm (the size of a proton) for seven extra dimensions.

A direct consequence of such compact extra dimensions in which gravity is allowed to propagate is a modification of Newton's law at the distances comparable with R_c: the gravitational potential would fall off as $1/r^{n+1}$ for $r \lesssim R_c$. This immediately rules out the possibility of a single extra dimension, as the very existence of our solar system requires the potential to fall off as $1/r$ at the distances comparable with the size of planetary orbits. However, as of 1998, any higher number of large extra dimensions have not been ruled out by gravitational measurements, as Newton's law has not been tested to the distances smaller than about 1 mm. As amazing as it sounds, large extra dimensions with the size as macroscopic as ~1 mm were perfectly consistent with the host of experimental measurements at the time when the original idea has appeared!

Note that the solution to the hierarchy problem via large extra dimensions is not quite rigorous. In a sense, the hierarchy of energy scales is traded for the hierarchy of distance scales from the "natural" electroweak symmetry breaking range of 1 TeV^{-1} ~ 10^{-19} m to the much larger size of compact extra dimensions. Nevertheless, this hierarchy could be significantly smaller than the hierarchy of energy scales, or perhaps even be stabilized by some topological means. Strictly speaking, the idea of large extra dimensions is really a paradigm, which can be used to build more or less realistic models, rather than a model by itself.

Since large extra dimensions as a possible solution for the hierarchy problem were suggested, numerous attempts to either find them or to rule out the model have been carried out. They include measurements of gravity at sub-millimeter distances (Hoyle et al. 2001, 2004; Adelberger 2002; Chiaverini et al. 2003; Long et al. 2003), studies of various astrophysical and cosmological implications (Cullen & Perelstein 1999; Hall & Smith 1999; Barger et al. 1999; Fairbairn 2001; Hanhart et al. 2001a,b; Hannestad & Raffelt 2002; Fairbairn & Griffiths 2002), and numerous collider searches for virtual and real graviton effects (Abbott et al. 2001; Bourilkov 2001; Affolder et al. 2002, 2004; Abazov et al. 2003, 2005b; Adloff et al. 2003; Chekanov et al. 2004; LEP Exotica Working Group 2004; Abulencia et al. 2006). For a detailed review of the existing constraints and sensitivity of future experiments, the reader is referred to Hewett & Spiropulu (2002) and Landsberg (2004). The host of experimental measurements to date have largely disfavored the case of two large extra dimensions. However, for three or more extra dimensions, the lower limits on the fundamental Planck scale are only ~ 1 TeV, i.e., most of its allowed range has not been probed experimentally yet.

The LHC experiments will extend the reach for the fundamental Planck scale to 6–9 TeV in the variety of measurements probing both direct graviton emission and virtual graviton effects. For expected sensitivity, see Vacavant & Hinchliffe (2001); Kabachenko et al. (2001) and CMS Collaboration (2006).

5. The Randall-Sundrum model

A significantly different and perhaps more rigorous solution to the hierarchy problem is offered in the Randall-Sundrum (RS) model (Randall & Sundrum 1999a,b) with non-factorizable geometry and a single compact extra dimension. This is achieved by placing two 3-dimensional branes with equal and opposite tensions at the fixed points of the S_1/Z_2 orbifold in the five-dimensional anti-deSitter space-time (AdS_5). The metric of the AdS_5 space is given by $ds^2 = \exp(-2kR_c|\varphi|)\eta_{\mu\nu}dx^\mu dx^\nu - R_c^2 d\varphi^2$, where $0 \leq |\varphi| \leq \pi$ is the coordinate along the compact dimension of radius R_c, k is the curvature of the AdS_5 space, often referred to as the warp factor, x^μ are the conventional (3+1)-space-time coordinates, and $\eta^{\mu\nu}$ is the metric tensor of Minkowski space-time. The positive-tension (Planck) brane is placed at $\varphi = 0$, while the second, negative-tension standard-model brane is placed at $\varphi = \pi$. In this framework, gravity originates on the Planck brane and the graviton wave function is exponentially suppressed away from the brane along the fifth dimension due to the warp factor. Consequently, the $O(M_{Pl})$ operators generated on the Planck brane yield low-energy effects on the standard-model brane with a typical scale of $\Lambda_\pi = \overline{M}_{Pl}\exp(-\pi kR_c)$, where $\overline{M}_{Pl} \equiv M_{Pl}/\sqrt{8\pi}$ is the reduced Planck mass. The hierarchy between the Planck and electroweak scales is solved if Λ_π is ~ 1 TeV, which can be achieved with little fine tuning by requiring $kR_c \approx 12$. It has been shown (Goldberger & Wise 1999) that the size of the extra dimension can be stabilized by the presence of a bulk scalar field (the radion). Consequently, the hierarchy problem in the Randall-Sundrum model is solved naturally for $k \sim 10/R_c \sim (10^{-2}$–$10^{-1})\overline{M}_{Pl}$, since $\overline{M}_{Pl}$ is the only fundamental scale in this model and both k and $1/R_c$ are of the same order as this scale.

It is convenient to introduce a dimensionless parameter $\tilde{k} \equiv k/\overline{M}_{Pl}$, which defines the strength of coupling between the graviton and the standard model fields. Theoretically preferred value of $\tilde{k}$ is between 0.01 and 0.1. For larger coupling theory becomes non-perturbative; if the coupling is too small, an undesirably high amount of fine tuning is still required to solve the hierarchy problem.

In the simplest form of the Randall-Sundrum model only gravitons are allowed to propagate in the fifth dimension; in the extensions of the model, other particles are allowed to "live" in the bulk (Davoudiasl et al. 2000, 2001). In both cases, since gravitons are allowed to propagate in the bulk, from the point of view of a 4D-observer on the

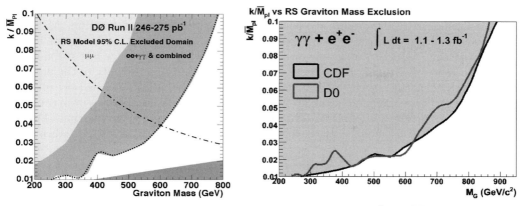

FIGURE 2. Limits on the Randall-Sundrum model parameters M_1 and $\tilde{k} = k/\overline{M}_{\rm Pl}$ from the Tevatron experiments (Abazov et al. 2005a; Abulencia et al. 2005; Aaltonen et al. 2007a,b). Left: pioneer analysis from DØ. The light-shaded area has been excluded in the dimuon channel; the medium-shaded area shows the exclusion obtained in the dielectron and diphoton channels; the dotted line corresponds to the combination of all three channels. The area below the dashed-dotted line is excluded from the precision electroweak data (see Davoudiasl et al. 2000, 2001). The dark-shaded area in the lower right-hand corner corresponds to $\Lambda_\pi > 10$ TeV, which requires undesirably high amount of fine tuning. Right: current preliminary results from the CDF and DØ experiments based on four times the original statistics.

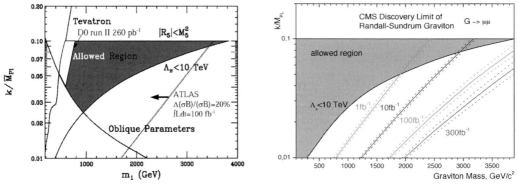

FIGURE 3. Projected 5σ discovery sensitivity for Randall-Sundrum gravitons, as a function of parameters M_1 and $\tilde{k} = k/\overline{M}_{\rm Pl}$ for the LHC experiments. Left: ATLAS sensitivity in the dielectron channel; right: CMS sensitivity in the dimuon channel; bands represent uncertainties due to systematics.

standard model brane, they acquire Kaluza-Klein modes with the masses given by the subsequent zeroes x_i of the Bessel function J_1 $(J_1(x_i) = 0)$: $M_i = kx_i e^{-\pi k R_c} = \tilde{k} x_i \Lambda_\pi$. The mass of the first excited mode is then $M_1 \approx 3.83 \tilde{k} \Lambda_\pi \sim 1$ TeV. The zeroth Kaluza-Klein mode of the graviton remains massless and couples to the standard model fields with the gravitational strength, $1/M_{\rm Pl}$, while the excited graviton modes couple with the strength of $1/\Lambda_\pi$. Coupling $\tilde{k}$ governs both the graviton width and the production cross section at colliders (both are proportional to $\tilde{k}^2$).

Indirect limits on the Randall-Sundrum model parameters come from precision electroweak data (dominated by the constraint on the S parameter; Davoudiasl et al. 2000, 2001). The only dedicated searches for Randall-Sundrum gravitons at colliders so far have been accomplished by the DØ and CDF experiments at the Fermilab Tevatron and set direct limits (Abazov et al. 2005a; Abulencia et al. 2005; Aaltonen et al. 2007a,b) on $\tilde{k}$ as a function of the mass of the first Kaluza-Klein mode of graviton, M_1. Both direct and indirect limits are summarized in Figure 2. As seen from the figure, for masses $M_1 \approx 3.83\tilde{k}\Lambda_\pi \gtrsim 850$ GeV most of the theoretically preferred range for $\tilde{k}$ is still allowed. Note also that quadrupling the statistics did little to help the sensitivity of the analysis, as it is limited by the available Tevatron energy and has essentially reached the ultimate sensitivity the Tevatron could provide. The LHC will be able to probe the entire preferred parameter range in the Randall-Sundrum model with just ~ 10 fb^{-1} in multiple channels (Allanach et al. 2000, 2002; Danheim 2007; CMS Collaboration 2006), as indicated in Figure 3.

Recently, several extensions of the simplest Randall-Sundrum model have been suggested (Davoudiasl et al. 2000, 2001; Agashe et al. 2004; Fitzpatrick et al. 2007), which attempt to solve fermion mass and flavor problems by placing fermions in the bulk. In these models, typically the top quark is the closest to the SM brane where the Higgs field is localized, thus explaining its large mass. In this scenario, the decay of graviton to bosons and light fermions is suppressed, and the dominating decay mode is $t\bar{t}$. This is a challenging channel, as for M_1 above ~ 2 TeV the top quarks from the decay are highly boosted and reconstructed as a single "fat" jet in their dominant three-jet decay mode. This is an experimentally challenging signature, requiring detailed experimental studies, which are currently under way. In this scenario, the discovery of the Kaluza-Klein excitations of the graviton may be significantly harder than in the simple scenario described above.

6. Mini black holes in large extra dimensions

As was pointed out several years ago (Argyres et al. 1998; Banks & Fischler 1999; Emparan et al. 2000), an exciting consequence of low-scale quantum gravity is the possibility of producing black holes (BH) at particle accelerators. More recently, this phenomenon has been quantified (Dimopoulos & Landsberg 2001; Giddings & Thomas 2002) for the case of particle collisions at LHC, resulting in a mesmerizing and unexpected prediction that future colliders would produce mini black holes at enormous rates (e.g., ~ 1 Hz at the LHC for $M_D = 1$ TeV), thus qualifying for black-hole factories. With the citation index of the original papers (Dimopoulos & Landsberg 2001; Giddings & Thomas 2002) well over three hundred, the production of mini black holes in the lab became one of the most actively studied and rapidly evolving subjects in the phenomenology of models with low-scale gravity.

In general relativity, black holes are well understood if their mass M_{BH} far exceeds the Planck scale. Consequently, in the model with large extra dimensions, general relativity would give an accurate description of black-hole properties when its mass is much greater than the fundamental (multi-dimensional) Planck scale $M_D \sim 1$ TeV. As its mass decreases and approaches M_D, the black hole becomes a quantum gravity object with unknown and presumably complex properties.

In this section, following Dimopoulos & Landsberg (2001), we will ignore this obstacle† and estimate the properties of light black holes by simple semiclassical arguments, strictly valid only for $M_{\mathrm{BH}} \gg M_D$. We expect this to be an adequate approximation, since the important experimental signatures rely on two simple qualitative properties: the absence

---

† Some of the properties of the "stringy" subplanckian "precursors" of black holes are discussed in Dimopoulos & Emparan (2002) and later in this review.

of small couplings and the "democratic" nature of black-hole decays, both of which may survive as average properties of the light descendants of black holes.

As we expect unknown quantum gravity effects to play an increasingly important role for the black-hole mass approaching the fundamental Planck scale, following the prescription of Dimopoulos & Landsberg (2001), we do not consider black-hole masses below the Planck scale. It is expected that the black-hole production rapidly turns on, once the relevant energy threshold $\sim M_D$ is crossed. At lower energies, we expect black-hole production to be exponentially suppressed due to the string excitations or other quantum effects.

We will first focus on the production in particle collisions and subsequent decay of small Schwarzschild black holes with the size much less than the compactification radius of extra dimensions. In this case, the standard Schwarzschild solution found for a flat $(3 + n)$-dimensional metric fully applies. The expression for the Schwarzschild radius R_S of such a black hole in $(3 + n)$ spacial dimensions is well known (Myers & Perry 1986):

$$R_S(M_{BH}) = \frac{1}{\sqrt{\pi}M_D}\left[\frac{M_{BH}}{M_D}\frac{8\Gamma\left(\frac{n+3}{2}\right)}{n+2}\right]^{\frac{1}{n+1}}. \tag{6.1}$$

Given $M_D \sim 1$ TeV and taking into account the fact that black-hole masses accessible at the next generation of particle colliders and in ultrahigh-energy cosmic-ray collisions are at most a few TeV, we note that the Schwarzschild radius of such black holes is $\sim 1/M_D$, i.e., indeed much smaller than the size of large extra dimensions even when their number approaches six or seven (the preferred number of extra dimensions expected in string theory). We also note that for $M_{BH} \sim M_D$ the Schwarzschild radius does not depend significantly on the number of extra dimensions n.

Given the current lower constraints on the fundamental Planck scale in the model with large extra dimensions of ≈ 1 TeV (Abbott et al. 2001; Bourilkov 2001; Affolder et al. 2002, 2004; Abazov et al. 2003, 2005b; Adloff et al. 2003; Chekanov et al. 2004; LEP Exotica Working Group 2004; Abulencia et al. 2006), the black holes that we may be able to study at colliders and in cosmic rays will be barely transplanckian. Hence, the unknown quantum corrections to their classical properties are expected to be large; therefore it is reasonable to focus only on the most robust properties of these mini black holes that are expected to be affected the least by unknown quantum gravity corrections. Consequently, when discussing production and decay of black holes, we do not consider the effects of spin and other black-hole quantum numbers, as well as gray-body factors. For detailed review of those the reader is referred to, e.g., Landsberg (2006).

6.1. Black-hole production in particle collisions

For two colliding partons with the center-of-mass energy $\sqrt{\hat{s}} = M_{BH}$, semiclassical reasoning suggests that if the impact parameter of the collision is less than the Schwarzschild radius R_S corresponding to this energy, a black hole with the mass M_{BH} is formed. Therefore the total cross section of black-hole production in particle collisions can be estimated from pure geometrical arguments and is of order πR_S^2.

Soon after the original calculations of Dimopoulos & Landsberg (2001) and Giddings & Thomas (2002) appeared, it was suggested in Voloshin (2001, 2002) that the geometrical cross section is in fact exponentially suppressed, based on the Gibbons-Hawking action argument (Gibbons & Hawking 1977). Subsequent detailed studies performed in simple string-theory models (Dimopoulos & Emparan 2002), using full general relativity calculations (Eardley & Giddings 2002; Yoshino & Nambu 2002, 2003), or a path-integral approach (Hsu 2003; Solodukhin 2002; Bilke et al. 2002) did not confirm this finding and

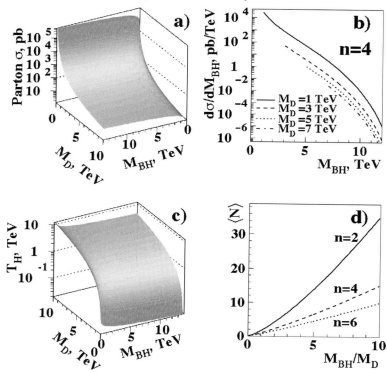

FIGURE 4. a) Parton-level and b) differential production cross section of black holes in pp collisions at the LHC. Also shown: c) Hawking temperature and d) average particle multiplicity in black-hole decays. Adapted from Dimopoulos & Landsberg (2001).

proved that the geometrical cross section is modified only by a numeric factor of order one. Further, a flaw in the Gibbons-Hawking action argument of Voloshin (2001, 2002) was found by Jevicki & Thaler (2002): the use of this action implies that the black hole has already been formed, so describing the evolution of the two colliding particles before they cross the event horizon and form the black hole via Gibbons-Hawking action is not justified. By now there is broad agreement that the production cross section is not significantly suppressed compared to a simple geometrical approximation, which we will consequently use through this talk.

Using the expression (6.1) for the Schwarzschild radius (Myers & Perry 1986), we derive the following parton-level black-hole production cross section (Dimopoulos & Landsberg 2001):

$$\sigma(M_{\mathrm{BH}}) \approx \pi R_S^2 = \frac{1}{M_D^{\,2}} \left[\frac{M_{\mathrm{BH}}}{M_D} \left(\frac{8\Gamma\left(\frac{n+3}{2}\right)}{n+2} \right) \right]^{\frac{2}{n+1}} . \qquad (6.2)$$

This cross section, as a function of the fundamental Planck scale and the black-hole mass, is shown in Figure 4a.

In order to obtain the production cross section in pp collisions at the LHC, we use the parton luminosity approach (Eichten et al. 1984; Dimopoulos & Landsberg 2001;

Giddings & Thomas 2002):

$$\frac{d\sigma(pp \to \mathrm{BH} + X)}{dM_{\mathrm{BH}}} = \frac{dL}{dM_{\mathrm{BH}}} \hat{\sigma}(ab \to \mathrm{BH}) \Big|_{\hat{s}=M_{\mathrm{BH}}^2} \quad ,$$

where the parton luminosity dL/dM_{BH} is defined as the sum over all the types of initial partons:

$$\frac{dL}{dM_{\mathrm{BH}}} = \frac{2M_{\mathrm{BH}}}{s} \sum_{a,b} \int_{M_{\mathrm{BH}}^2/s}^{1} \frac{dx_a}{x_a} f_a(x_a) f_b\left(\frac{M_{\mathrm{BH}}^2}{sx_a}\right) \quad ,$$

and $f_i(x_i)$ are the parton distribution functions (PDFs). We used the MRSD$-'$ Martin et al. (1993) PDFs with the Q^2 scale taken to be equal to M_{BH}, which is within the allowed range of these PDFs for up to the kinematic limit at the LHC. The dependence of the cross section on the choice of PDF is only ∼10%. The total production cross section for $M_{\mathrm{BH}} > M_D$ at the LHC, obtained from the above equation, ranges between 15 nb and 1 pb for the Planck scale between 1 TeV and 5 TeV, and varies by ∼10% for n between 2 and 7. The differential cross section is shown in Figure 4b.

Black holes can also be produced in fixed-target ultra-high-energy cosmic-ray collisions in the upper atmosphere, if the center-of-mass energy exceeds production threshold. The original idea was suggested in Feng & Shapere (2002) and studied by a number of authors. For a review, see, e.g., Landsberg (2006).

6.2. *Black-hole evaporation*

In general relativity, black-hole evaporation is expected to occur in three distinct stages: "balding," spin-down, and Hawking evaporation. During the first stage, the black hole loses its multipole momenta and quantum numbers via emission of gauge bosons until it reaches the Kerr solution for a spinning black hole; at the second stage it gets rid of the residual angular momentum and becomes a Schwarzschild black hole; and at the last stage it decays via emission of black-body radiation (Hawking 1975) with a characteristic Hawking temperature:

$$T_H = M_D \left(\frac{M_D}{M_{\mathrm{BH}}} \frac{n+2}{8\Gamma\left(\frac{n+3}{2}\right)} \right)^{\frac{1}{n+1}} \frac{n+1}{4\sqrt{\pi}} = \frac{n+1}{4\pi R_S} \tag{6.3}$$

of ∼1 TeV (Dimopoulos & Landsberg 2001; Giddings & Thomas 2002). The dependence of the Hawking temperature on the fundamental Planck scale and the black-hole mass is shown in Figure 4c.

Note that if a certain quantum number (e.g., $B - L$) is gauged, it will be conserved in the process of black-hole evaporation. Since the majority of black holes at the LHC are produced in quark-quark collisions, one would expect many of them to have the baryon number and fractional electric charge. Consequently, the details of black-hole evaporation process will allow us to determine if these quantum numbers are truly conserved.

In quantum gravity it is expected that there is a fourth, Planckian stage of black-hole evaporation, which is reached when the mass of the evaporating black hole approaches the Planck scale. The details of the Planckian stage are completely unknown, as they are governed by the effects of quantum gravity, which should be dominant at such low black-hole masses. Some authors speculate that the Planckian stage terminates with a formation of a stable or semi-stable black-hole remnant with the mass $\sim M_{\mathrm{Pl}}$. Others argue that the evaporation proceeds until the entire mass of the black hole is radiated. The truth is that no unambiguous predictions about the Planckian regime are possible, given our lack of knowledge of quantum gravity.

The average multiplicity of particles produced in the process of black-hole evaporation is given by:

$$\langle N \rangle = \left\langle \frac{M_{\text{BH}}}{E} \right\rangle \quad,$$

where E is the energy spectrum of the decay products. In order to find $\langle N \rangle$, we note that evaporation is a black-body radiation process, with the energy flux per unit of time given by Planck's formula:

$$\frac{df}{dx} \sim \frac{x^3}{e^x + c} \quad, \tag{6.4}$$

where $x \equiv E/T_H$, and c is a constant, which depends on the quantum statistics of the decay products ($c = -1$ for bosons, $+1$ for fermions, and 0 for Boltzmann statistics).

The spectrum of the decay products in the massless particle approximation is given by:

$$\frac{dN}{dE} \sim \frac{1}{E}\frac{df}{dE} \sim \frac{x^2}{e^x + c} \quad.$$

For averaging the multiplicity, we use the average of the distribution in the inverse particle energy:

$$\left\langle \frac{1}{E} \right\rangle = \frac{1}{T_H} \frac{\int_0^\infty dx \frac{1}{x}\frac{x^2}{e^x+c}}{\int_0^\infty dx \frac{x^2}{e^x+c}} = a/T_H \quad, \tag{6.5}$$

where a is a dimensionless constant that depends on the type of produced particles and numerically equals 0.68 for bosons, 0.46 for fermions, and $1/2$ for Boltzmann statistics. Since a mixture of fermions and bosons is produced in the black-hole decay, we can approximate the average by using Boltzmann statistics, which gives the following formula for the average multiplicity: $\langle N \rangle \approx M_{\text{BH}}/2T_H$. Using expression (6.3) for Hawking temperature, we obtain:

$$\langle N \rangle = \frac{2\sqrt{\pi}}{n+1}\left(\frac{M_{\text{BH}}}{M_D}\right)^{\frac{n+2}{n+1}}\left(\frac{8\Gamma\left(\frac{n+3}{2}\right)}{n+2}\right)^{\frac{1}{n+1}} \quad, \tag{6.6}$$

which corresponds to production of about a half-a-dozen particles for typical black-hole masses accessible at the LHC, see Figure 4d.

Naïvely, one would expect that a large fraction of Hawking radiation is emitted in the form of gravitons, escaping in the bulk space. However, as was shown in Emparan et al. (2000), this is not the case, since the wavelength $\lambda = 2\pi/T_H$ corresponding to the Hawking temperature is larger than the size of the black hole. Therefore, the black hole acts as a point-radiator and consequently emits mostly s-waves. Since the s-wave emission is sensitive only to the radial coordinate, bulk radiation per graviton degree of freedom is the same as radiation of any standard model degree of freedom on the brane. While many angular degrees of freedom are available in the bulk space, the s-wave emission cannot take advantage of them, thus suppressing bulk graviton component. Since there are many more particles on the brane than in the bulk space, this has the crucial consequence that the black hole mainly decays to visible standard model particles.

Since the gravitational coupling is flavor-blind, a black hole emits all the ≈ 120 standard model particle and antiparticle degrees of freedom with roughly equal probability. Accounting for the color and spin and ignoring the graviton emission, we expect $\approx 75\%$ of particles produced in black-hole decays to be quarks and gluons, $\approx 10\%$ charged leptons, $\approx 5\%$ neutrinos, and $\approx 5\%$ photons or W/Z bosons, each carrying hundreds of GeV of energy. Similarly, if new particles exist with masses $\lesssim 100$ GeV, they would be produced

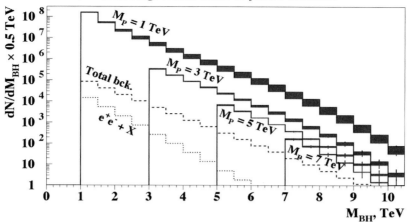

FIGURE 5. Number of black holes produced at the LHC in the electron- or photon-decay channels (assuming 15% efficiency for an electron or a photon tag), with 100 fb^{-1} of integrated luminosity, as a function of the black-hole mass. The shaded regions correspond to the variation in the number of events for n between 2 and 7. The dashed line shows the total standard model background (from inclusive $Z(ee)$ and direct photon production). The dotted line corresponds to the $Z(ee) + X$ background alone. Adapted from Dimopoulos & Landsberg (2001).

in the decays of black holes with the probability similar to that for the standard model species. For example, a sufficiently light Higgs boson is expected to be emitted with $\sim$1% probability. This has exciting consequences for searches for new physics at the LHC and beyond, as the production cross section for any new particle via this mechanism is large and depends only weakly on the particle mass, in contrast with an exponential dependence characteristic of direct production.

A relatively large fraction of prompt and energetic photons, electrons, and muons expected in the high-multiplicity black-hole decays would make it possible to select pure samples of black holes, which are also easy to trigger (Dimopoulos & Landsberg 2001; Giddings & Thomas 2002). At the same time, only a small fraction of particles produced in the black-hole decays are undetectable gravitons and neutrinos, so most of the black-hole mass is radiated in the form of visible energy, making it easy to detect.

It has been argued (Anchordoqui & Goldberg 2003) that the fragmentation of quarks and jets emitted in the black-hole evaporation might be significantly altered by the presence of a dense and hot QCD plasma ("chromosphere") around the event horizon. If this argument is correct, one would expect much a softer hadronic component in the black-hole events. However, we would like to point out that one would still have a significant number of energetic jets due to the decay of weakly interacting W/Z and Higgs bosons, as well as tau leptons, emitted in the process of black-hole evaporation and penetrating the chromosphere before decaying into jetty final states. In any case, tagging of the black-hole events by the presence of an energetic lepton or a photon and large total energy deposited in the detector is a fairly model-independent approach.

In Figure 5 we show the number of black-hole events tagged by the presence of an energetic electron or photon among the decay products in 100 fb^{-1} of data collected at the LHC, along with estimated backgrounds, as a function of the black-hole mass (Dimopoulos & Landsberg 2001). It is clear that very clean and large samples of black holes can be produced at the LHC up to the Planck scale of $\sim$5 TeV. Note that the black-hole discovery potential at the LHC is maximized in the $e/\mu + X$ channels, where background is much smaller than that in the $\gamma + X$ channel (see Figure 5). The reach of a simple counting experiment extends up to $M_D \approx 9$ TeV ($n = 2$–7), for which one would expect to see a handful of black-hole events with negligible background.

The lifetime of a black hole can be estimated using Stefan's law of thermal radiation. Since black-hole evaporation occurs primarily in three spatial dimensions, the canonical 3-dimensional Stefan's law applies, and therefore the power dissipated by the Hawking radiation per unit area of the event horizon is $p = \sigma T_H^4$, where σ is the Stefan-Boltzmann constant and T_H is the Hawking temperature. Since the effective evaporation area of a black hole is the area of a 3D-sphere with radius R_S and Stefan's constant in natural units is $\sigma = \pi^2/60 \sim 1$, dropping numeric factors of order unity we obtain the following expression for the total power dissipated by a black hole: $P \sim R_S^2 T_H^4 \sim R_S^{-2}$.

The black-hole lifetime τ then can be estimated as: $\tau \sim M_{\mathrm{BH}}/P \sim M_{\mathrm{BH}} R_S^2$, and using (6.1) we find:

$$\tau \sim \frac{1}{M_D} \left(\frac{M_{\mathrm{BH}}}{M_D} \right)^{\frac{n+3}{n+1}} . \tag{6.7}$$

Therefore, a typical lifetime of a mini black hole is $\sim 10^{-27}$–10^{-26} s. A multi-TeV black hole would have a relatively narrow width ~ 100 GeV, i.e., similar to, e.g., a W' or Z' resonance of a similar mass. This is not surprising, as the strength of gravity governing the black-hole evaporation rate is similar in the model with large extra dimensions to that of electroweak force responsible for the W' or Z' decay rates.

While the detection of Hawking radiation from rapidly evaporating TeV-scale black holes remains one of the most clear and well-studied signatures at the LHC, several authors have studied complementary signatures as well. Among those most notable are studies of the suppression of dijet production at the invariant masses exceeding the fundamental Planck scale (Banks & Fischler 1999; Lonnblad et al. 2005; Stöcker 2006), anomalous inclusive hadron production due to products of black-hole decays (Mocioiu et al. 2003), and production of black holes in heavy ion collisions at the LHC (Chamblin & Novak 2002; Chamblin et al. 2004a).

Recently, there have been attempts to estimate the decrease of the black-hole production cross section due to inelasticity in parton collisions. The reason for the inelasticity is the emission of gravitational waves during the formation of black holes, some of which may not be trapped within the event horizon and escape. For discussion of this effect, see Anchordoqui et al. (2004).

6.3. *Testing Hawking radiation in decays of black holes*

A sensitive test of properties of Hawking radiation can be performed by measuring the relationship between the mass of the black hole (reconstructed from the total energy of all the visible decay products) and its Hawking temperature (measured from the energy spectrum of the emitted particles). One can use the measured M_{BH} vs. T_H dependence to determine both the fundamental Planck scale M_D and the dimensionality of space n. This is a multidimensional equivalent of the Wien's law of thermal radiation. It is particularly interesting that the dimensionality of extra space can be determined in a largely model-independent way via taking a logarithm of both parts of (6.3) for Hawking temperature:

$$\log_{10}(T_H/1 \text{ TeV}) = -\frac{1}{n+1} \log_{10}(M_{\mathrm{BH}}/1 \text{ TeV}) + \text{const} ,$$

where the constant does not depend on the black-hole mass, but only on the M_D and on detailed properties of the bulk space, such as relative size of extra dimensions (Dimopoulos & Landsberg 2001). Therefore, the slope of a straight-line fit to the $\log_{10}(T_H/1 \text{ TeV})$ vs. $\log_{10}(M_{\mathrm{BH}}/1 \text{ TeV})$ data offers a direct way of determining the dimensionality of space.

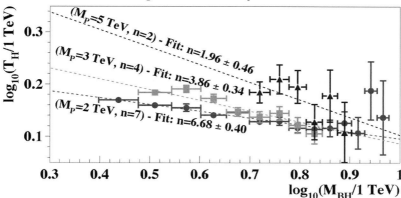

FIGURE 6. Determination of the dimensionality of space via Wien's displacement law at the LHC with 100 fb^{-1} of data. Adapted from Dimopoulos & Landsberg (2001).

The reach of this method at the LHC is illustrated in Figure 6 and discussed in detail in Dimopoulos & Landsberg (2001). Note that the determination of the dimensionality of space by this method is fundamentally different from other ways of determining n, e.g., by studying a monojet signature or a virtual graviton exchange processes, also predicted in the models with large extra dimensions. The latter always depend on the volume of extra space, and therefore cannot provide a direct way of measuring n without making assumptions about the relative size of large extra dimensions. The former depends only on the area of the event horizon of a black hole, which is not sensitive to the size of large extra dimensions.

An interesting possibility studied in Dimopoulos & Emparan (2002) is production of a precursor of a black hole, i.e., a long and jagged, highly exited string state, dubbed as a "string ball" due to its folding in a ball-like object via a random walk. As shown in Dimopoulos & Emparan (2002), there are three characteristic string ball production regimes, which depend on the mass of the string ball M, the string scale $M_S < M_D$, and the string coupling $g_s < 1$. For $M_S < M < M_S/g_s$, the production cross section increases $\propto M^2$, until it reaches saturation at $M \sim M_S/g_s$ and stays the same up to the string ball mass $\sim M_S/g_s^2$, at which point a black hole is formed and the production cross section agrees with that from Dimopoulos & Landsberg (2001).

A string ball has properties similar to those of a black hole, except that its evaporation temperature, known as Hagedorn temperature (Hagedorn 1965), is constant: $T_S = M_S/(2\sqrt{2}\pi)$. Thus, the correlation between the temperature of the characteristic spectrum and the string-ball mass may reveal the transition from the Hagedorn to Hawking regime, which can be used to estimate M_S and g_s. Another possibility is a production of higher-dimensional objects, e.g., black p-branes, rather than spherically symmetric black holes ($p = 0$) (Ahn et al. 2003; Jain et al. 2003; Anchordoqui et al. 2002a). For a detailed review see, e.g., Cheung (2002).

6.4. *Simulation of black-hole production and decay*

In order to study properties of black holes at colliders, it is important to have tools capable of simulating their production and decay. Currently, there are three Monte Carlo generators capable of doing this: TRUENOIR (Landsberg 2001), CHARYBDIS (Harris et al. 2003), and CATFISH (Cavaglià 2006).

TRUENOIR (Landsberg 2001) is a plug-in module for PYTHIA (Sjöstrand et al. 2001) Monte Carlo package. It simulates the production and decay of black holes as described

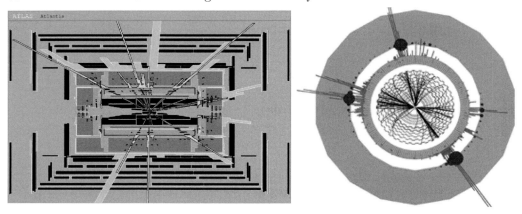

FIGURE 7. Left: a typical black-hole event, generated with CHARYBDIS, as seen by the ATLAS detector (courtesy of ATLAS-Japan group). Right: a typical black-hole event, generated with TRUENOIR, as seen by the CMS detector (courtesy of Albert De Roeck).

in Dimopoulos & Landsberg (2001), assuming conservation of the individual baryon and lepton numbers, as well as the electric and color charges. It does not include any gray-body factors for the black-hole decay and assumes that the evaporation takes place in a single, Schwarzschild phase, thus ignoring any spin effects. While the deficiency of these assumptions is obvious, the logic behind this simple approach is that, given the uncertainties related to unknown quantum corrections, an incremental improvement from taking into account classically calculated corrections is minor.

CHARYBDIS (Harris et al. 2003) is interfaced with the HERWIG (Corcella et al. 2001) Monte Carlo package. The authors of CHARYBDIS incorporated classically calculated gray-body factors and also gave the user flexibility of defining when the Schwarzschild stage terminates and a stable Planckian remnant of the black hole is formed.

Both generators have been successfully interfaced with the LHC detector simulation packages. Figure 7 shows simulations of a black-hole event in the ATLAS and CMS detectors, using the CHARYBDIS and TRUENOIR Monte Carlo packages, respectively. A number of dedicated studies (Akchurin et al. 2004; Harris et al. 2005; Tanaka et al. 2005; Koch et al. 2005) of the black-hole decay signatures at the LHC have been performed using these Monte Carlo generators.

The new event generator CATFISH (Cavaglià 2006) includes additional effects, such as back-reaction, inelasticity in black-hole formation, and superradiance. Note that all these improvements are entirely classically calculated, which makes their use very limited, as most of black holes will be produced with the masses close to the threshold, where quantum effects are expected to be overwhelming.

6.5. *Discovering new particles in black-hole decays*

As was mentioned earlier, new particles with the mass ∼100 GeV would be produced in the process of black-hole evaporation with a relatively large probability: ∼1% times the number of their quantum degrees of freedom. Consequently, it may be advantageous to look for new particles among the decay products of black holes in large samples accessible at the LHC and other future colliders.

As an example, following Landsberg (2002), we demonstrate the discovery potential of a black-hole sample to be collected at the LHC for a light Higgs boson. We pick the Higgs boson mass of 130 GeV, which is still allowed in low-scale supersymmetry models, but makes it quite hard to discover Higgs via direct means either at the Fermilab Tevatron

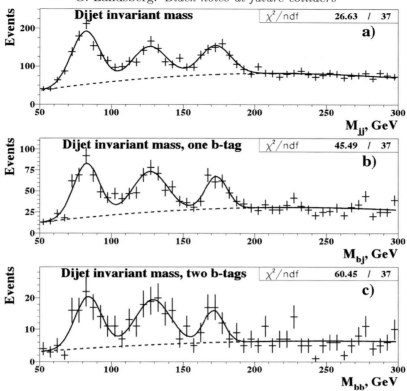

FIGURE 8. Dijet invariant mass observed in black-hole decays with a prompt lepton or photon tag in ≈ 3 pb^{-1} of the LHC data, for $M_D = 1$ TeV and $n = 3$: (a) all jet combinations; (b) jet combinations with at least one of the jets tagged as a b-jet; (c) jet combinations with both jets tagged as b-jets. The solid line is a fit to a sum of three Gaussians and a polynomial background (shown with the dashed line). The three peaks correspond to the W/Z bosons, the Higgs boson, and the top quark (see text). From Landsberg (2002).

(Carena et al. 2000) or at the LHC. We consider the decay of the Higgs boson into a pair of jets (with the branching fraction of 67%), dominated by the $b\bar{b}$ final state (57%), with an additional 10% contribution from the $c\bar{c}$, gg, and hadronic $\tau\tau$ final states.

We model the production and decay of the black holes with the TRUENOIR Monte Carlo generator (Landsberg 2001). We used a 1% probability to emit the Higgs particle in the black-hole decay. We reconstruct final-state particles within the acceptance of a typical LHC detector and smear their energies with the expected resolutions. We select the black-hole events by requiring an energetic electron, photon, or a muon, as well as total multiplicity of energetic objects (jets, electrons, photons, or muons) of at least four.

The simplest way to look for the Higgs boson in the black-hole decays is to use the dijet invariant mass spectrum for all possible combinations of jets found among the final-state products in the above sample. This spectrum is shown in Figure 8 for $M_D = 1$ TeV and $n = 3$. The three panes correspond to all jet combinations (with the average of approximately four jet combinations per event), combinations with at least one b-tagged jet, and combinations with both jets b-tagged. (We used typical tagging efficiency and mistag probabilities of an LHC detector to simulate b-tags.)

The most prominent feature in all three plots is the presence of three peaks with the masses around 85, 130, and 175 GeV. The first peak is due to the hadronic decays of the W and Z bosons produced in the black-hole decay either directly or in top-quark decays. (The resolution of a typical LHC detector does not allow to resolve the W and Z in the dijet decay mode.) The second peak is due to the $h \to jj$ decays, and the third peak is due to the $t \to Wb \to jjb$ decays, where the top quark is highly boosted. In this case, one of the jets from the W decay sometimes overlaps with the prompt b-jet from the top-quark decay, and thus the two are reconstructed as a single jet; when combined with the second jet from the W decay, this gives a dijet invariant mass peak at the top quark mass. The data set shown in Figure 8 corresponds to 50K black-hole events, which, given the 15 nb production cross section for $M_D = 1$ TeV and $n = 3$, is equivalent to the integrated luminosity of 3 pb^{-1}, or less than an hour of the LHC operation at the nominal instantaneous luminosity. The significance of the Higgs signal shown in Figure 8a is 6.7σ, even without b-tagging involved.

With this method, a 5σ discovery of the 130 GeV Higgs boson may be possible with $\mathcal{L} \approx 2$ pb^{-1} (first day), 100 pb^{-1} (first week), 1 fb^{-1} (first month), 10 fb^{-1} (first year), and 100 fb^{-1} (one year at the nominal luminosity) for the fundamental Planck scale of 1, 2, 3, 4, and 5 TeV, respectively, even with an incomplete and poorly calibrated detector. If the Planck scale is below $\lesssim 4$ TeV, the integrated luminosity required is significantly lower than that for the Higgs discovery in direct production.

While this study was done for a particular value of the Higgs boson mass, the dependence of the new approach on the Higgs mass is small. Moreover, this approach is applicable to searches for other new particles with the masses ~ 100 GeV, e.g., low-scale supersymmetry (Chamblin et al. 2004b; Nayak & Smith 2006). Light slepton or top squark searches via this technique may be particularly fruitful. Very similar conclusions apply not only to black holes, but to intermediate quantum states, such as string balls (Dimopoulos & Emparan 2002), which have similar production cross section and decay modes as black holes. In this case, the relevant mass scale is not the Planck scale, but the string scale, which determines the Hagedorn evaporation temperature.

Large samples of black holes accessible at the LHC can be used even to study some of the properties of known particles, see, e.g., Uehara (2002a,b).

7. Randall-Sundrum black holes

Mini black holes can be also produced in TeV particle collisions in the Randall-Sundrum model (Anchordoqui et al. 2002b; Rizzo 2005a,b,c, 2006a,b). In this case, the warp-factor-suppressed Planck scale, $\Lambda_\pi \sim 1$ TeV, plays the role of the fundamental Planck scale in the model with large extra dimensions.

The event horizon of the Randall-Sundrum black holes has a pancake shape with the radius in the fifth dimension suppressed compared to the radius R_S on the standard model brane by the warp factor $e^{-\pi k R_c}$. Thus, for $R_S e^{-\pi k R_c} \ll \pi R_c$, the black hole can be considered "small" and has properties similar to that in the $n = 1$ (5D) large extra dimensions scenario, if the effects of the curvature of the AdS space at the standard model brane are ignored.

In order to derive properties of the Randall-Sundrum black holes, it is convenient to introduce the fundamental 5D Planck scale M, which enters the Lagrangian of the Randall-Sundrum model. The relationship between the reduced 4-dimensional Planck scale $\overline{M}_{\text{Pl}}$ and M is as follows:

$$\overline{M}_{\text{Pl}}^2 = \frac{M^3}{k}(1 - e^{-2\pi k R_c}) \approx M^3/k \quad .$$

Since k is 0.01–0.1 $\times \overline{M}_{\text{Pl}}$, $M = 0.2$–$0.5\overline{M}_{\text{Pl}}$, i.e., both the 5D and 4D Planck scales are of the same order. Since the curvature of the slice of the AdS space is given by $k^2/M^2 \sim$

$\tilde{k}^2 \ll 1$ (Rizzo 2005a,b,c, 2006a,b), one can indeed ignore higher-order curvature effects and consider Randall-Sundrum black holes as if they were black holes in flat Minkowski space.

The Schwarzschild radius of a black hole of mass M_{BH} is given by Argyres et al. (1998); Myers & Perry (1986); Rizzo (2005a,b, 2006b)†:

$$R_S = \frac{1}{\pi M e^{-\pi k R_c}} \sqrt{\frac{M_{\mathrm{BH}}}{3 M e^{-\pi k R_c}}} \quad .$$

Taking into account $M^3 \approx k \overline{M}_{\mathrm{Pl}}^2 = \Lambda_\pi^2 k e^{2\pi k R_c}$, we get:

$$R_S \approx \frac{1}{\sqrt{3}\pi\Lambda_\pi} \sqrt{\frac{M_{\mathrm{BH}}}{\tilde{k}\Lambda_\pi}} \quad . \tag{7.1}$$

Since the expression under the square root is ~ 10 for a typical range of $M_{\mathrm{BH}}/\Lambda_\pi = O(1)$ and $k/\overline{M}_{\mathrm{Pl}} = O(0.01)$, we find that a typical Schwarzschild radius of the Randall-Sundrum black hole is $R_S \sim 1/\Lambda_\pi \sim 1 \text{ TeV}^{-1}$, similar to that for the black holes in models with large extra dimensions. Indeed, using (6.1) for the ADD model with $n = 1$, we get:

$$R_S(\mathrm{ADD}, 5\mathrm{D}) = \frac{1}{\sqrt{\pi}M_D} \sqrt{\frac{8M_{\mathrm{BH}}}{3M_D}} \quad ,$$

which turns into (7.1) for $\Lambda_\pi = M_D$ and $\tilde{k} = 1/8\pi \approx 0.04$.

Moreover, it is easy to see that such a black hole is still small from the point of view of the 5th dimension, as the condition of the black-hole "smallness" mentioned above can be expressed as:

$$R_S \ll \frac{\pi R_c}{\exp(-\pi k R_c)} = \frac{k\pi R_c}{\frac{k}{\overline{M}_{\mathrm{Pl}}} \overline{M}_{\mathrm{Pl}} \exp(-\pi k R_c)} = \frac{k\pi R_c}{\Lambda_\pi \tilde{k}} \sim \frac{36}{\Lambda_\pi \tilde{k}} \quad .$$

Given that $\tilde{k}$ is between 0.01 and 0.1, the inequality becomes:

$$R_S \ll \frac{360\text{--}3600}{\Lambda_\pi} \quad ,$$

which is clearly satisfied for $R_S \sim 1/\Lambda_\pi$. In fact, one would need to produce a black hole with the mass $\sim 10^6$ TeV to exceed this limit. Such energy is achievable neither at any foreseen collider nor in fixed-target collisions of ultra-high-energy particles from cosmic accelerators.

The Hawking temperature of the Randall-Sundrum black hole can be found from expression (6.3) for a black hole in models with large extra dimensions by requiring $n = 1$, i.e.,

$$T_H = \frac{1}{2\pi R_S} \quad , \tag{7.2}$$

which, given $R_S \sim 1/\Lambda_\pi$, makes it very similar to that for the case of large extra dimensions. Consequently, for the preferred range of model parameters, both the production cross section and the decay properties of a Randall-Sundrum black hole are very similar to those in models with large extra dimensions. In fact, all the Monte Carlo generators

---

† Note that this expression differs from the analogous expression (9) in Anchordoqui et al. (2002b) by a $\sqrt{2}$ factor; the difference stems from the fact that the mass parameter M used in Anchordoqui et al. (2002b) is different from the true 5D Planck scale in the Lagrangian of the Randall-Sundrum model, which we refer to as M in this talk.

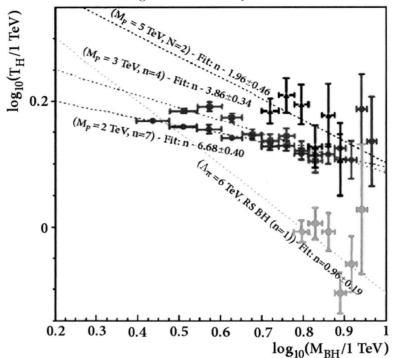

FIGURE 9. Comparison of the Randall-Sundrum ($n = 1$) and ADD black holes via Wien's displacement law at the LHC with 100 fb^{-1} of data. $\Lambda_\pi = 6$ TeV and $\tilde{k} = 1/8\pi$ have been assumed for the Randall-Sundrum case.

reviewed above can be used to simulate Randall-Sundrum black holes by setting $n = 1$ and

$$M_D = \Lambda_\pi \sqrt[3]{8\pi\tilde{k}} \approx 0.765 M_1 \tilde{k}^{-2/3} \ .$$

Note that since the Randall-Sundrum black holes look like the black holes in the case of large extra dimensions with $n = 1$, one could infer this fact by analyzing the correlation between the Hawking temperature and black-hole mass, as it was done for the ADD black holes (see Section 6.3). The result of such a test is shown in Figure 9, which compares the ADD and Randall-Sundrum cases. The slope of the straight-line fit in this double-logarithmic plot can be used to distinguish Randall-Sundrum black holes from their ADD counterparts.

Figure 10 shows the number of black holes produced in 100 fb^{-1} of LHC data and tagged via a presence of an energetic electron or photon among the final-state particles, as a function of Randall-Sundrum model parameters. Here we assume that the production threshold for the black hole is $M_{\rm BH} > \Lambda_\pi$. As seen from this plot, one would expect a few black holes to be produced even at the boundaries of the preferred parameter space of the Randall-Sundrum model.

Note that $M_{\rm BH} > \Lambda_\pi$ condition may be too optimistic. Another relevant parameter in the Randall-Sundrum model is the five-dimensional Planck scale reduced by the warp factor: $\tilde{M} = M \exp(-\pi k R_c)$. The value of $\tilde{M}$ is numerically similar to Λ_π but differs by the $\tilde{k}^{-1/3}$ factor:

$$\tilde{M} \approx \Lambda_\pi \left(\frac{k}{\overline{M}_{\rm Pl}} \right)^{-\frac{1}{3}} \ .$$

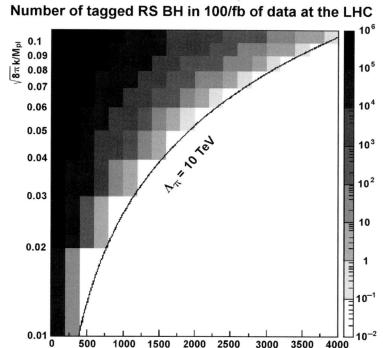

Number of tagged RS BH in 100/fb of data at the LHC

FIGURE 10. The number of Randall-Sundrum black holes produced at the LHC and tagged via a prompt electron or photon in the final state in 100 fb^{-1} of data. Production threshold is assumed to be $M_{\mathrm{BH}} > \Lambda_\pi$.

For the $\tilde{k}$ range between 0.01 and 0.1, this factor ranges between 2 and 5. Consequently, if one assumes that the black-hole production threshold is, in fact, $\tilde{M}$ and not Λ_π, production rate will be significantly lower and it may not be possible to observe Randall-Sundrum black holes at the LHC with $\sim$100 fb^{-1} of data (Meade & Randall 2008). If this is the case, the existence of warped extra dimensions is likely to be established first by looking for Kaluza-Klein modes of the graviton. The proposed LHC upgrade—SuperLHC, with 10 times the luminosity—may have a better shot at black holes in the Randall-Sundrum model if the production threshold significantly exceeds Λ_π.

Stojkovic (2005) has suggested that black holes in Randall-Sundrum model and in large extra dimensions can nevertheless be distinguished by the different dynamics of an early stage of black-hole evaporation due to the fact that the angular momentum of a Randall-Sundrum black hole, unlike that for a black hole in large extra dimensions, cannot have any bulk component (due to the existence of a discrete Z_2 orbifold symmetry). Thus, bulk evaporation for a Randall-Sundrum black hole is suppressed, compared to that for a black hole of similar mass in models with large extra dimensions. Since it is argued that the bulk component of Hawking radiation of gravitons for a black hole in large extra dimensions may be significant during the early stages of its evaporation, it is suggested that the black-hole evaporation may result in less missing energy in the Randall-Sundrum scenario. Of course, there are other ways of distinguishing the two cases, particularly the excitation of narrow graviton TeV-scale resonances in Drell-Yan and diboson production in the Randall-Sundrum case (Davoudiasl et al. 2000, 2001), versus an overall enhancement of the high end of the Drell-Yan and diboson mass spectra in the case of large extra dimensions.

Recently, there have been studies of the modification of black-hole properties due to Gauss-Bonnet or Lovelock terms added to the Einstein-Hilbert action. These modifications affect properties of black holes in models with either large or warped extra dimension(s). Additional terms could naturally introduce a minimum threshold on the black-hole mass. For detailed studies of modifications related to these higher-order terms, see Rizzo (2005a,b,c, 2006a,b).

8. Conclusions

To conclude, black-hole production at the LHC and in cosmic rays may be one of the early signatures of TeV-scale quantum gravity in the models with large or warped extra dimensions. It has three advantages:

(*a*) Generally large cross section, as no small dimensionless coupling constants, analogous to α, suppress the production of black holes; this leads to potentially enormous rates.

(*b*) Hard, prompt, charged leptons and photons, as thermal decays are flavor-blind; this signature has practically vanishing standard model background.

(*c*) Little missing energy, as most of the black-hole evaporation products are detectable; this facilitates the determination of the mass and the temperature of the black hole, and may lead to a test of Hawking radiation.

If low-scale gravity is realized in nature, the production and detailed studies of black holes in the lab are just a few years away. Large samples of black holes accessible by the LHC may allow for precision determination of the parameters of bulk space, and may even result in the discovery of new particles in black-hole evaporation. Limited samples of black-hole events may be observed in ultra-high-energy cosmic-ray experiments, even before the LHC turns on.

A discovery of mini black holes in particle collisions would mark an exciting transition for astroparticle physics: its true unification with cosmology—the "Grand Unification" to strive for.

I would like to thank the organizers of the ST ScI 2007 Spring Symposium on Black Holes for their kind invitation and for an excellent meeting. I am indebted to my coauthor on the original black-hole paper, Savas Dimopoulos, for a number of stimulating discussions and support. Many thanks to Lisa Randall and Tom Rizzo for several enlightening discussions of the properties of Randall-Sundrum black holes. I would also like to thank Lisa for sharing her paper in preparation with me. This work has been partially supported by the U.S. Department of Energy under Grant No. DE-FG02-91ER40688 and by the National Science Foundation under the CAREER Award PHY-0239367.

REFERENCES

AALTONEN, T., ET AL. (CDF COLLABORATION) 2007a *Phys. Rev. Lett.* **99**, 171801.
AALTONEN, T., ET AL. (CDF COLLABORATION) 2007b *Phys. Rev. Lett.* **99**, 171802.
ABAZOV, V. M., ET AL. (DØ COLLABORATION) 2003 *Phys. Rev. Lett.* **90**, 251802.
ABAZOV, V. M., ET AL. (DØ COLLABORATION) 2005a *Phys. Rev. Lett.* **95**, 091801.
ABAZOV, V. M., ET AL. (DØ COLLABORATION) 2005b *Phys. Rev. Lett.* **95**, 161602.
ABBOTT, B., ET AL. (DØ COLLABORATION) 2001 *Phys. Rev. Lett.* **86**, 1156.
ABULENCIA, A., ET AL. (CDF COLLABORATION) 2005 *Phys. Rev. Lett.* **95**, 252001.
ABULENCIA, A., ET AL. (CDF COLLABORATION) 2006 *Phys. Rev. Lett.* **97**, 171802.

ADELBERGER, E. G. (EÖT-WASH COLLABORATION) 2002. In *Proceedings of the Second Meeting on CPT and Lorentz Symmetry* (ed. V. Alan Kostelecky). p. 9. World Scientific.

ADLOFF, C., ET AL. (H1 COLLABORATION) 2003 *Phys. Lett. B* **568**, 35.

AFFOLDER, T., ET AL. (CDF COLLABORATION) 2002 *Phys. Rev. Lett.* **89**, 281801.

AFFOLDER, T., ET AL. (CDF COLLABORATION) 2004 *Phys. Rev. Lett.* **92**, 121802.

AGASHE, K., ET AL. 2003 *J. High Energy Phys.* **08**, 050.

AHN, E. J., CAVAGLIA, A., & OLINTO, A. V. 2003 *Phys. Lett. B* **551**, 1.

AKCHURIN, N., ET AL. 2004 *CMS Internal Note* CMS IN–2004/027.

ALLANACH, B. C., ET AL. 2000 *J. High Energy Phys.* **09**, 019.

ALLANACH, B. C., ET AL. 2002 *J. High Energy Phys.* **12**, 039.

ANCHORDOQUI, L. A., FENG, J. L., & GOLDBERG, H. 2002a *Phys. Lett. B* **535**, 302.

ANCHORDOQUI, L. A., GOLDBERG, H., & SHAPERE, A. D. 2002b *Phys. Rev. D* **66**, 024033.

ANCHORDOQUI, L. A., ET AL. 2004 *Phys. Lett. B* **594**, 363.

ANCHORDOQUI, L. & GOLDBERG, H. 2003 *Phys. Rev. D* **67**, 064010.

ANTONIADIS, I., ET AL. 1998 *Phys. Lett. B* **436**, 257.

ARGYRES, P. C., DIMOPOULOS, S., & MARCH-RUSSELL, J. 1998 *Phys. Lett. B* **441**, 96.

ARKANI-HAMED, N., DIMOPOULOS, S., & DVALI, G. 1998 *Phys. Lett. B* **429**, 263.

ARKANI-HAMED, N., DIMOPOULOS, S., & DVALI, G. 1999 *Phys. Rev. D* **59**, 086004.

BANKS, T. & FISCHLER, W. 1999; arXiv:hep-th/9906038v1.

BARGER, V. D., ET AL. 1999 *Phys. Lett. B* **461**, 34.

BILKE, S., LIPARTIA, E., & MAUL, M. 2002; arXiv:hep-ph/0204040v2.

BOURILKOV, D. 2001. In *Proc. 15th Les Rencontres de Physique de la Vallée d'Aoste: Results and Perspective in Particle Physics* (ed. Mario Greco). Frascati Physics Series, Vol. 22.

CARENA, M., ET AL. 2000; arXiv:hep-ph/0010338v2.

CAVAGLIÁ, M. 2006; http://www.phy.olemiss.edu/GR/catfish/.

CHAMBLIN, A., COOPER, F., & NAYAK, G. C. 2004a *Phys. Rev. D* **69**, 065010.

CHAMBLIN, A., COOPER, F., & NAYAK, G. C. 2004b *Phys. Rev. D* **70**, 075018.

CHAMBLIN, A. & NAYAK, G. C. 2002 *Phys. Rev. D* **66**, 091901.

CHEKANOV, S., ET AL. (ZEUS COLLABORATION) 2004 *Phys. Lett. B* **591**, 23.

CHEUNG, K. 2002 *Phys. Rev. D* **66**, 036007.

CHIAVERINI, J., ET AL. 2003 *Phys. Rev. Lett.* **90**, 151101.

CMS COLLABORATION 2006 *CMS Physics TDR 8.2 Volume II: Physics Performance*, CERN/LHCC 2006-021. CERN.

CORCELLA, G., ET AL. 2001 *J. High Energy Phys.* **01**, 010.

CULLEN, S. & PERELSTEIN, M. 1999 *Phys. Rev. Lett.* **83**, 268.

DANHEIM, D. 2007. In *SUSY06: the 14th International Conference on Supersymmetry and the Unification of Fundamental Interactions* (ed. J. L. Feng). AIP Conf. Proc. 903. AIP.

DAVOUDIASL, H., HEWETT, J. L., & RIZZO, T. G. 2000 *Phys. Rev. Lett.* **84**, 2080.

DAVOUDIASL, H., HEWETT, J. L., & RIZZO, T. G. 2001 *Phys. Rev. D* **63**, 075004.

DIMOPOULOS, S. & EMPARAN, R. 2002 *Phys. Lett. B* **526**, 393.

DIMOPOULOS, S. & LANDSBERG, G. 2001 *Phys. Rev. Lett.* **87**, 161602.

EARDLEY, D. M. & GIDDINGS, S. B. 2002 *Phys. Rev. D* **66**, 044011.

EICHTEN, E., ET AL. 1984 *Rev. Mod. Phys.* **56**, 579.

EINSTEIN, A. 1907 *Jahrbuch der Radioaktivität und Elektronik* **4**, 411.

EINSTEIN, A. 1911 *Annalen der Physik* **35**, 898.

EINSTEIN, A. 1916 *Annalen der Physik* **49**, 769.

EINSTEIN, A. & GROSSMAN, M. 1913 *Z. F. Mathematik und Physik* **62**, 225.

EMPARAN, R., HOROWITZ, G. T., & MYERS, R. C. 2000 *Phys. Rev. Lett.* **85**, 499.

FAIRBAIRN, M. 2001 *Phys. Lett. B* **508**, 335.

FAIRBAIRN, M. & GRIFFITHS, L. M. 2002 *J. High Energy Phys.* **02**, 024.

FENG, J. L. & SHAPERE, A. D. 2002 *Phys. Rev. Lett.* **88**, 021303.

FITZPATRICK, A. L., KAPLAN, J., & RANDALL, L. 2007 *J. High Energy Phys.* **09**, 013.

GIBBONS, G. W. & HAWKING, S. W. 1977 *Phys. Rev. D* **15**, 2752.

GIDDINGS, S. B. & THOMAS, S. 2002 *Phys. Rev. D* **65**, 056010.

GOLDBERGER, W. D. & WISE, M. B. 1999 *Phys. Rev. Lett.* **83**, 4922.

HAGEDORN, R. 1965 *Nuovo Cimento Suppl.* **3**, 147.

HALL, L. & SMITH, D. 1999 *Phys. Rev. D* **60**, 085008.

HANHART, C., ET AL. 2001a *Phys. Lett. B* **509**, 1.

HANHART, C., ET AL. 2001b *Nucl. Phys. B* **595**, 335.

HANNESTAD, S. & RAFFELT, G. G. 2002 *Phys. Rev. Lett.* **88**, 071301.

HARRIS, C. M., RICHARDSON, P., & WEBBER, B. R. 2003 *J. High Energy Phys.* **08**, 033.

HARRIS, C. M., ET AL. 2005 *J. High Energy Phys.* **05**, 053.

HAWKING, S. W. 1975 *Commun. Math. Phys.* **43**, 199.

HEWETT, J. L. & SPIROPULU, M. 2002 *Ann. Rev. Nucl. Part. Sci.* **52**, 397.

HOYLE, C. D., ET AL. (EÖT-WASH COLLABORATION) 2001 *Phys. Rev. Lett.* **86**, 1418.

HOYLE, C. D., ET AL. 2004 *Phys. Rev. D* **70**, 042004.

HSU, S. D. H. 2003 *Phys. Lett. B* **555**, 92.

JAIN, P., ET AL. 2003 *Int. J. Mod. Phys. D* **12**, 1593.

JEVICKI, A. & THALER, J. 2002 *Phys. Rev. D* **66**, 024041.

KABACHENKO, V., MIAGKOV, A., & ZENIN, A. 2001 *ATLAS Internal Note* ATL-PHYS-2001-12.

KALUZA, TH. 1921 *Sitzungsber. Preuss. Akad. Wiss. Phys. Math. Klasse*, p. 996.

KLEIN, O. 1926a *Z. F. Physik* **37**, 895.

KLEIN, O. 1926b *Nature* **118**, 516.

KOCH, B., BLEICHER, M., & HOSSENFELDER, S. 2005 *J. High Energy Phys.* **10**, 053.

LANDSBERG, G. 2001 http://hep.brown.edu/users/Greg/TrueNoir/index.htm/.

LANDSBERG, G. 2002 *Phys. Rev. Lett.* **88**, 181801.

LANDSBERG, G. 2004. In *Proc. of the 32nd SLAC Summer Institute on Particle Physics* (eds. J. Hewett, J. Jaros, T. Kamae, & C. Prescott). http://www.slac.stanford.edu/econf/C040802/papers/MOT006.PDF/; arXiv:hep-ex/0412028v2.

LANDSBERG, G. 2006 *J. Phys. G* **32**, R337.

LEP EXOTICA WORKING GROUP (ALEPH, DELPHI, L3, AND OPAL COLLABORATIONS) 2004 *CERN Note LEP Exotica WG 2004–03* http://lepexotica.web.cern.ch/LEPEXOTICA/notes/2004-03/ed_note_final.ps.gz and references therein.

LONG, J. C., ET AL. 2003 *Nature* **421**, 922.

LONNBLAD, L., SJODAHL, M., & AKESSON, T. 2005 *J. High Energy Phys.* **09**, 019.

MALDACENA, J. 1998 *Adv. Theor. Math. Phys.* **2**, 231.

MARTIN, A. D., ROBERTS, R. G., & STIRLING, W. J. 1993 *Phys. Lett. B* **306**, 145; Erratum *Phys. Lett. B* **309**, 492.

MEADE, P. & RANDALL, L. 2008 *J. High Energy Phys.* **05**, 003.

MOCIOIU, I., NARA, Y., & SARCEVIC, I. 2003 *Phys. Lett. B* **557**, 87.

MYERS, R. C. & PERRY, M. J. 1986 *Ann. Phys.* **172**, 304.

NAYAK, G. C. & SMITH, J. 2006 *Phys. Rev. D* **74**, 014007.

RANDALL, L. & SUNDRUM, R. 1999a *Phys. Rev. Lett.* **83**, 3370.

RANDALL, L. & SUNDRUM, R. 1999b *Phys. Rev. Lett.* **83**, 4690.

RIEMANN, G. F. B. 1868 *Abh. Ges. Wiss. Gött.* **13**, 1.

RIZZO, T. G. 2005a *J. High Energy Phys.* **01**, 028.

RIZZO, T. G. 2005b SLAC PUB 11534; arXiv:hep-ph/0510420v4.

RIZZO, T. G. 2005c *J. High Energy Phys.* **06**, 079.

RIZZO, T. G. 2006a *Class. Quant. Grav.* **23**, 4263.

RIZZO, T. G. 2006b SLAC PUB 11666; arXiv:hep-ph/0603242v5.

SJÖSTRAND, T., ET AL. 2001 *Comput. Phys. Commun.* **135**, 238. (We used PYTHIA v6.157).

SOLODUKHIN, S. N. 2002 *Phys. Lett. B* **533**, 153.

STÖCKER, H. 2006 *Int. J. Mod. Phys. D,* **16**, 185.

STOJKOVIC, D. 2005 *Phys. Rev. Lett.* **94**, 011603.

SUSSKIND, L. 2005 *The Cosmic Landscape: String Theory and the Illusion of Intelligent Design.* Little, Brown.

TANAKA, J., ET AL. 2005 *Eur. Phys. J. C* **41**, 19.

UEHARA, Y. 2002a; arXiv:hep-ph/0205122v2.

UEHARA, Y. 2002b; arXiv:hep-ph/0205199v1.

VACAVANT, L. & HINCHLIFFE, I. 2001 *J. Phys. G* **27**, 1839.

VOLOSHIN, M. B. 2001 *Phys. Lett. B* **518**, 137.

VOLOSHIN, M. B. 2002 *Phys. Lett. B* **524**, 376.

YOSHINO, H. & NAMBU, Y. 2002 *Phys. Rev. D* **66**, 065004.

YOSHINO, H. & NAMBU, Y. 2003 *Phys. Rev. D* **67**, 024009.

Black holes in globular clusters

By STEPHEN L. W. MCMILLAN

Department of Physics, Drexel University, Philadelphia, PA 19104, USA

Dynamical evolution in star clusters naturally creates an environment in which interactions among massive stars, binaries, and compact remnants are common. Young clusters may temporarily contain a significant population of stellar black holes, and close encounters and physical collisions among stars in dense cluster cores may lead to the formation of very massive stars and high-mass black holes via runaway merging. Numerical simulations suggest runaway masses in the range commonly cited for intermediate-mass black holes. While our understanding of black hole formation and retention has improved greatly in recent years, substantial uncertainties remain in both the physics of the runaway merger process and the evolution of very massive stars. Direct and indirect observational evidence have been reported for massive black holes in globular clusters, although here too interpretations remain controversial. I examine critically some details of the processes possibly leading to massive black holes in present-day globular clusters, and discuss some observational constraints on the various theoretical scenarios.

1. Introduction

Black holes are natural products of stellar evolution in massive stars, and may also result from dynamical interactions in dense stellar systems, such as star clusters and galactic nuclei. They can significantly influence the dynamics of their parent cluster, and may also have important observational consequences, via their x-ray emission, the production of gravitational radiation, and their effect on the structural properties of the system in which they reside.

Globular clusters offer particularly rich environments for the production of black holes in statistically significant numbers. Direct evidence for black holes in globulars is scarce, although several independent lines of investigation now hint at their presence. Despite the lack of firm observational support, the past three decades have seen many theoretical studies of the formation and dynamics of stellar- and intermediate-mass black holes in star clusters.

In this review I discuss the observational evidence for black holes (and, in particular, intermediate-mass black holes) in globular clusters. I then turn to recent theoretical insights into their formation in dense star clusters, and their subsequent dynamics. Finally, I describe in some detail the modeling techniques used to simulate dense star clusters, and present some promising new directions for future simulations of these complex systems.

2. Observations of stellar-mass black holes in star clusters

Given the numerous theoretical studies of the formation and dynamical consequences of black holes (BH) in star clusters, and the overwhelming weight of opinion on the inevitability of BHs as consequences of stellar evolution, there is remarkably little observational evidence for stellar-mass (i.e., less than a few tens of solar masses) BHs in globular clusters. There appears to be no evidence for such low-mass BHs in the Galactic globular cluster population, and only one firm observation of a stellar-mass BH in an extragalactic globular cluster.

The sole extragalactic BH is reported by Maccarone et al. (2007), who report a bright x-ray source in a cluster in NGC 4472, a bright elliptical galaxy in the Virgo cluster. Its observed x-ray luminosity of $\sim 4 \times 10^{39}$ erg s^{-1} is the Eddington luminosity for a 35 $M_\odot$

object, and it shows substantial variability on a time scale of hours, perhaps indicating a considerably larger mass, so even this candidate may actually fall into the intermediate-mass black hole (IMBH) range. Interestingly, the parent cluster is itself quite bright ($V = 21$, corresponding to $L \sim 7.5 \times 10^5 \, L_\odot$ at a distance of 16 Mpc) and lies far (30 kpc) from the center of the host galaxy, perhaps placing it in the same category as the leading candidates for IMBHs in the Milky Way (ω Centauri: $L \approx 1.1 \times 10^6 \, L_\odot$; Harris 1996) and M31 (G1: $L \sim 2.1 \times 10^6 L_\odot$; Meylan et al. 2001), as discussed in Section 3 below.

3. Observational evidence for intermediate-mass black holes

Oddly, given the considerable theoretical uncertainty about how such objects might form, the observational evidence for IMBHs, while incomplete, is rather stronger than that for stellar-mass BHs. Several lines of reasoning support the assertion that IMBHs exist in globular clusters: (1) observations of ultraluminous x-ray sources (ULXs), (2) dynamical modeling, and (3) studies of cluster structure.

3.1. *Ultraluminous x-ray sources*

The bright x-ray source M82 X–1 (Matsumoto & Tsuru 1999; Matsumoto et al. 2001; Kaaret et al. 2001) is the strongest ULX candidate for an IMBH. With a peak luminosity of more than 10^{41} erg s^{-1}, it is too bright to be an ordinary x-ray binary—while its location 200 pc from the center of M82 argues against a supermassive black hole. This luminosity is consistent with an accreting compact object of at least 350 solar masses, possibly an IMBH. The discovery of 54.4 ± 0.9 mHz quasi-periodic oscillations (Strohmayer & Mushotsky 2003) supports this view.

An intriguing aspect of M82 X–1 is its apparent association with the young dense cluster MGG–11 (McCrady et al. 2003). Figure 1 (from Portegies Zwart et al. 2004) shows superimposed near-IR (*HST*) and x-ray (*Chandra*) images of the region containing M82 X–1 and MGG–11. (The offset between the infrared cluster and the x-ray source is consistent with the absolute pointing accuracies of the two telescopes.) The cluster age is between 7 and 12 Myr. Portegies Zwart et al. (2004) found that such an association would be consistent with the scenario described in Section 4 for runaway stellar growth in a dense cluster.

Numerous authors have noted that high x-ray luminosity is by no means conclusive evidence of an IMBH (see Miller & Colbert 2004 for a review of some alternative possibilities). Soria (2006, 2007) points out that most ULXs may in fact be consistent with the high-luminosity tail of the x-ray binary luminosity function, and that ULXs are generally not associated with young star clusters. Nevertheless, the runaway collision scenario has arguably become the "standard" mechanism for IMBH formation in clusters, against which others are assessed.

3.2. *Dynamical modeling of cluster velocity structure*

Currently the most definitive statements about IMBHs in globular clusters have come from Gebhardt and collaborators (Gerssen et al. 2002; Gebhardt, Rich & Ho 2002, 2005; Noyola, Gebhardt, & Bergmann 2006, based on axisymmetric, three-integral dynamical models of cluster potentials, using full line-of-sight velocity information to constrain the model fits. IMBH masses have been reported for the Galactic globular clusters M15 and ω Centauri, and for the cluster G1 in M31, although these results have not been without controversy.

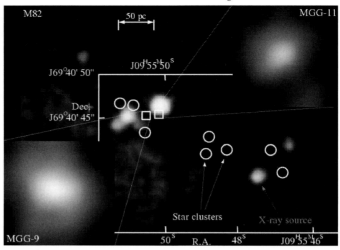

FIGURE 1. Combined *HST* and *Chandra* images of the region containing M82–X1 and MGG–11. The x-ray observations of Matsumoto et al. (2001) are shown; M82 X–1 is near the center of the image. The star clusters from Table 3 of McCrady et al. (2003), are indicated by circles. The positions of the clusters MGG–9 and MGG–11 studied by Portegies Zwart et al. 2004 are indicated by squares. Infrared images from the McCrady et al. observations are presented in the upper right (MGG–11) and lower left (MGG–9) corners. The quasi-periodic oscillator is not shown because of its low (7 arcsecond) positional accuracy, but its position is consistent with the x-ray source in MGG–11 (from Portegies Zwart et al. 2004).

3.2.1. *The Milky Way globular cluster M15*

Gerssen et al. (2002) reported dynamical evidence for a $4 \pm 2 \times 10^3 \, M_\odot$ IMBH in M15. This result was criticized by Baumgardt et al. (2003a), who pointed out that a "standard" dynamically evolved cluster model would yield similar results, and was marred by a crucially mislabeled figure in an earlier study of M15's dynamics. Subsequently the estimate of the IMBH mass was reduced to $2 \pm 2 \times 10^3 \, M_\odot$ (Gerssen et al. 2003). It now seems clear, from a variety of different methods, that the core of this highly centrally concentrated cluster contains on the order of $1000 \, M_\odot$ of non-luminous matter. In addition to the dynamical models of Gerssen et al., these methods also include Fokker–Planck (Dull et al. 1997) and N-body (Baumgardt et al. 2003a) cluster simulations, as well as earlier studies of pulsar accelerations (Phinney 1993).

The key issue, of course, is the interpretation of this non-luminous component. The evolutionary models make it clear that, given the age and likely history of M15, the dark material most plausibly consists of stellar remnants—neutron stars and/or heavy white dwarfs—providing a much more natural explanation of the invisible mass. The models to not rule out an IMBH at the ~500–$1000 \, M_\odot$ level, but present no compelling reason to conclude that one exists. We will see in Section 3.3.2 that evolved, concentrated systems like M15 are in fact probably not the best places to look for IMBHs.

3.2.2. *Globular clusters G1 (M31) and ω Centauri (Milky Way)*

Much firmer evidence for IMBHs is found in G1 and ω Centauri, the largest globular clusters in M31 and the Milky Way, respectively. Both clusters are large enough that their relaxation times are long, rendering it unlikely that the dynamics of stellar evolution products could mimic the effect of a central IMBH, as in M15.

In G1, a cluster with mass $\sim1.5 \times 10^7 \, M_\odot$, situated some 40 kpc from the center of M31 (Meylan et al. 2001), the dynamical models yield a black hole mass of $1.8 \pm 0.5 \times$

$10^4 \, M_\odot$ (Gebhardt, Rich, & Ho 2002, 2005). The initial results were also questioned by Baumgardt et al. (2003b), in part because the $0.013''$-radius sphere of influence of the supposed IMBH lies well inside the central pixel of the *HST* image, but also because plausible N-body models without black holes could reproduce the cluster's observed surface density and velocity dispersion profiles quite well. However, Gebhardt, Rich, & Ho (2005) have argued that the use of the full velocity distribution, and not just its lowest moments, is essential in order to properly define the cluster potential, and that the N-body simulations simply lack the resolution necessary for them to be meaningfully compared with the observational data.

Support for the Gebhardt et al. result has come from the recent detection of x-ray emission from G1 (Pooley & Rappaport 2006). The observed luminosity $L_X \sim 2 \times 10^{36} \, \text{erg s}^{-1}$ is consistent with Bondi-Hoyle accretion of intracluster gas onto an IMBH in the relevant mass range. Further support comes from radio observations of G1 (Ulvestad, Greene, & Ho 2007) which imply a radio–to–x-ray flux ratio consistent with a $2 \times 10^4 \, M_\odot$ black hole (Merloni, Heinz, & di Matteo 2003).

In ω Centauri, a globular cluster with mass $\sim 5 \times 10^6 \, M_\odot$ lying ~ 6 kpc from the center of the Milky Way, Noyola, Gebhardt, & Bergmann (2006) report an IMBH mass of $4 \pm 1 \times 10^4 \, M_\odot$.

Interestingly, both G1 and ω Centauri (and, in fact also M15) lie near the low-mass extension of the "M-σ" relation for active galactic nuclei (Merritt & Ferrarese 2001; Tremaine et al. 2002), reinforcing the suspicion that G1 and ω Centauri are in fact not "real" globular clusters, but rather are the cores of dwarf spheroidal systems stripped by the tidal fields of M31 and the Milky Way (Meylan et al. 2001, Freeman 1993). This possibility does not explain the origin of the IMBHs, but it obviously moves the question into a very different arena.

3.3. *Indirect evidence for central black holes*

Detailed dynamical simulations of IMBHs in clusters extend back to the 1970s, with the pioneering work of Bahcall and Wolf (1976) and the Monte-Carlo studies of Shapiro and co-workers (e.g., Marchant & Shapiro 1980). These early investigations were motivated in part by the steep central density and velocity dispersion profiles observed in M15 and the suspicion that central black holes might be responsible. More recently, Baumgardt and collaborators (2004, 2005) have carried out N-body simulations of clusters containing central black holes having masses between 0.1 and 10% of the total cluster mass.

Two important results of these studies are directly relevant here. The first is that the stellar density and velocity dispersion distributions in the vicinity of a black hole follow distinctive power-law "cusps" that may be directly observable, or (for lower-mass IMBHs) at least have detectable consequences for the observed structure of the cluster core. The second is that black holes are efficient heating sources to stellar systems, and this fact has ramifications for the entire cluster.

3.3.1. *Weak cusps*

Noyola and Gebhardt (2006) find that a surprisingly large fraction ($\sim 25\%$) of globular clusters hitherto thought to have "classical" cores in fact have shallow power-law surface-brightness profiles in their central regions. The detection of these weak cusps is due in large part to high-resolution *HST* observations of the innermost arcsecond of these systems. These observations remain somewhat controversial, but are in good agreement with theoretical simulations by Baumgardt, Makino, & Hut (2005) of clusters containing central black holes comprising ~ 0.1–1% of the total cluster mass, and have been interpreted as indirect indicators of IMBHs in these clusters.

Ongoing detailed dynamical studies by Noyola et al. (2008) of selected clusters from their earlier study, including NGC 2808, 47 Tucanae, and the "weak cusp" systems M54 and M80 should shed much further light on the dynamics of these intriguing systems.

3.3.2. *Cluster structural Parameters*

Although dense stellar systems are among the most promising environments for the formation of IMBHs (see Section 4), they may not be the best place to look for evidence of massive black holes. Baumgardt, Makino, & Hut (2005) and Heggie et al. (2007) have pointed out that core-collapse clusters like M15 are probably the least likely to harbor IMBHs. Rather, dynamical heating by even a modest IMBH is likely to lead to a cluster containing a fairly extended core. The reasoning is straightforward:

• Within the sphere of influence of the IMBH (radius $R_g \sim GM_{\mathrm{BH}}/v_c^2$, where v_c is the velocity dispersion in the cluster core), a Bahcall-Wolf (1976) cusp steadily transports stars inward by two-body relaxation.

• The outward energy flux in the cusp is independent of radius, so this inward diffusion implies a net core heating rate of mv_c^2/t_{rc} per star within Rg, where m is the mean stellar mass and t_{rc} is the core relaxation time. The total heating rate—the net outward heat flux from the core, in steady state—then is $\rho_c R_g^3 v_c^2/t_{\mathrm{rc}}$, where ρ_c is the core mass density.

• The great disparity in time scales means that the core heat flux due to any central energy source must always come into equilibrium with the much more slowly varying demands of the cluster half-mass radius. The heat flux across this radius is $\sim Mv_h^2/t_{\mathrm{rh}}$, where M is the total cluster mass and subscript "h" refers to the half-mass radius.

• Equating the fluxes at the core and the half-mass radius determines the equilibrium core to half-mass radius ratio R_c/R_h. Calibrating to simulations, Heggie et al. conclude that

$$\frac{R_c}{R_h} \sim 0.7 \left(\frac{M_{\mathrm{BH}}}{M} \right)^{3/4} \quad .$$

Trenti suggests that the imprint of this process can be seen in his "isolated and relaxed" sample of Galactic clusters having relaxation times less than 1 Gyr, a half-mass to tidal radius ratio $R_h/R_t < 0.1$, and an orbital ellipticity of less than 0.1. Roughly half of the clusters in this sample have core radii substantially larger than would be expected on the basis of simple stellar dynamics and binary heating.

4. Formation of intermediate-mass black holes

Accepting without further debate the still sketchy observational evidence for IMBHs in globular clusters, I now turn to the question of how such black holes might have formed. The leading possibilities are (1) they are primordial, the result of stellar evolution in supermassive population III stars, (2) they are the result of runaway stellar collisions in young dense star clusters, and (3) they are the result of mergers of BHs over the lifetime of a cluster. I focus here on the latter two possibilities, since the formation mechanisms involved are arguably much more relevant to the physics of dense stellar environments such as globular clusters, and lead most naturally to my final topic, modeling the detailed evolution of such systems.

The dynamics of dense stellar systems inevitably leads to conditions favorable to repeated stellar collisions, and hydrodynamic simulations indicate that, when massive stars collide in clusters, they are very likely to merge. The likelihood of multiple stellar collisions was first demonstrated in N-body simulations of dense clusters (Portegies Zwart

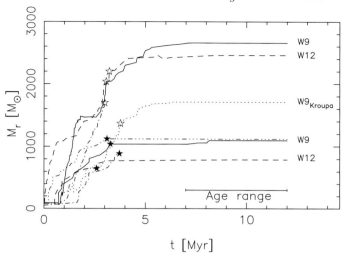

FIGURE 2. Growth in mass of the runaway star for simulations performed with *Starlab* (Portegies Zwart et al. 2001) and NBODY4 (Aarseth 1999; Baumgardt & Makino 2003). The choice of initial concentration is labeled by the dimensionless central potential, where W12(9) implies a King parameter $W_0 = 12(9)$. The upper curves are for NBODY4; the lower ones for *Starlab*. The star symbols indicate the moment when a supernova occurs, typically around 3 Myr.

et al. 1999, 2004, 2005; Portegies Zwart & McMillan 2002; McMillan & Portegies Zwart 2007), and later also in Monte-Carlo models (Freitag, Gürkan, & Rasio 2006).

The central question then becomes "What next?" Unfortunately, while the dynamical processes leading to collision runaways are simple and well known, there are several prominent "missing links" in the chain of reasoning starting from a young stellar system and ending with a massive black hole, all involving key aspects of the physics of massive stars.

4.1. *Mass segregation and runaway mergers*

The dynamics of runaway mergers is straightforward:

• Massive stars sink to the cluster core. The mass segregation time scale is $\sim t_r \langle m \rangle / m$, where t_r is the local relaxation time, m is the mass of the star, and $\langle m \rangle$ is the mean stellar mass.

• The result, for a realistic (Kroupa 2001) stellar mass spectrum, is a "core collapse" of sorts, leading to the formation of a dense sub-core of massive stars in a time $t_{cc} \sim 0.2 t_r$. Here, t_r can be the half-mass relaxation time for a small system, or the core relaxation time for a larger one, as discussed below.

• High densities in the core lead to collisions and mergers, which naturally involve the most massive stars.

• The process runs away, with one collision product growing rapidly in mass and radius and outstripping the competition. The resultant merger (and, we assume, black hole) mass is in the IMBH range.

• For a runaway to occur, the collision process must complete before the first supernovae occur, that is, within 3–5 Myr. This requires short relaxation times, or, equivalently, high cluster densities.

Figure 2 shows a typical set of results, obtained by Portegies Zwart et al. (2004) in their *N*-body study of the M82 clusters discussed earlier. The different runs represent a broad range of initial conditions, with and without initial binaries and with both Salpeter

(1955) and Kroupa (2001) mass functions, for systems of 128k–585k stars. The runaway masses in NBODY4 are generally larger than those in *Starlab*, since NBODY4 adopts systematically larger stellar radii and hence collision cross sections, for the massive stars considered here. Comparable results (in both time scale and in total runaway mass) have been obtained in Monte-Carlo simulations (e.g., Freitag, Gürkan, & Rasio 2006).

According to dynamical models, a runaway will always occur if a central core of massive stars ($m \gtrsim 20\,M_\odot$) can form by mass segregation before the occurrence of the first supernova. In small systems—having $t_{\rm rh} \lesssim 25$ Myr—essentially all the massive stars can reach the center in the time available. The merger mass fraction for a typical (Kroupa 2001) mass function is ~ 0.01 (Portegies Zwart & McMillan 2002). In large systems, only a fraction of the massive stars can reach the center before exploding as supernovae. Freitag, Gürkan, and Rasio (2006) report a somewhat smaller mass fraction for the IMBH in their Monte-Carlo models. McMillan and Portegies Zwart (2007) find that, for $t_{\rm rh} \gtrsim 25$, the merger fraction is expected to scale as

$$t_{\rm rh}^{-1/2} \quad .$$

Gürkan, Fregeau, and Rasio (2006) have reported the intriguing possibility of forming *binary* IMBHs, resulting from independent collision runaways in a sufficiently large system. Binary IMBH formation has not been seen in any N-body simulations to date.

4.2. *Evolution of the merger product*

Freitag, Gürkan, and Rasio (2006; also Lombardi 2007) find that direct mass loss due to the collision itself is generally unimportant, amounting to less than $\sim 10\%$ of the total mass in most cases. However, the problem of computing the evolution of a possibly rapidly rotating collision product, from its initial non-equilibrium state back to the anomalous main sequence and beyond, presents significant challenges (see, e.g., Sills et al. 2003), although important inroads are now being made, as discussed in Section 5 below.

Perhaps the greatest uncertainty in the runaway process has to do with mass loss from supermassive stars. Massive main-sequence stars are well known to have very strong winds, and these compete with collisions in determining the mass of the final object (e.g., Vanbeveren et al. 2009; Belkus, Van Bever, & Vanbeveren 2007). Portegies Zwart et al. (1999) recognized that the specifics of the mass-loss prescription completely control the outcome of the collision runaway. Applying most of the mass-loss late, at the terminal-age main sequence, permits the process to build massive stars, and this prescription has (unfortunately) been adopted in most dynamical models to date. A more realistic treatment of mass loss (based on Langer et al. 1994) can significantly reduce the mass of the final collision product, while early mass loss (probably unrealistically so) can shut the process down completely.

By way of calibration, it is worth pointing out that dynamical models typically predict the accretion of $\sim 10^3\,M_\odot$ of material in $\sim 10^6$ yr, for a net accretion rate of $\sim 10^{-3}\,M_\odot$ yr^{-1}. This is comparable to the mass-loss rates reported for several massive stars, but substantially less than the largest rates known, e.g., the $\sim 0.1\,M_\odot$ yr^{-1} inferred during the decade around the "great outburst" of Eta Carinae in 1843 (Morris et al. 1999). Belkus, Van Bever, and Vanbeveren (2007) find that massive (300–1000 $M_\odot$) solar-metallicity stars end their lives as relatively low-mass (40–50 $M_\odot$) black holes, but comparably massive stars in low-metallicity clusters may give rise to IMBHs in the 150–200 $M_\odot$ range. Yungelson et al. (2008) find a final black hole mass of $\sim 150\,M_\odot$ for a star with initial mass 1000 $M_\odot$.

These studies suggest that it might be difficult to retain enough mass to form an IMBH, but recently, Suzuki et al. (2007) have performed simulations of collisionally merged stars, using a more realistic mass-loss model than in most earlier dynamical studies, and find that, because of the extended envelopes of the merged systems, most collisions are expected to occur early, during the Kelvin-Helmholtz contraction phase, before mass loss by stellar winds become significant. They conclude that stellar mass loss does not prevent the formation of massive stars with masses up to $\sim 1000\,M_\odot$. Obviously, any statements about collision runaways and their outcomes must be tempered by our relative ignorance of the proper mass-loss prescription and rate.

Finally, even if mass were not a factor, the assumption that a $1000\,M_\odot$ star will form a $1000\,M_\odot$ black hole is also largely a matter of conjecture.

4.3. *Connection with globular clusters*

Consider a compact, centrally concentrated young star cluster, with a central density high enough for the runaway collision scenario to operate. It is easily shown that, for a $\sim 10^6\,M_\odot$ system, the constraint on the relaxation time presented in Section 4.1 requires a mean density of $\sim 5 \times 10^7\,M_\odot\,\mathrm{pc}^{-3}$, corresponding to a characteristic radius of ~ 0.2 pc.

One might reasonably wonder what such a highly concentrated, dense system has to do with the relatively low-density globular clusters seen today. Can runaway collisional processes, shortly after their formation, account for the IMBHs that may exist in the Galactic globular cluster system? Based on numerous simulations of many aspects of the problem, we can construct a plausible scenario in which collisionally formed IMBHs might now reside in globular clusters.

After the IMBH has formed, numerous processes combine to expand the cluster overall (that is, its half-mass radius) and also to reduce its central concentration (as measured by, say, the ratio R_h/R_c). If the cluster is not significantly mass segregated at birth, the dynamical effects of segregation by massive stars and their remnants will cause the concentration to decrease significantly, and will also result in a modest overall expansion of the cluster (Merritt et al. 2004; Mackey et al. 2007). At the same time, mass loss due to stellar evolution drives further overall expansion, by a factor of 2–3 (Takahashi & Portegies Zwart 2000; Mackey et al. 2007). Vesperini, McMillan, and Portegies Zwart (2008) find that mass loss from the cores of initially mass-segregated clusters is particularly effective in reducing the central concentration. Finally, heating due to the central IMBH (Section 3.3.2) again reduces the concentration and causes significant expansion over the lifetime of the cluster; the half-mass radii of the model clusters considered by Baumgardt, Makino, and Hut (2005) expanded by factors of 5–7.

Taken together, these (admittedly approximate) figures suggest that the collision runaway scenario could indeed account for an IMBH in a typical globular cluster—if one is ever confirmed!

4.4. *Systems of black holes*

Rapid collisions in a sufficiently dense cluster may lead to the prompt formation of an IMBH. Let us now briefly consider the opposite limit, in which the relaxation time is long enough that the massive stars form stellar-mass black holes "in place," before significant dynamical evolution can occur.

The evolution of black-hole systems in low-density clusters has been studied by a number of authors. Kulkarni, Hut, and McMillan (1993) and Sigurdsson and Hernquist (1993) considered the dynamics of a population of black holes in an idealized star cluster. They found that mass segregation rapidly transports the black holes to the cluster core, where the black hole subsystem evolves rapidly, forming binaries that ultimately eject most of the black holes from the cluster. Both papers concluded that few observable stellar-mass black holes would be expected in the Galactic globular-cluster system today.

Subsequently, Portegies Zwart and McMillan (2000) performed N-body simulations, and generally verified these findings, but also concluded that the ejected black-hole binaries could be significant sources of gravitational radiation in the LIGO band.

These studies were idealized in several important ways. They were not dynamically self-consistent (the N-body simulations combined many small-N simulations to achieve good statistics), they did not include a spectrum of black hole masses, and they did not include post-Newtonian relativistic effects (all used the Peters and Mathews [1963] formula for gravitational radiation energy losses). Miller and Hamilton (2002) pointed out that a sufficiently massive black hole, at the top end of the stellar black hole mass spectrum, or perhaps the result of a "failed" runaway merger, could survive ejection by dynamical interactions, and proposed a mechanism in which black holes in binary systems merge by the emission of gravitational radiation to form successively more massive objects, ultimately resulting in an IMBH. It is unclear whether the large gravitational radiation recoil velocities found in recent numerical simulations of black-hole mergers would still allow such a process to operate.

O'Leary et al. (2006) carried out Monte-Carlo simulations of a population of black holes (roughly 50% binaries) taken from a population synthesis calculation starting from realistic distributions of stellar and binary properties. They included post-Newtonian effects in their model, and found that, although modest IMBHs ($\sim 100\,M_\odot$ up to a maximum of $\sim 600\,M_\odot$) did form, in most cases the recoil speed substantially exceeded the escape speed from the cluster, and that the most likely outcome was evaporation of the entire black-hole subsystem. Their inferred LIGO detection rate was consistent with the range given previously by Portegies Zwart and McMillan (2000).

Recently, Mackey et al. (2007) have reported a set of fully self-consistent (Newtonian) dynamical simulations of young dense clusters. They find that the combination of stellar mass loss and dynamical heating has a significant dynamical impact on the cluster, causing the core (and, to a lesser extent, the cluster as a whole) to expand in a manner strikingly similar to the core radius–age relation observed in the young clusters in the LMC (Mackey & Gilmore 2003). However, because of the expansion, the black-hole interaction rate drops sharply at late times, and a substantial population of black holes remains after 10 Gyr. If, as seems plausible, this general dynamical result scales to systems comparable in size to the Milky Way globular clusters, it suggests that the absence of evidence of stellar-mass black holes is not evidence of absence, but simply reflects their low current interaction rate with other cluster members.

5. Modeling dense stellar systems

The proper treatment of collision runaways—and the central question of whether or not they can ultimately lead to the formation of an IMBH—is a complex problem involving several branches of astrophysics, whose interfaces have traditionally been handled in at best a rudimentary fashion. In this section I discuss some strengths and weaknesses of the modeling techniques currently in use, and how they might be improved.

5.1. *The state of the art*

The chief uncertainties in our understanding of the collision runaway process lie in the details of stellar-mass loss and stellar evolution. Unfortunately, most large-scale simulations to date have concentrated on the dynamics of the collisions, and on how an IMBH in a cluster core affects the global properties of the cluster.

The reason for this somewhat asymmetric application of resources lies in the structure of the codes currently available for the simulation of dense stellar systems. The state-of-the-art programs in this area are the various "kitchen sink" packages that combine treatments of dynamics, stellar and binary evolution, and stellar hydrodynamics within a single simulation. Of these, the most widely used are the N-body codes NBODY (Aarseth 2003) and KIRA (e.g., Portegies Zwart et al. 2001), and the Monte-Carlo codes developed by Freitag (see Freitag, Rasio, & Baumgardt 2006) and by Rasio and coworkers (e.g., Fregeau et al. 2003). These codes differ principally in their treatment of the large-scale dynamics, employing conceptually similar approaches to stellar and binary evolution and collisions. They typically contain the following elements:

• The N-body codes incorporate detailed descriptions of stellar dynamics at all levels, using direct integration of the individual (Newtonian) stellar equations of motion.

• Monte-Carlo codes use an orbit-averaged or other approximate description of stellar orbits on large scales, limiting them to spherical symmetry, dynamical equilibrium, and global dynamical processes occurring on relaxation time scales.

• Both N-body and Monte-Carlo codes rely on specialized procedures to follow the small-scale dynamics, allowing them to resolve close stellar encounters and multiple interactions. These may include post-Newtonian terms in the interactions between compact objects.

• The current codes employ approximate treatments of stellar evolution, generally derived from look-up tables based on the detailed evolutionary models of Eggleton, Fitchett, and Tout (1989) and/or Hurley, Pols, and Tout (2000). They also employ extensive semi-analytic and heuristic rule-based treatments of binary evolution, conceptually similar in principle, but significantly different in detail.

• Most of the current codes implement collisions in the simple "sticky-sphere" approximation, where stars are taken to collide (and merge) if they approach within the sum of their effective radii. The effective radii may be calibrated using hydrodynamical simulations, and mass loss may be included in some approximate way. Freitag's code uses a more sophisticated approach, interpolating encounter outcomes from a pre-computed grid of SPH simulations (Freitag & Benz 2005).

5.2. *Shortcomings of current models*

These programs have been very successfully applied to the dynamics of dense stellar systems, and are responsible for much of our detailed insight into the collision runaway problem, but their complex internal structure often renders them hard to modify or extend, and makes experimentation difficult. Choosing a particular program amounts to selecting a specific menu of treatments of all aspects of the simulation.

For example, if one elects to use Aarseth's NBODY4 or NBODY6, this choice necessarily also entails use of the SSE and BSE stellar and binary evolution packages (Hurley, Pols, & Tout 2000) and a simple sticky-spheres treatment of collisions. Use of KIRA implies also using the SeBa stellar and binary-evolution package (Portegies Zwart & Verbunt 1996), again with sticky-sphere collisions. The binary evolution portions of the Monte-Carlo codes rely on the "StarTrack" binary population synthesis code (Belczynski, Kalogera, & Bulik 2002), and so on.

The basic problem with this approach is clear to any user who has tried to modify these packages to add new functionality, or who wishes, for example, to install a new stellar-evolution prescription to study how the details of the adopted approach affect the outcome of the evolution. These large codes are monolithic in design, internally complex in structure, and usually contain non-obvious dependencies on specific treatments of the internal physics, making them unforgiving of novice attempts to alter them and intolerant of changes to their core algorithms.

The realization that dense stellar systems entail not just interactions among stars, but also among modelers and their programs, was a major motivating factor in the inception of the MODEST initiative.† Short for MOdeling DEnse STellar systems, MODEST is a loosely knit collection of various groups working on all aspects of the theory and observations of star clusters, including stellar dynamics, stellar evolution, stellar hydrodynamics, and cluster formation. MODEST has hosted some 20 meetings over the past 5 years, providing an invaluable forum for discussion and collaboration among researchers in this field.

One important aspect of MODEST is the attempt to provide a software framework for large-scale simulations of dense stellar systems, within which existing codes for dynamics, stellar evolution, and hydrodynamics can be easily coupled. In this regard, MODEST has become a platform for a focused attempt to address some of the shortcomings found in current simulation packages. I highlight here the problems most urgently in need of attention:

• Stellar evolution is handled by interpolation from precomputed tracks, making it incapable of handling the "new" stars formed by collisions with any degree of precision. In all cases, following a collision, stars are simply "rejuvenated"—merged, mixed, and restarted on a new evolutionary track selected from the precomputed grid.

• Stellar collisions are at best handled by lookup from precomputed models, and more usually by prescriptions that are checked against detailed calculations after the fact (e.g., Lombardi 2007). In neither case is the resulting dynamical simulation self-consistent.

• The program structure makes it difficult to incorporate new physical processes into any of the component modules—stellar dynamics, stellar evolution, or stellar collisions. The result of these deficiencies is inconsistency and the inability to make detailed comparisons of specific algorithms, rendering the structure and state of runaway collision products little more than conjecture.

5.3. *The MUSE Project*

A key goal of MODEST is the incorporation of "live" treatments of stellar and binary evolution and ultimately stellar hydrodynamics directly into kitchen-sink N-body simulations. Such an undertaking is essential if one wishes to model the evolution of a dense stellar system, in which stellar collisions may be commonplace events, creating wholly new channels for stars to evolve and allowing the formation of stellar species completely inaccessible by standard stellar and binary evolutionary pathways.

A pioneering effort to couple stellar dynamics and stellar evolution was reported by Church (2006), who combined Aarseth's NBODY6 with a version of Eggleton's EV (Eggleton 2006) in a single program. This heroic programming accomplishment hard-coded the two programs into a single application, providing proof of concept that two such disparate modules could in fact be successfully merged.

The MODEST approach to code integration, called "MUSE," for MUltiscale MUltiphysics Scientific Environment, adopts a rather different approach, with the intention of providing a modular framework within which programs written by many different authors, and in many different languages, can interoperate with as little intrusion as possible into the internal operation of each program.

In a typical kitchen-sink code, the program simply cycles back and forth between invocations of the dynamics subroutine, which moves particles around the system and the stellar-evolution subroutine, which advances the internal states of individual stars.

† http://www.manybody.org/modest

Collisions and close encounters between stars cannot be scheduled, and instead are handled as they are detected within the dynamics module.

The MUSE approach differs from this model in a number of important ways. As discussed in more detail in Portegies Zwart et al. (2008), the key elements of MUSE are as follows.

- All basic operations are handled by program modules, conceptually similar to subroutines in a traditional code, except that they are not necessarily written in the same language, or by the same author. The intent is to allow experts to share their knowledge without first having to become well versed in the algorithms or programming style of the others.
- Rather than handling collisions and multiple encounters within the dynamics module, all distinct physical processes—large-scale dynamics, stellar and binary evolution, stellar interactions, dynamics of multiple systems, etc.—are treated as peers by the top-level scheduling loop. This complicates the scheduling, but greatly increases the modularity of the resultant program.
- Each module handles its own detailed internal data (probably the largest single programming task for the author to perform in order to make the module compliant with MUSE, since this will often involve data management operations alien to the original standalone program). Normally these data are not needed by, and are invisible to, the rest of the program.
- Each module provides standard interface functions, defined by the MUSE community, to provide specific pieces of information (stellar mass, position, radius, temperature, etc.) on demand. The essential idea here is that the amount of data that must move around the system to specify (say) a stellar collision is much less than the amount of internal data needed to describe a star or a binary in detail.
- See from outside, each module is a "black box," with completely specified functionality, largely hidden data, and a well-defined interface to the rest of the system. This approach is already tried and tested, and should be familiar to users of GRAPE- and GPU-accelerated applications (Makino et al. 1997, Portegies Zwart, Belleman, & Geldof 2007). Figure 3 illustrates this programming model, in the case of the stellar evolution module. The actual implementation of stellar evolution might be as simple as a "main-sequence to giant to remnant" cartoon, or as complex as a full-featured stellar-evolution code. The interface is the same in either case.
- The top-level loop in the traditional program is replaced by a loop written in a "glue" language, which manages all the modules in the simulation. We have chosen PYTHON as the glue language for MUSE because of its sophisticated object-oriented programming features, large user base, and extensive scientific and numerical libraries. The interfaces to Fortran (77, 90, 95), C, and C++ are constructed using the utilities F2PY and SWIG.
- The conceptually simpler module structure puts all physical processes on the same footing, but places a larger burden on the scheduler, since it now has to handle a range of tasks requested by specific modules (such as merging two stars, removing the original stars from the simulation, and transmitting the merger product data to the stellar and dynamics modules), in addition to pushing particles and evolving stars and binaries. PYTHON is particularly well suited to such challenging management operations.

The MUSE model has many advantages. By providing well-defined interfaces between modules, it allows a researcher (or a student) to quickly build real scientific applications using state-of-the-art techniques, without first having to become an expert in the many details of each module. Equally important, the MUSE structure allows users to write generic scripts that do not depend on the details of the algorithms used, providing, for

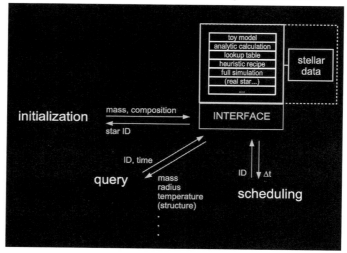

FIGURE 3. Schematic diagram of the stellar module in MUSE. As with all MUSE modules, the interaction between the module and the rest of the system is defined by interface functions which (1) initialize and identify a star, (2) return specific pieces of information about the star, and (3) provide a time scale for use by the scheduler. The internal implementation of the module may vary widely, and all implementation-specific local data on all stars are managed locally, invisible to the other modules in the simulation. The interface provides a uniform programming environment, yet allows great diversity in actual implementation.

the first time, "plug and play" functionality that allows different combinations of modules to be combined and compared.

The following (somewhat schematic) PYTHON fragment implements an N-body code with stellar evolution and stellar mergers included. Switching from one treatment (e.g., a hermite integrator with simple Eggleton, Fitchett, & Tout stellar lookup and sticky sphere collisions) to another (say, an implementation of Aarseth's NBODY with full stellar evolution and an entropy-sorting stellar merger scheme [Lombardi et al. 2003]) is as simple as changing only the first three lines of the program.

```
from gravity.hermite0.muse_dynamics import Hermite as dyn
from stellar.EFT89.muse_stellar import EFT89 as star
from collisions.sticky_spheres.muse_collisions \
        import StickySpheres as coll


    .
    .    (initialization)
    .


while time < t_max:
  time += delta_time
  while dyn.get_time() < time:
    id1 = dyn.evolve(time)
    if id1 > 0:                          # id1>0 is a colliding primary
      id2 = dyn.find_colliding_secondary(id1)
      evolve_stars(dyn.get_time())    # update stellar evolution
      collide_stellar_pair(id1, id2)  # merge the two stars
    evolve_stars(time)
```

```
print "end at t = ", time,   ", Nstars= ", star.get_number()
```

More details on the current state of MUSE, including specifications of the interface functions and the growing list of dynamics, stellar and binary evolution, multiple interactions, and collision modules may be found at the project web site, http://muse.li/.

6. Summary and Conclusions

There is growing, but arguably still inconclusive, evidence that some globular star clusters harbor massive black holes in their cores. Perhaps the strongest cases come from dynamical studies of G1 in M31 and ω Centauri in the Milky Way, but even here the conclusions are tainted to some degree by the suspicion that these massive clusters may actually be the stripped cores of dwarf spheroidal galaxies, in which case they shed little light on the physics of young star clusters or the collisional processes that may give rise to IMBHs in stellar systems. Indirect evidence based on cluster structural parameters and central density and velocity profiles is almost as compelling, and follow-up studies of the central velocity structure of "real" globular clusters have the potential to revolutionize the debate on this subject.

From a dynamical modeler's perspective, at least, the possibility that collisions in sufficiently dense young clusters might lead to runaway mergers offers the most interesting path to IMBHs in globular clusters. Extensive dynamical simulations leave no doubt that stellar collisions occur in sufficiently dense systems, and simple arguments lead to the inevitable conclusion that there is easily enough mass potentially available in these systems to produce objects with masses exceeding $1000\,M_\odot$. In many ways, this process provides a "natural" mechanism for IMBH formation, but critical aspects of the evolution of the merger product—specifically, the role of stellar winds and the end result of the evolution of very massive stars—remain poorly understood.

The connection between young, compact, and centrally concentrated clusters needed to form IMBHs and the old globular clusters we see today is tortuous, but plausible arguments can be made that early mass segregation, followed by stellar mass loss, and then heating by a central IMBH could expand an initially compact cluster into a globular-cluster-sized system by the present day.

Current studies of the dynamics of dense stellar systems, and specifically the runaway merger problem, are hampered by the structure of the "kitchen-sink" codes used to carry out the simulations. Existing packages are extremely capable but also very complex, inhibiting experimentation and imposing artificial constraints on the scientific questions that can be addressed. New modeling techniques that allow independently written modules to be freely combined and compared promise major advances in the near future.

This work has been suppported by NASA grants NNG04GL50G and NNX07AG95G, and by NSF grant AST-0708299.

REFERENCES

AARSETH, S. J. 1999 *PASP* **111**, 1333.

AARSETH, S. J. 2003 *Gravitational N-Body Simulations*. Cambridge University Press.

BAHCALL, J. N. & WOLF, R. A. 1976 *ApJ* **209**, 214.

BAUMGARDT, H., HUT, P., MAKINO, J., MCMILLAN, S. L. W., & PORTEGIES ZWART, S. F. 2003a *ApJ* **582**, L21.

BAUMGARDT, H. & MAKINO, J. 2003 *MNRAS* **340**, 227.

BAUMGARDT, H., MAKINO, J., & EBISUZAKI, T. 2004 *ApJ* **613**, 1133.

BAUMGARDT, H., MAKINO, J., & HUT, P. 2005 *ApJ* **620**, 238.

BAUMGARDT, H., MAKINO, J., HUT, P., MCMILLAN, S. L. W., & PORTEGIES ZWART, S. F. 2003b *ApJ* **589**, L25.

BELCZYNSKI, K., KALOGERA, V., & BULIK, T. 2002 *ApJ* **572**, 407.

BELKUS, VAN BEVER, J., & VANBEVEREN, D. 2007 *ApJ* **659**, 1576.

CHURCH, R. 2006 Ph.D. thesis, Cambridge University.

DULL, J. D., COHN, H. N., LUGGER, P. M., MURPHY, B. W., SEITZER, P. O., CALLANAN, P. J., RUTTEN, R. G. M., & CHARLES, P. A. 1997 *ApJ* **481**, 267.

EGGLETON, P. P. 2006 *Evolutionary Processes in Binary and Multiple Stars*. Cambridge University Press.

EGGLETON, P. P., FITCHETT, M. J., & TOUT, C. A. 1989 *ApJ* **347**, 998.

FREEMAN, K. C. 1993. In *Galactic Bulges* (eds. H. Dejonghe & H. J. Habing). IAU Symposium 153, p. 263. Kluwer.

FREGEAU, J. M., GÜRKAN, M. A., JOSHI, K. J., & RASIO, F. A. 2003 *ApJ* **593**, 772.

FREITAG, M. & BENZ, W. 2005 *MNRAS* **358**, 1133.

FREITAG, M., GÜRKAN, M. A., & RASIO, F. A. 2006 *MNRAS*, **368**, 141.

FREITAG, M., RASIO, F. A., & BAUMGARDT, H. 2006 *MNRAS* **368**, 121.

GEBHARDT, K., RICH, R. M. R., & HO, L. C. 2002 *ApJ* **578**, L41.

GEBHARDT, K., RICH, R. M. R., & HO, L. C. 2005 *ApJ* **634**, 1093.

GERSSEN, J., VAN DER MAREL, R. P., GEBHARDT, K., GUHATHAKURTA, P., PETERSON, R., & PRYOR, C. 2002 *AJ* **124**, 327

GERSSEN, J., VAN DER MAREL, R. P., GEBHARDT, K., GUHATHAKURTA, P., PETERSON, R., & PRYOR, C. 2003 *AJ* **125**, 376.

GÜRKAN, M. A., FREGEAU, J. M., & RASIO, F. A. 2006 *ApJ* **640**, 39.

HARRIS, W. E. 1996 *AJ* **112**, 1487.

HEGGIE, D. C., HUT, P., MINISHIGE, S., MAKINO, J., & BAUMGARDT, H. 2007 *PASJ* **59**, 507.

HURLEY, J. R., POLS, O. R., & TOUT, C. A. 2000, *MNRAS* **315**, 543.

KAARET, P., PRESTWICH, A. H., ZEZAS, A., MURRAY, S. S., KIM, D.-W., KILGARD, R. E., SCHLEGEL, E. M., & WARD, M. J. 2001 *MNRAS* **321**, L29.

KROUPA, P. 2001 *MNRAS* **322**, 231.

KULKARNI, S. R., HUT, P., & MCMILLAN, S. L. W. 1993 *Nature* **364**, 421.

LANGER, N., HAMANN, W-R., LENNON, M., NAJARRO, F., PAULDRACH, A. W. A., & PULS, J. 1994 *A&A* **290**, 819.

LOMBARDI, J. C. 2007 MODEST-8 workshop, Bonn

LOMBARDI, J. C., THRALL, A. P., DENEVA, J. S., FLEMING, S. W., GRABOWSKI, P. E. 2003 *MNRAS* **345**, 762.

MACCARONE, T. J., KUNDU, A., ZEPF, S. E., & RHODE, K. L. 2007 *Nature* **445**, 183.

MACKEY A. D. & GILMORE, G. F. 2003 *MNRAS* **338**, 120.

MACKEY, A. D., WILKINSON, M. I., DAVIES, M. B., & GILMORE, G. F. 2007 *MNRAS* **379**, 40.

MAKINO, J., TAIJI, M., EBISUZAKI, T., & SUGIMOTO, D. 1997 *ApJ* **480**, 432.

MARCHANT, A. B. & SHAPIRO, S. L. 1980 *ApJ* **239**, 685.

MATSUMOTO, H. & TSURU, T. G. 1999 *PASJ* **51**, 321.

MATSUMOTO, H., TSURU, T. G., KOYAMA, K., AWAKI, H., CANIZARES, C. R., KAWAI, N., MATSUSHITA, S., & KAWABE, R. 2001 *ApJ* **547**, L25.

MCCRADY, N., GILBERT, A. M., & GRAHAM, J. R. 2003 *ApJ* **596**, 240.

MCMILLAN, S. L. W. & PORTEGIES ZWART, S. F. 2007. In *Massive Stars in Interactive Binaries* (eds. N. St.-Louis & A. F. J. Moffat). ASP Conf. Ser. 367, p. 697. Astronomical Society of the Pacific.

MERLONI, A., HEINZ, S., & DI MATTEO, T. 2003 *MNRAS* **345**, 1057.

MERRIT, D. & FERRARESE, L. 2001 *ApJ* **547**, 140.

MERRIT, D., PIATEK, S., PORTEGIES ZWART, S. F., & HEMSENDORF, M. 2004 *ApJ* **608**, 25.

MEYLAN, G., SARAJEDINI, A., JABLONKA, P., DJORGOVSKI, S. G., BRIDGES, T., & RICH, R. M. 2001 *AJ* **122**, 830.

MILLER, M. C. & COLBERT, E. J. M. 2004 *Int. J. Mod. Phys. D* **13**, 1.

MILLER, M. C. & HAMILTON, D. P. 2002 *MNRAS* **330**, 232.

MORRIS, P. W., WATERS, L. B. F. M., BARLOW, M. J., LIM, T., DE KOTER, A., VOORS, R. H. M., COX, P., DE GRAAUW, TH., HENNING, TH., HONY, S., LAMERS, H. J. G. L. M., MUTSCHKE, H., & TRAMS, N. R. 1999 *Nature* **402**, 502.

NOYOLA, E. & GEBHARDT, K. 2006 *AJ* **132**, 447.

NOYOLA, E., GEBHARDT, K. J., & BERGMANN, M. 2006. In *New Horizons in Astronomy: Frank N. Bash Symposium* (ed. S. J. Kannappan, S. Redfield, J. E. Kessler-Silacci, M. Landriau, & N. Drory). ASP Conf. Ser. 352, p. 269. Astronomical Society of the Pacific.

NOYOLA, E., GEBHARDT, K. J., & KISSLER-PATIG, M. 2008, in preparation.

O'LEARY, R. M., RASIO, F. A., FREGEAU, J. M., IVANOVA, N., O'SHAUGHNESSY, R. 2006 *ApJ* **637**, 937.

PETERS, P. C. & MATHEWS, J. 1963 *Phys. Rev. D* **131**, 345.

PHINNEY, E. S. 1993. In *Structure and Dynamics of Globular Clusters* (eds. S. Djorgovski & G. Meylan). p. 141. Astronomical Society of the Pacific.

POOLEY, D. & RAPPAPORT, S. 2006 *ApJ* **644**, 45.

PORTEGIES ZWART, S. F., BAUMGARDT, H., HUT, P., MAKINO, J., & MCMILLAN, S. L. W. 2004 *Nature* **428**, 724.

PORTEGIES ZWART, S. F., BELLMANN, R. G., & GELDOF, P. M. 2007 *New Astronomy* **12**, 641.

PORTEGIES ZWART, S. F., DEWI, J., & MACCARONE, T. 2005 *Ap&SS* **300**, 247.

PORTEGIES ZWART, S. F., MAKINO, J., MCMILLAN, S. L. W., & HUT, P. 1999 *A&A* **348**, 117.

PORTEGIES ZWART, S. F. & MCMILLAN, S. L. W. 2000 *ApJ* **528**, 17.

PORTEGIES ZWART, S. F. & MCMILLAN, S. L. W. 2002 *ApJ* **576**, 899.

PORTEGIES ZWART, S. F., MCMILLAN, S. L. W., HUT, P., & MAKINO, J., ET AL. 2001 *MNRAS* **321**, 199.

PORTEGIES ZWART, S. F., MCMILLAN, S. L. W., O'NUALLÁIN, B., HEGGIE, D. C., LOMBARDI, J., HUT, P., BANERJEE, S., BELKUS, H., FRAGOS, T., FREGEAU, J., FUJII, M., GABUROV, E., HARFST, S., IZZARD, R., JURIĆ, M., JUSTHAM, S., TEUBEN, P., VAN BEVER, J., YARON, O., & ZEMP, M. 2008. In *Simulation of Multiphysics Multiscale Systems* (ed. R. Bellman). Computational Science–ICCS 2008, p. 207. Springer.

PORTEGIES ZWART, S. F. & VERBUNT, F. 1996 *A&A* **309**, 179.

SALPETER, E. E. 1955 *ApJ* **121**, 161.

SIGURDSSON, S. & HERNQUIST 1993 *Nature*, **364**, 423.

SILLS, A., DEITERS, S., EGGLETON, P., FREITAG, M., GIERSZ, M., HEGGIE, D., HURLEY, J., HUT, P., IVANOVA, N., KLESSEN, R. S., KROUPA, P., LOMBARDI, J. C., MCMILLAN, S. L. W., PORTEGIES ZWART, S. F., ZINNECKER, H. 2003 *New Astronomy* **8**, 605.

SORIA, R. 2006. In *Populations of High Energy Sources in Galaxies* (eds. E. J. A. Meurs & G. Fabbiano). IAU Symp. 230, p. 473. Cambridge University Press.

SORIA, R. 2007 *Ap&SS* **311**, 213.

STROHMAYER, T. E. & MUSHOTSKY, R. F. 2003, *ApJ* **586**, L61.

SUZUKI, T. K., NAKASATO, N., BAUMGARDT, H., IBUKIYAMA, A., MAKINO, J., & EBISUZAKI, T. 2007 *ApJ* **668**, 435.

TAKAHASHI, K. & PORTEGIES ZWART, S. F. 2000 *ApJ* **535**, 759.

TREMAINE, S., GEBHARDT, K., BENDER, R., BOWER, G., DRESSLER, A., FABER, S. M., FILLIPENKO, A. V., GREEN, R., GRILLMAIR, C., HO, L. C., KORMENDY, J., LAUER, T., MAGORRIAN, J., PINKNEY, J., & RICHSTONE, D. 2002 *ApJ* **574**, 740.

ULVESTAD, J. S., GREENE, J., & HO, L. C. 2007 *ApJ* **661**, 151.

VANBEVEREN, D., BELKUS, H., VAN BEVER, J., & MENNEKENS, N. 2009 *Ap&SS* **324**, 271.

VESPERINI, E., MCMILLAN, S. L. W., & PORTEGIES ZWART, S. F. 2008. In *Dynamical Evolution of Dense Stellar Systems* (eds. E. Vesperini, M. Giersz, & A. Sills). IAU Symp. 246, p. 181. Cambridge University Press.

YUNGELSON, L. R., VAN DEN HEUVEL, E. P. J., VINK, J. S., PORTEGIES ZWART, S. F., & DE KOTER, A. 2008 *A&A* **477**, 223.

Evolution of massive black holes

By MARTA VOLONTERI

University of Michigan, Ann Arbor, MI 48109, USA

Supermassive black holes are nowadays believed to reside in most local galaxies. Accretion of gas and black-hole mergers play a fundamental role in determining the two parameters defining a black hole: mass and spin. I briefly review here some of the physical processes that are conducive to the evolution of the massive black-hole population. I'll discuss black-hole formation processes that are likely to place at early cosmic epochs, and how massive black holes evolve in a hierarchical universe. The mass of the black holes that we detect today in nearby galaxy has mostly been accumulated by accretion of gas. While black-hole–black-hole mergers do not contribute substantially to the final mass of massive black holes, they influence the occupancy of galaxy centers by black hole, owing to the chance of merging black holes being kicked from their dwellings due to the "gravitational recoil." Similarly, accretion leaves a deeper imprint on the distribution of black-hole spins than black-hole mergers do. The differences in accretion histories for black holes hosted in elliptical or disk galaxies may reflect on different spin distributions.

1. Introduction

Black holes (BHs), as physical entities, span the full range of masses, from tiny BHs predicted by string theory, to monsters weighing by themselves almost as much as a dwarf galaxy (massive black holes, MBHs). Notwithstanding the several orders of magnitude difference between the smallest and the largest BH known, we believe that all of them can be described by only three parameters: mass, spin, and charge. Astrophysical BHs are even simpler system, as charge can be neglected as well. The interaction between astrophysical BHs and their environment is where complexity enters the game. I will focus here on the formation and evolution of MBHs with masses above thousands of solar masses, and how we believe their evolution is symbiotic with that of their host.

Let's start by recalling that MBHs in galaxy centers are far from being really "black." We can easily trace their presence, as they are the engines powering the luminous quasars that have been detected up to high redshift. Nowadays we can detect the dead remnants of this bright past activity in neighboring galaxies. It is indeed well established that the centers of most local galaxies host MBHs with masses in the range $M_{\rm BH} \sim 10^6$–$10^9 \, M_\odot$ (e.g., Ferrarese & Merritt 2000; Kormendy & Gebhardt 2001; Richstone et al. 1998). The MBH population may extend down to the smallest masses. For example, the dwarf Seyfert 1 galaxy POX 52 is thought to contain a BH of mass $M_{\rm BH} \sim 10^5 \, M_\odot$ (Barth et al. 2004). At the other end, however, the Sloan Digital Sky survey detected luminous quasars at very high redshift, $z > 6$. Follow-up observations confirmed that at least some of these quasars are powered by supermassive black holes with masses $\simeq 10^9 \, M_\odot$ (Barth et al. 2003; Willott et al. 2005). We are therefore left with the task of explaining the presence of very large MBHs when the universe is less than 1 Gyr old, and of much smaller BHs lurking in 13 Gyr-old galaxies.

2. Massive black holes in a hierarchical universe

The demography of massive black holes in the local universe has been clarified in the last ten years by studies of the central regions of relatively nearby galaxies (mainly with quiescent nuclei). The mass of MBHs detected in neighboring galaxies scales with the

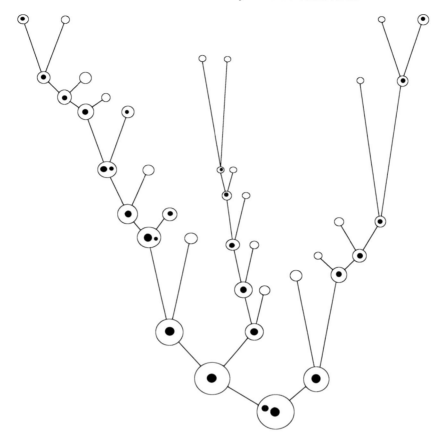

FIGURE 1. A diagram of the assembly of a galaxy and its central black hole in cold dark-matter cosmology. Time increases from top to bottom, and every junction between two "branches" of the merger tree marks a galaxy merger. The central MBHs are indicated by black dots (not to scale). In this case, the final galaxy is assembled from the merger of twenty smaller galaxies, containing a total of four seed black holes, resulting in four mergers of binary black holes.

bulge luminosity—or stellar velocity dispersion—of their host galaxy (Ferrarese & Merritt 2000; Gebhardt et al. 2000; Tremaine et al. 2002), suggests a single mechanism for assembling black holes and forming spheroids in galaxy halos. The evidence is therefore in favor of co-evolution between galaxies, black holes and quasars. In the currently favored cold dark-matter cosmogonies (Spergel et al. 2007), present-day galaxies have been assembled via a series of mergers from small-mass building blocks which form at early cosmic times. In this paradigm, galaxies experience multiple mergers during their lifetime. If most galaxies host BHs in their center, and a local galaxy has been made up by multiple mergers, then a black-hole binary is a natural evolutionary stage. During each galaxy merger event, the central black hole already present in each galaxy would be dragged to the center of the newly formed galaxy via dynamical friction (≈ 0.1–10 pc), and then if/when they get close (≈ 0.01–0.001 pc) the black-hole binary would coalesce via emission of gravitational radiation (Figure 1). The gap between the point where dynamical friction ceases to be efficient ($a_h \simeq Gm_2/(4\sigma_*^2)$) and the emission of gravitational waves takes over—at binary separations about two orders of magnitude smaller—could be the bottleneck of the merger process.

In gas-poor systems, the sub-parsec evolution of the binary while gravitational radiation emission is still negligible may be largely determined by three-body interactions with background stars (Begelman et al. 1980), and by capturing the stars that pass within a distance of the order of the binary semi-major axis and ejecting them at much higher velocities (Quinlan 1996; Milosavljević & Merritt 2001; Sesana et al. 2007a). Dark-matter particles will be ejected by decaying binaries in the same way as the stars, i.e., through the gravitational slingshot. The hardening of the binary modifies the density profile, removing mass interior to the binary orbit, depleting the galaxy core of stars and dark matter, and slowing down further decay. We can use a toy model to understand the typical timescales (Volonteri et al. 2003). If we assume that the stellar mass removal scours a core of radius r_c and constant density $\rho_c \equiv \rho_*(r_c)$ into a pre-existing isothermal sphere, the total mass ejected as the binary shrinks from a_h to $a < a_h$ can be written as

$$\mathcal{M}_{\rm ej} = \frac{4}{3} \frac{\sigma_*^2(r_c)}{G} \ , \tag{2.1}$$

where m_2 is the least massive BH in the binary, and σ_* is the velocity dispersion of the stellar system. The core radius then grows as

$$r_c(t) \approx \frac{3}{4\sigma_*^2} G(m_1 + m_2) \int_{a(t)}^{a_h} \frac{1}{a} \, da \ . \tag{2.2}$$

The binary separation quickly falls below r_c and subsequent evolution is slowed down due to the declining stellar density, with a hardening time $t_h = |a/\dot{a}| = 2\pi r_c(t)^2/(H\sigma_* a)$ that becomes increasingly long as the binary shrinks and $r_c(t)$ increases.

In gas-rich systems, however, the orbital evolution of the central MBH is likely dominated by dynamical friction against the surrounding gaseous medium. The available simulations (Escala et al. 2004; Dotti et al. 2006; Mayer et al. 2008) show that the binary can shrink to about parsec or slightly sub-parsec scale by dynamical friction against the gas, depending on the gas thermodynamics. The interaction between a BH binary and an accretion disk can also lead to a very efficient transport of angular momentum, and drive the secondary BH to the regime where emission of gravitational radiation dominates on short timescales, comparable to the viscous timescale (Armitage & Natarajan 2005; Gould & Rix 2000).

The viscous timescale depends on the properties of the accretion disk and of the binary:

$$t_{\rm vis} = 0.1 \, {\rm Gyr} \, a_{\rm pc}^{3/2} \left(\frac{H}{R}\right)_{0.1}^{-2} \alpha_{0.1}^{-1} \left(\frac{m_1}{10^4 \, M_\odot}\right)^{-1/2} \ , \tag{2.3}$$

where $a_{\rm pc}$ is the initial separation of the binary when the secondary MBH starts interacting with the accretion disk in units of parsec, (h/r) is the aspect ratio of the accretion disk, $h/r = 0.1$ above, α is the Shakura & Sunyaev viscosity parameter, $\alpha = 0.1$ above, and m_1 is the mass of the primary MBH, in solar masses. The emission of gravitational waves takes over the viscous timescales at a separation (Armitage & Natarajan 2005):

$$a_{\rm GW} = 10^{-8} \, {\rm pc} \left(\frac{H}{R}\right)_{0.1}^{-16/5} \alpha_{0.1}^{-8/5} q_{0.1}^{3/5} \left(\frac{m_1}{10^4 \, M_\odot}\right) \ , \tag{2.4}$$

where $q = m_2/m_1 \lesssim 1$, is the binary mass ratio. The timescale for coalescence by emission of gravitational waves from $a_{\rm GW}$ is much shorter than the Hubble time:

$$t_{\rm gr} = 2.3 \times 10^6 \, {\rm yr} \left(\frac{a(t)}{0.1 \, {\rm pc}}\right)^4 \left(\frac{m_1 m_2 (m1 + m2)}{2 \times 10^{18} \, M_\odot}\right) \ . \tag{2.5}$$

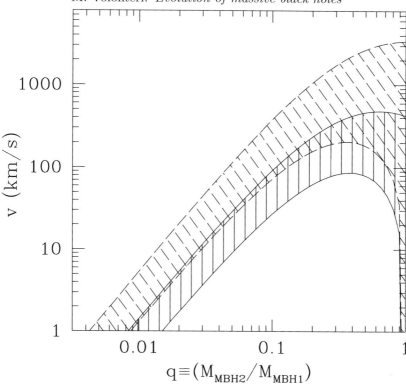

FIGURE 2. Recoil velocity of spinning black holes as a function of binary mass ratio, q. Solid curves: spin axis aligned (or anti-aligned) with the orbital angular momentum (Baker et al. 2007). Dashed curves: spin axis in the orbital plane (Campanelli et al. 2007). For every mass ratio we plot the combination of spins, $\hat{a}_1$ and $\hat{a}_2$, which minimizes (lower curves) or maximizes (upper curves) the recoil velocity (from Volonteri 2007).

Hence, the physical processes driving the evolution of MBH binaries are likely *redshift and environmental* dependent. Fast mergers are probably common in high-redshift young galaxies, and sluggish binaries might reach an impasse in low-redshift, gas-poor spheroids.

Somehow counter-intuitively fast MBH mergers at very high redshift can bring an overall damage to the growth of the MBH population, rather than contribute to the build-up of more massive holes. This is due to the so-called "gravitational recoil" (a.k.a. rocket, kick). When the members of a black-hole binary coalesce, the center of mass of the coalescing system recoils due to the non-zero net linear momentum carried away by gravitational waves in the coalescence. This recoil could be so violent that the merged hole breaks loose from shallow potential wells, especially in small-mass pregalactic building blocks (Figure 2).

Comparing the recoil velocity to the escape velocity from their hosts, Volonteri (2007) find that the fraction of "lost" binaries is very high (>50–90%) at $z > 10$, but it decreases at later times due to a combination of (i) the mass ratio distribution becoming shallower and, (ii) the hierarchical growth of the hosts (Figure 3). Schnittman (2007) shows in a very elegant way that, even for large recoils, the very hierarchical nature of structure evolution ensures that a substantial fraction of galaxies retain their BHs if evolution proceeds over a long series of mergers (see also Menou et al. 2001). Since, especially at high redshift, binaries represent the exception rather than the rule, the possible ejection

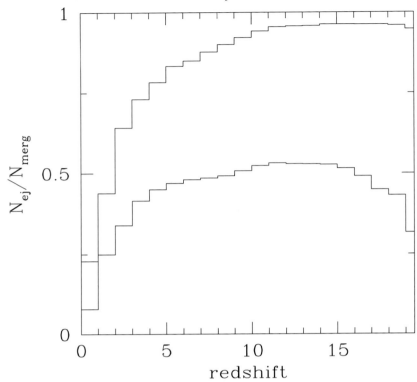

FIGURE 3. Fraction of ejected binaries as a function of redshift for Schwarzschild MBHs (lower histogram) and spinning MBHs (upper histogram), assuming isotropic orbits. A binary is defined as ejected if the recoil velocity is larger than the escape velocity from the host.

of most binaries before $z \simeq 5$ is not the threat to the evolution of the MBH population that has been detected in nearby galaxies.

Although the gravitational recoil does not do damage to the evolution of the MBH population that we locally observe, it can be dangerous in very special cases. Haiman (2004) points out that the recoil can be indeed threatening the growth of MBHs that are believed to be powering the luminous quasars at $z \simeq 6$ detected in the Sloan survey (e.g., Fan et al. 2001). In fact, in such a biased volume, the density of halos where MBH formation can be efficient (either by direct collapse, or via Pop III stars) is highly enhanced. The net result is an higher fraction of binary systems, and binarity is especially common for the central galaxy of the main halo. While the "average" MBH experiences at most one merger in its lifetime, an MBH hosted in a rare, exceptionally massive halo can experience up to a few tens of mergers, and the probability of ejecting the central MBH, halting its growth, is 50–80% at $z > 6$ (Volonteri & Rees 2006). This implies that MBHs at high redshift do not primarily grow via mergers.

3. Scenarios for massive black-hole formation

A single big galaxy can be traced back to the stage when it was split up in hundreds of smaller components with individual internal velocity dispersions as low as 20 km s^{-1}. Did black holes form with the same efficiency in small galaxies (with shallow potential wells), or did their formation have to await the buildup of substantial galaxies with deeper potential wells?

The formation of massive black holes is far less understood than those of their light, stellar-mass counterparts. The "flowing chart" presented by Rees (1978) still stands as a guideline for the possible paths leading to the formation of massive BH seeds in the center of galactic structures. The first possibility is the direct formation of a BH from a collapsing gas cloud (Haehnelt & Rees 1993; Loeb & Rasio 1994; Eisenstein & Loeb 1995; Bromm & Loeb 2003; Koushiappas et al. 2004; Begelman et al. 2006; Lodato & Natarajan 2006). In the most common situations, rotational support can halt the collapse before the densities required for MBH formation are reached. Halos, and their baryonic cores, in fact possess angular momentum, J, believed to be acquired by tidal torques created by interactions with neighboring halos. This can be quantified through the so-called spin parameter, which represents the degree of rotational support available in a gravitational system:

$$\lambda_{\rm spin} \equiv J \frac{|E|^{1/2}}{G M_h^{5/2}} \quad ,$$

where E and M_h are the total energy and mass of the halo.

Let $f_{\rm gas}$ be the gas fraction of a proto-galaxy mass, and f_d the fraction of the gas which can cool; a mass $M = f_d f_{\rm gas} M_h$ would then settle into a rotationally supported disk (Mo et al. 1998; Oh & Haiman 2002) with a scale radius $\simeq \lambda_{\rm spin} r_{\rm vir}$, where $r_{\rm vir}$ is the virial radius of the proto-galaxy. Spin parameters found in numerical simulations are distributed lognormally in $\lambda_{\rm spin}$, with mean $\bar{\lambda}_{\rm spin} = 0.04$ and standard deviation $\sigma_\lambda = 0.5$ (e.g., Bullock et al. 2001; van den Bosch et al. 2002). The tidally induced angular momentum would therefore be enough to provide centrifugal support at a distance $\simeq 20$ pc from the center, and halt collapse. Additional mechanisms inducing transport of angular momentum are needed to further condense the gas.

The loss of angular momentum can be driven either by (turbulent) viscosity or by global dynamical instabilities, such as the "bars-within-bars" mechanism (Shlosman et al. 1989; Begelman et al. 2006). The gas can therefore condense to form a central massive object, either a supermassive star, which eventually becomes subject to post-Newtonian gravitational instability and forms a seed BH, or via a low-entropy star-like configuration where a small black hole forms in the core and grows by accreting the surrounding envelope (Begelman, Rossi, & Armitage 2008). The masses of the seeds predicted by different models vary, but they are typically in the range $M_{\rm BH} \sim 10^4$–$10^6 \, M_\odot$.

Alternatively, the seeds of MBHs can be associated with the remnants of the first generation of stars, formed out of zero-metallicity gas. The first stars are believed to form at $z \gtrsim 10$ in halos which represent high-σ peaks of the primordial density field. With the absence of metals, the main coolant is molecular hydrogen, which is rather inefficient. This inefficient cooling might lead to a very top-heavy initial stellar-mass function, and in particular, to the production of very massive stars with masses $> 100 \, M_\odot$ (Carr, Bond, & Arnett 1984). If very massive stars form above $260 \, M_\odot$, they would rapidly collapse to massive BHs with little mass loss (Fryer et al. 2001), i.e., leaving behind seed BHs with masses $M_{\rm BH} \sim 10^2$–$10^3 \, M_\odot$ (Madau & Rees 2001; Volonteri et al. 2003).

3.1. *Observational tests of MBH-formation scenarios*

What are the possible observational tests of MBH-formation scenarios? Detection of gravitational waves from seeds merging at the redshift of formation (Sesana et al. 2007b) is probably one of the best ways to discriminate among formation mechanisms. The planned *Laser Interferometer Space Antenna* (*LISA*) in principle will be sensitive to

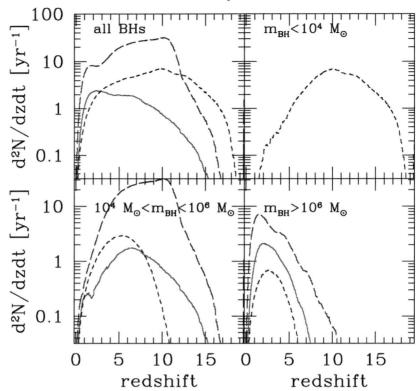

FIGURE 4. Predicted rate of MBH binary coalescences per unit redshift, in different MBH mass intervals: solid line - MBH seeds from Pop III stars; long-dashed line - MBH seeds from direct collapse (Koushiappas et al. 2004); short-dashed line - MBH seeds from direct collapse (Begelman, Volonteri, & Rees 2006). Adapted from Sesana et al. 2007b.

gravitational waves from binary MBHs with masses in the range 10^3–10^6 $M_\odot$, basically at any redshift of interest. A large fraction of coalescences will be directly observable by *LISA*, and on the basis of the detection rate, constraints can be put on the MBH-formation process. Different theoretical models for the formation of MBH seeds and dynamical evolution of the binaries predict merger rates that largely vary one from the other (Figure 4; Sesana et al. 2007b).

The imprint of different formation scenarios can also be sought in observations at lower redshifts (Volonteri, Lodato, & Natarajan 2008). Since during the quasar epoch MBHs increase their mass by a large factor, signatures of the seed-formation mechanisms are likely more evident at *earlier epochs*. Figure 5 compares the integrated comoving mass density in MBHs to the expectations from Soltan-type arguments (F. Haardt, private communication), assuming that quasars are powered by radiatively efficient flows (for details, see Yu & Tremaine 2002; Elvis et al. 2002; Marconi et al. 2004). The curves differ only with respect to the MBH-formation scenario. We either assume that seeds are Population III remnants (black curve), or that seeds are formed via direct collapse with different efficiencies (Lodato & Natarajan 2006). While during and after the quasar epoch the mass densities in our theoretical models differ by less than a factor of 2, at $z > 3$ the differences become more pronounced. The comoving mass density, an integral constraint, is reasonably well determined out to $z = 3$, but is still poorly known at higher

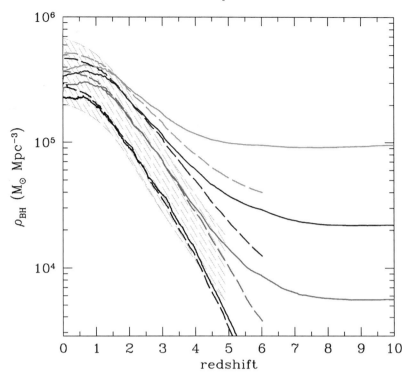

FIGURE 5. Integrated black-hole mass density as a function of redshift. Solid lines: total mass density locked into nuclear black holes. Dashed lines: integrated mass density *accreted* by black holes. Models based on BH remnants of Population III stars (lowest curve), models based on direct collapse (Lodato & Natarajan 2006), with different efficiencies. Shaded area: constraints from Soltan-type arguments, where we have varied the radiative efficiency from a lower limit of 6% (applicable to Schwarzschild MBHs, upper envelope of the shaded area), to about 20% (Wang et al. 2006).

redshifts. The increasing area and depth of high-redshift survey, especially in x-rays, will increase the strength of our constraints (Salvaterra 2007).

In our neighborhood, the best diagnostic of MBH-formation mechanisms would be the measure of MBH masses in low-luminosity galaxies. This can be understood in terms of the cosmological bias. The progenitors of massive galaxies (or clusters of galaxies) have both a high probability of hosting MBH seeds (cf. Madau & Rees 2001), and a high probability that the central MBH is not "pristine," that is, it has increased its mass by accretion, or it has experienced mergers and dynamical interactions. In the case of low-bias systems, such as isolated dwarf galaxies, very few of the high-z progenitors have the deep potential wells needed for gas retention and cooling, a prerequisite for MBH formation. The signature of the efficiency of the formation of MBH seeds will consequently be stronger in isolated dwarf galaxies. Hence, MBH-formation models are distinguishable at the low mass end of the BH-mass function, while at the high-mass end the effect of initial seeds appears to be sub-dominant. The clearest signature of massive seeds, compared to Population III remnants, would be a lower limit of the order of the typical mass of seeds to the mass of MBHs in galaxy centers, as shown in Figure 6. Additionally, the fraction of galaxies without an MBH increases with decreasing halo masses at $z = 0$. A larger fraction of low mass halos are devoid of central black holes

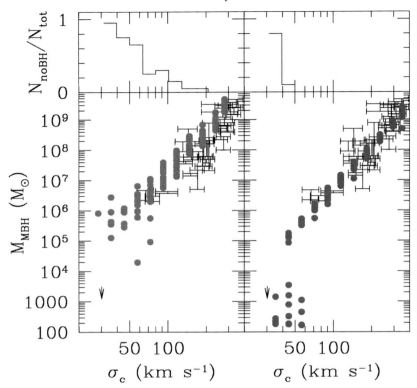

FIGURE 6. The M_{bh}−velocity dispersion (σ_c) relation at $z = 0$. Every circle represents the central MBH in a halo of given σ_c. Observational data are marked by their quoted error bars, both in σ_c, and in M_{bh} (Tremaine et al. 2002). Left panel: direct collapse seeds, Population III star seeds. Top panels: fraction of galaxies at a given velocity dispersion which *do not* host a central MBH. Adapted from Volonteri, Lodato, & Natarajan (2008).

for lower seed formation efficiencies. While current data in the low-mass regime is scant, future campaigns with the Giant Magellan Telescope or the *James Webb Space Telescope* (*JWST*) are likely to probe this region of parameter space with significantly higher sensitivity.

4. Accretion: mass growth and quasar activity

Accretion is inevitable during the "active" phase of a galactic nucleus. Observations tell us that AGN are widespread in both the local and early universe. All the information that we have gathered on the evolution of MBHs is indeed due to studies of AGN, as we have to await for *LISA* to be able to "observe" quiescent MBHs in the distant universe. A key issue is then the relative importance of mergers and accretion in the build-up of the black holes, in dependence of the host properties (mass, redshift, environment).

The accretion of mass at the Eddington rate would cause a black-hole mass to increase in time as

$$M(t) = M(0) \exp\left(\frac{1-\epsilon}{\epsilon} \frac{t}{t_{Edd}}\right), \tag{4.1}$$

where $t_{Edd} = 0.45\,\mathrm{Gyr}$ and ϵ is the radiative efficiency. The classic argument of Soltan (1982) compares the total mass of black holes today with the total radiative output by known quasars, by integration over redshift and luminosity of the luminosity function of quasars (Yu & Tremaine 2002; Elvis et al. 2002; Marconi et al. 2004). The total energy density can then be converted into the total mass density accreted by black holes during the active phase, by assuming a mass-to-energy conversion efficiency, ϵ (Aller & Richstone 2002; Merloni et al. 2004; Elvis et al. 2002; Marconi et al. 2004). The similarity of the total mass in MBHs today and the total mass accreted by MBHs implies that the last 2–3 e-folds of the mass is grown via radiatively efficient accretion, rather than accumulated through mergers or radiatively inefficient accretion. However, most of the 'e-folds' (corresponding to a relatively small amount of mass, say the first 10% of mass) could be gained rapidly via, e.g., radiatively inefficient accretion. This argument is particularly important at early times.

The Sloan Digital Sky Survey detected luminous quasars at very high redshift, $z > 6$, when the universe was less than 1 Gyr old. Follow-up observations confirmed that at least some of these quasars are powered by supermassive black holes with masses $\simeq 10^9 \ M_\odot$ (Barth et al. 2003; Willott et al. 2005). Given a seed mass $M(0)$ at $z = 50$ or less, the higher the efficiency, the longer it takes for the MBH to grow in mass by, say, 10 e-foldings. If accretion is radiatively efficient, via a geometrically thin disk, the alignment of an MBH with the angular momentum of the accretion disk tends to efficiently spin holes up (see Section 5), and radiative efficiencies can therefore approach 30–40%. With such a high efficiency, $\epsilon = 0.3$, it can take longer than 2 Gyr for the seeds to grow up to a billion solar masses.

Let us consider the extremely rare, high-redshift (say, $z > 15$) metal-free halos with virial temperatures $T_{\rm vir} > 10^4$ K, where gas can cool even in the absence of H_2 via neutral hydrogen atomic lines. The baryons can therefore collapse until angular momentum becomes important. Afterward, gas settles into a rotationally supported dense disk at the center of the halo (Mo et al. 1998; Oh & Haiman 2002). This gas can supply fuel for accretion onto an MBH within it. Estimating the mass accreted by the MBH within the Bondi-Hoyle formalism, the accretion rate is initially largely above the Eddington limit (Figure 7; Volonteri & Rees 2005). When the supply is super-critical the excess radiation can be trapped, as radiation pressure cannot prevent the accretion rate from being super-critical, while the emergent luminosity is still Eddington limited in case of spherical or quasi-spherical configurations (Begelman 1979; Begelman & Meier 1982). In the spherical case, though this issue remains unclear, it still seems possible that when the inflow rate is super-critical, the radiative efficiency drops so that the hole can accept the material without greatly exceeding the Eddington luminosity. The efficiency could be low either because most radiation is trapped and advected inward, or because the flow adjusts so that the material can plunge in from an orbit with small binding energy (Abramowicz & Lasota 1980). The creation of a radiation-driven outflow, which can possibly stop the infall of material, is also a possibility. If radiatively inefficient, super-critical accretion requires metal-free conditions in exceedingly rare massive halos, rapid early growth, therefore, can happen for only a tiny fraction of MBH seeds. These MBHs are powering the most luminous high-redshift quasars, and later on will be found in the most biased halos. The global MBH population, instead, evolves at a more quiet and slow pace.

5. Probing the other hair of astrophysical black holes

Astrophysical BHs are characterized by just two parameters, mass and spin, which measure the angular momentum of the hole. The spin is typically expressed via the

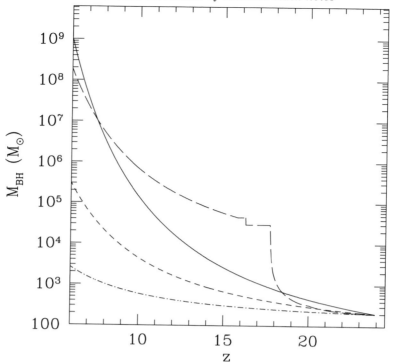

FIGURE 7. Growth of an MBH mass under different assumptions for the accretion rate and efficiency. Eddington limited accretion: $\epsilon = 0.1$ (solid line), $\epsilon = 0.2$ (short-dashed line), $\epsilon = 0.4$ (dot-dashed line). Radiatively inefficient super-critical accretion, as in Volonteri & Rees 2005 (long-dashed line).

dimensionless parameter $\hat{a} \equiv J_h/J_{max} = c\,J_h/G\,M_{BH}^2$, where J_h is the angular momentum of the black hole.

The spin of a hole affects the efficiency of the "classical" accretion processes themselves; the value of $\hat{a}$ in a Kerr BH also determines how much energy is, in principle, extractable from the hole itself. Assuming that relativistic jets are powered by rotating black holes through the Blandford-Znajek mechanism, the so-called "spin paradigm" asserts that powerful relativistic jets are produced in AGN with fast-rotating black holes (Blandford et al. 1990).

Spin-up is a natural consequence of prolonged disk-mode accretion: any hole that has, for instance, doubled its mass by capturing material with constant angular-momentum axis would end up spinning rapidly, close to the maximum allowed value (Bardeen 1970; Thorne 1974). However, when an accretion disk does not lie in the equatorial plane of the BH, that is, when the angular momentum of the accretion disk is misaligned with respect to the direction of J_h, accretion of counter-rotating material can cause the spin-down of MBHs, at least under particular conditions. A misaligned disk is subject to the Lense-Thirring precession, which tends to align the inner parts of the disk with the angular momentum of the black hole, causing the inclination angle between the angular momentum vectors to decrease with decreasing distance from the MBH, forcing the inner parts of the accretion disk to rotate in the equatorial plane of the MBH (Bardeen & Petterson 1975). The orbits can be co-rotating or counter-rotating depending the value of J_d/J_h.

King et al. (2005) indeed argue that accretion of counter-rotating material is more common than accretion of co-rotating material. The counter-alignment condition depends on the ratio $0.5\,J_d/J_h$, where J_h and J_d are the angular momenta of the hole and of the disk, to be compared with the cosine of the inclination angle, ϕ. If $\cos\phi < -0.5\,J_d/J_h$, the counter-alignment condition is satisfied.

However, sustained accretion from a twisted disk would align the MBH spin (and the innermost equatorial disk) with the angular momentum vector of the disk at large radii (Scheuer & Feiler 1996). If the disk was initially counter-rotating with respect to the MBH, a complete flip-over would eventually occur, and then accretion of co-rotating material would act to spin up the MBH (Bardeen 1970). Early work by Moderski et al. (1998) concluded that the Bardeen-Petterson effect can be neglected because the alignment time (10^7 years; Rees 1978) is longer than the duration of a single accretion event. Later, however, a series of papers revised the alignment timescale, suggesting that it could be much shorter (Scheuer & Feiler 1996; Natarajan & Pringle 1998). This framework was investigated by Volonteri et al. (2005), who argue that the lifetime of quasars is long enough that angular-momentum coupling between black holes and accretion disks through the Bardeen-Petterson effect effectively forces the innermost region of accretion disks to align with black-hole spins (possibly through spin flips), and hence all AGN black holes should have large spins.

In this context, Volonteri et al. (2007) have explored the dependence of the alignment timescale in a Shakura & Sunyaev (1973) disk on: viscosity ν_2,† black-hole mass M_{BH}, Eddington ratio f_{Edd}, accreted mass Δm.

Within this simple picture, the timescale for disk-BH alignment can be estimated as

$$t_{\mathrm{align}} \simeq \frac{J_h}{J_d(R_w)}\, t_{\mathrm{acc}}(R_w) \ , \tag{5.1}$$

where t_{acc} is simply the accretion timescale, $t_{\mathrm{acc}} = R_w^2/\nu_1$, and R_w marks the transition between (inner) alignment and (outer) misalignment. R_w corresponds to the location in the disk where the timescale for radial diffusion of the warp is comparable to the local Lense-Thirring precession timescale (Scheuer & Feiler 1996; Natarajan & Pringle 1998). The warp radius scales with the Schwarzschild radius of the BH as:

$$\frac{R_w}{R_s} = 3.6 \times 10^3 \hat{a}^{5/8} \left(\frac{M_{\mathrm{BH}}}{10^8\,\mathrm{M_\odot}}\right)^{1/8} f_{\mathrm{Edd}}^{-1/4} \left(\frac{\nu_2}{\nu_1}\right)^{-5/8} \alpha^{-1/2} \ . \tag{5.2}$$

Defining the mass accreted during t_{align} as $m_{\mathrm{align}} = t_{\mathrm{align}}\dot{M}$, one gets:

$$m_{\mathrm{align}} \simeq M_{\mathrm{BH}}\, \hat{a} \left(\frac{R_s}{R_w}\right)^{1/2} \ . \tag{5.3}$$

Therefore, for all plausible assumptions, $m_{\mathrm{align}} \ll M_{\mathrm{BH}}$, and a series of many randomly oriented accretion events with accreted mass $\Delta m \ll m_{\mathrm{align}}$ should result in black-hole spin oscillating around zero. For the opposite case of $\Delta m \gg m_{\mathrm{align}}$ the black hole will be spun-up to large positive spins; for $\Delta m \sim M_{\mathrm{BH}}$ the hole will be spun-up to $\hat{a} \sim 1$.

† The viscosity characterizing the alignment of the disk can be different from the accretion-driving viscosity, ν_1, which is responsible for the transfer of the component of the angular momentum parallel to the spin of the disk. The relation between ν_1 and ν_2 is the main uncertainty of the problem, assuming of course that such two-viscosity description is adequate at all. Describing ν_1 by the Shakura–Sunyaev parameter α, one can show that the regime in which $H/R < \alpha \ll 1$ (H being the disk thickness) one has $\nu_1/\nu_2 \approx \alpha^2$ (Papaloizou & Pringle 1983). However, for high accretion rates $\alpha \ll 1$ might not be appropriate, and in such a case, ν_1 is comparable to ν_2 (Kumar & Pringle 1985).

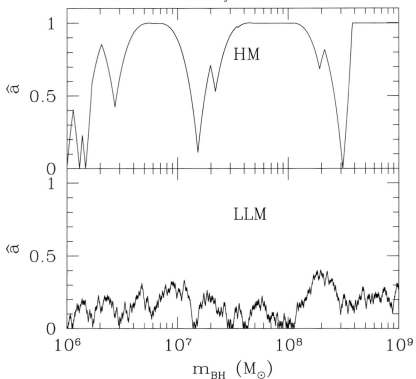

FIGURE 8. Evolution of an MBH spin during a series of accretion episodes lasting for a Hubble time. Initial mass $m_{BH} = 10^6 \, M_\odot$, initial spin $\hat{a} = 10^{-3}$. Lower panel: the accreted mass at every accretion episode is constrained to be less than 0.01 of the MBH mass (LLM). Upper panel: the accreted mass is randomly extracted in the range 0.01–10 times the MBH mass (HM). Adapted from Volonteri, Sikora, & Lasota 2007.

However, both semi-analytical models of the cosmic MBH evolution (Volonteri et al. 2005) and simulations of merger-driven accretion (di Matteo et al. 2005) show that most MBHs increase their mass by an amount $\Delta m \gg m_{\text{align}}$ if the evolution of the LF of quasars is kept as a constraint. These high-Δm values are likely characteristic of the most luminous quasars and most massive black holes—especially at high redshift. We expect therefore that bright quasars at $z > 3$ will have large spins (upper panel in Figure 8). High spins in bright quasars are also indicated by the high radiative efficiency of quasars, as deduced from observations applying the Soltan argument (Soltan 1982; Wang et al. 2006, and references therein).

5.1. *MBH spins and galaxy morphology*

If the events powering quasars coincide with the formation of elliptical galaxies (di Matteo et al. 2005), we might expect that the MBH hosted by an elliptical galaxy had, as its last major accretion episode, a large increase in its mass. During this episode, the spin increased significantly as well, possibly up to very high values.

Black holes in spiral galaxies, on the other hand, probably had their last major merger (i.e., last major accretion episode), if any, at high redshift, so that enough time elapsed for the galaxy disk to reform. Moreover, several observations suggest that single accretion events last $\simeq 10^5$ years in Seyfert galaxies, while the total activity lifetime (based on the fraction of Seyfert disk galaxies) is 10^8–10^9 years (e.g., Kharb et al. 2006; Ho et al.

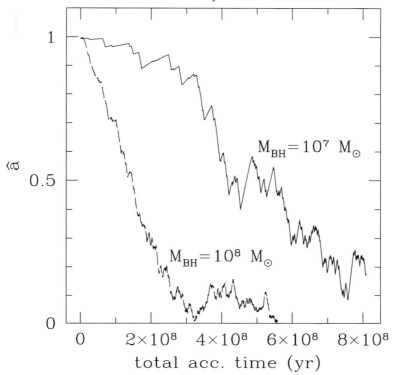

FIGURE 9. Evolution of MBH spins due to accretion of molecular cloud cores. We assume a lognormal distribution for the mass function of molecular clouds peaked at $\log(M_{MC}/M_\odot = 4$, with a dispersion of 0.75. The initial spin of the MBHs is 0.998. Upper curve: the initial MBH mass is 10^7 $M_\odot$, lower curve: the initial MBH mass is 10^8 $M_\odot$.

1997). This suggests that accretion events are very small and very 'compact.' Smaller MBHs, powering low-luminosity active galactic nuclei, likely grow by accreting smaller packets of material, such as tidally disrupted stars (for MBHs with mass $< 2 \times 10^6$ $M_\odot$, Milosavljević et al. 2006), or possibly molecular clouds (Hopkins & Hernquist 2006).

Compact self-gravitating cores of molecular clouds (MC) can occasionally reach sub-parsec regions. Although the rate of such events is uncertain, we can adopt the estimates of Kharb et al. (2006), and assume that about 10^4 of such events happen. We can further assume a lognormal distribution for the mass function of MC close to galaxy centers (based on the Milky Way case, e.g., Perets et al. 2007). We do not distinguish here giant MC and clumps, and, for illustrative purposes, we assume a single lognormal distribution peaked at $\log(M_{MC}/M_\odot) = 4$, with a dispersion of 0.75.

Figure 9 shows the possible effect that accretion of molecular clouds can have on spinning MBHs. The result is, on the whole, similar to that produced by minor mergers of black holes (Hughes & Blandford 2003), that is, a spin down in a random walk fashion.

In a gas-poor elliptical galaxy, however, substantial populations of molecular clouds are lacking (e.g., Sage et al. 2007), eliminating this channel of MBH feeding. Main sequence stars, however, linger in galaxy centers. Tidal disruption of stars is a feeding mechanism that has been proposed long ago (Hills 1975; Rees 1988). One expects disks formed by stellar debris to form with a random orientation. Stellar disruptions would therefore contribute to the spin-down of MBHs. The number of tidal disruptions of solar-type stars in an isothermal cusp per billion years can be written as:

$$N_* = 4 \times 10^5 \left(\frac{\sigma}{60\,\text{km s}^{-1}} \right) \left(\frac{M_{\text{BH}}}{10^6\,M_\odot} \right)^{-1} . \qquad (5.4)$$

Assuming that MBH masses scale with the velocity dispersion, σ, of the galaxy (we adopt here the Tremaine et al. 2002 scaling), we can derive the relative mass increase for an MBH in 1 billion years:

$$\frac{M_*}{M_{\text{BH}}} = 0.37 \left(\frac{M_{\text{BH}}}{10^6\,M_\odot} \right)^{-9/8} . \qquad (5.5)$$

The maximal level of spin down would occur assuming that all the tidal disruption events form counter-rotating disks, leading to retrograde accretion. Equation (5.5) shows that a small (say, $10^6\,M_\odot$) MBH starting at $\hat{a} = 0.998$ would be spun down completely. On the other hand, the spin of a larger (say, $10^7\,M_\odot$) MBH would not be drastically changed. This feeding channel is likely efficient in early-type disks which typically host faint bulges characterized by steep-density cusps, both inside (Bahcall & Wolf 1976) and outside (Faber et al. 1997) the sphere of influence of the BH. In this environment, the rate of stars which are tidally disrupted by MBHs (Hills 1975; Rees 1988) less massive than $10^8\,M_\odot$† is non-negligible (eq. 5.5; Milosavljević et al. 2006). The situation is different for giant ellipticals: the central density profile often displays a shallow core, and tidal disruption of stars is unlikely to play a dominant role.

Summarizing, the spin of MBHs in giant elliptical galaxies is likely dominated by massive accretion events which follow galaxy mergers. Both tidal disruption of stars and accretion of gaseous clouds is unlikely in shallow, stellar-dominated galaxy cores. Consequently, the spin rate stays high. In a galaxy displaying power-law (cuspy) brightness profiles, the rate of stellar tidal disruptions is much higher, and random small-mass accretion events contribute to spin MBHs down.

The different accretion histories in elliptical and disk galaxies seem to lead to a morphology-related bimodality of black-hole spin distribution in the centers of galaxies: central black holes in giant elliptical galaxies may have, on average, much larger spins than black holes in spiral/disk galaxies.

This result is in agreement with Sikora et al. (2007), who found that disk galaxies tend to be weaker radio sources with respect to elliptical hosts. Sikora et al. (2007) therefore proposed a revised "spin paradigm," which incorporates elements of the "accretion paradigm" (Ulvestad & Ho 2001; Merloni et al. 2003; Nipoti et al. 2005; Körding et al. 2006), according to which the radio-loudness is entirely related to the state of the accretion disks, similarly to radio-emission from x-ray binaries (Gallo et al. 2003), related to transitions between two different accretion modes (Livio et al. 2003). It should be emphasized that even if the production of powerful relativistic jets is conditioned by the presence of fast-rotating BHs, it also depends on the accretion rate and on the presence of disk MHD winds required to provide the initial collimation of the central Poynting-flux–dominated outflow.

To assess the role of spins in jet production, it is crucial to be able to measure MBH spins in AGN. An ingenious technique employs the Fe Kα line at 6.4 keV in the x-ray spectrum (2–10 keV), which is typically observed with broad asymmetric profiles indicative of a relativistic disk (Miller 2007). The iron line can, in principle, constrain the value of black-hole spins (Fabian et al. 1989; Laor 1991). The value of $\hat{a}$ affects the location

---

† For black-hole masses $\geqslant 2 \times 10^8\,M_\odot$ the Schwarzschild radius exceeds the tidal disruption radius for main-sequence stars.

of the inner radius of the accretion disk (corresponding to the innermost stable circular orbit in the standard picture), which in turn has a large impact on the shape of the line profile, because when the hole is rapidly rotating, the emission is concentrated closer in, and the line displays larger shifts. There is some evidence that this must be the case in some local Seyfert galaxies (Miniutti et al. 2004; Streblyanska et al. 2005). The assumption that the inner disk radius corresponds to the ISCO is not a trivial one, however, especially for thick disks (see Krolik 1999; Afshordi & Paczyński 2003).

Although observations of the iron line with the *Chandra* and *XMM* x-ray satellites are extending the studies of the innermost regions of MBHs, aiming at probing black-hole properties, interpretation of these studies is impeded by the inherent 'messiness' of gas dynamics. A clean probe would be a compact mass in a precessing and decaying orbit around a massive hole. The detection of gravitational waves from a stellar-mass BH—or even a white dwarf or neutron star—falling into a massive black hole (Extreme Mass Ratio Inspiral, or EMRI) can provide a unique tool to constrain the geometry of spacetime around BHs, and as a consequence, BH spins. Indeed the spin is a measurable parameter, with a very high accuracy, in the gravitational waves *LISA* signal (Barack & Cutler 2004; Berti et al. 2005, 2006; Lang & Hughes 2006; Vecchio 2004). Gravitational waves from an EMRI can be used to map the spacetime of the central massive dark object. The resulting 'map' can tell us if the standard picture for the central massive object, a Kerr BH described by general relativity, holds.

REFERENCES

ABRAMOWICZ, M. A. & LASOTA, J. P. 1980 *Acta Astronomica* **30**, 35.

AFSHORDI, N. & PACZYŃSKI, B. 2003 *ApJ* **592**, 354.

ALLER, M. C. & RICHSTONE, D. 2002 *AJ* **124**, 3035.

ARMITAGE, P. J. & NATARAJAN, P. 2005 *ApJ* **634**, 921.

BAHCALL, J. N. & WOLF, R. A. 1976 *ApJ* **209**, 214.

BAKER, J. G., BOGGS, W. D., CENTRELLA, J., KELLY, B. J., MCWILLIAMS, S. T., MILLER, M. C., & VAN METER, J. R. 2007 *ApJ* **668**, 1140.

BARACK, L. & CUTLER, C. 2004 *Phys. Rev. D* **69** (8), 082005.

BARDEEN, J. M. 1970 *Nature* **226**, 64.

BARDEEN, J. M. & PETTERSON, J. A. 1975 *ApJ* **195**, L65.

BARTH, A. J., HO, L. C., RUTLEDGE, R. E., & SARGENT, W. L. W. 2004 *ApJ* **607**, 90.

BARTH, A. J., MARTINI, P., NELSON, C. H., & HO, L. C. 2003 *ApJ* **594**, L95.

BEGELMAN, M. C. 1979 *MNRAS* **187**, 237.

BEGELMAN, M. C., BLANDFORD, R. D., & REES, M. J. 1980 *Nature* **287**, 307.

BEGELMAN, M. C. & MEIER, D. L. 1982 *ApJ* **253**, 873.

BEGELMAN, M. C., ROSSI, E. M., & ARMITAGE, P J. 2008 *MNRAS* **387**, 1649.

BEGELMAN, M. C., VOLONTERI, M., & REES, M. J. 2006 *MNRAS* **370**, 289.

BERTI, E., BUONANNO, A., & WILL, C. M. 2005 *Phys. Rev. D* **71** (8), 084025.

BERTI, E., CARDOSO, V., & WILL, C. M. 2006 *Phys. Rev. D* **73** (6), 064030.

BLANDFORD, R. D., NETZER, H., WOLTJER, L., COURVOISIER, T. J.-L., & MAYOR, M. 1990. In *Active Galactic Nuclei*. Saas-Fee Advanced Course 20. Lecture Notes 1990. Swiss Society for Astrophysics and Astronomy, XII. Springer-Verlag.

BROMM, V. & LOEB, A. 2003 *ApJ* **596**, 34.

BULLOCK, J. S., DEKEL, A., KOLATT, T. S., KRAVTSOV, A. V., KLYPIN, A. A., PORCIANI, C., & PRIMACK, J. R. 2001 *ApJ* **555**, 240.

CAMPANELLI, M., LOUSTO, C., ZLOCHOWER, Y., & MERRITT, D. 2007 *ApJ* **659**, L5.

CARR, B. J., BOND, J. R., & ARNETT, W. D. 1984 *ApJ* **277**, 445.

DI MATTEO, T., SPRINGEL, V., & HERNQUIST, L. 2005 *Nature* **433**, 604.

DOTTI, M., COLPI, M., & HAARDT, F. 2006 *MNRAS* **367**, 103.

EISENSTEIN, D. J. & LOEB, A. 1995 *ApJ* **443**, 11.

Elvis, M., Risaliti, G., & Zamorani, G. 2002 *ApJ* **565**, L75.

Escala, A., Larson, R. B., Coppi, P. S., & Mardones, D. 2004 *ApJ* **607**, 765.

Faber, S. M., Tremaine, S., Ajhar, E. A., Byun, Y.-I., Dressler, A., Gebhardt, K., Grillmair, C., Kormendy, J., Lauer, T. R., & Richstone, D. 1997 *AJ* **114**, 1771.

Fabian, A. C., Rees, M. J., Stella, L. & White, N. E. 1989 *MNRAS* **238**, 729.

Fan, X., Strauss, M. A., Schneider, D. P., Gunn, J. E., Lupton, R. H., Becker, R. H., Davis, M., Newman, J. A., Richards, G. T., White, R. L., Anderson, J. E., Annis, J., Bahcall, N. A., Brunner, R. J., Csabai, I., Hennessy, G. S., Hindsley, R. B., Fukugita, M., Kunszt, P. Z., Ivezić, Ž., Knapp, G. R., McKay, T. A., Munn, J. A., Pier, J. R., Szalay, A. S., & York, D. G. 2001 *AJ* **121**, 54.

Ferrarese, L. & Merritt, D. 2000 *ApJ* **539**, L9.

Fryer, C. L., Woosley, S. E., & Heger, A. 2001 *ApJ* **550**, 372.

Gallo, E., Fender, R. P., & Pooley, G. G. 2003 *MNRAS* **344**, 60.

Gebhardt, K., et al. 2000 *ApJ* **539**, L13.

Gould, A. & Rix, H.-W. 2000 *ApJ* **532**, L29.

Haehnelt, M. G. & Rees, M. J. 1993 *MNRAS* **263**, 168.

Haiman, Z. 2004 *ApJ* **613**, 36.

Hills, J. G. 1975 *Nature* **254**, 295.

Ho, L. C., Filippenko, A. V., & Sargent, W. L. W. 1997 *ApJ* **487**, 591.

Hopkins, P. F. & Hernquist, L. 2006 *ApJS* **166**, 1.

Hughes, S. A. & Blandford, R. D. 2003 *ApJ* **585**, L101.

Kharb, P., O'Dea, C. P., Baum, S. A., Colbert, E. J. M., & Xu, C. 2006 *ApJ* **652**, 177.

King, A. R., Lubow, S. H., Ogilvie, G. I., & Pringle, J. E. 2005 *MNRAS* **363**, 49.

Körding, E. G., Jester, S., & Fender, R. 2006 *MNRAS* **372**, 1366.

Kormendy, J. & Gebhardt, K. 2001. In *Relativistic Astrophysics: 20th Texas Symposium* (eds. J. C. Wheeler & H. Martel). AIP Conference Proceedings, vol. 586, p. 363. AIP.

Koushiappas, S. M., Bullock, J. S., & Dekel, A. 2004 *MNRAS* **354**, 292.

Krolik, J. H. 1999 *ApJ* **515**, L73.

Kumar, S. & Pringle, J. E. 1985 *MNRAS* **213**, 435.

Lang, R. N. & Hughes, S. A. 2006 *Phys. Rev. D* **74** (12), 122001.

Laor, A. 1991 *ApJ* **376**, 90.

Livio, M., Pringle, J. E., & King, A. R. 2003 *ApJ* **593**, 184.

Lodato, G. & Natarajan, P. 2006 *MNRAS* **371**, 1813.

Loeb, A. & Rasio, F. A. 1994 *ApJ* **432**, 52.

Madau, P. & Rees, M. J. 2001 *ApJ* **551**, L27.

Marconi, A., Risaliti, G., Gilli, R., Hunt, L. K., Maiolino, R., & Salvati, M. 2004 *MNRAS* **351**, 169.

Mayer, L., Kazantzidis, S., Madau, P., Colpi, M., Quinn, T., & Wadsley, J. 2008. In *Relativistic Astrophysics Legacy and Cosmology—Einstein's* (eds. B. Aschenbach, V. Burwitz, G. Hasinger, & B. Leibundgut). ESO Astrophysics Symposia, p. 152. Springer-Verlag.

Menou, K., Haiman, Z., & Narayanan, V. K. 2001 *ApJ* **558**, 535.

Merloni, A., Heinz, S., & di Matteo, T. 2003 *MNRAS* **345**, 1057.

Merloni, A., Rudnick, G., & Di Matteo, T. 2004 *MNRAS* **354**, L37.

Miller, J. M. 2007 *ARA&A* **45**, 441.

Milosavljević, M. & Merritt, D. 2001 *ApJ* **563**, 34.

Milosavljević, M., Merritt, D., & Ho, L. C. 2006 *ApJ* **652**, 120.

Miniutti, G., Fabian, A. C., & Miller, J. M. 2004 *MNRAS* **351**, 466.

Mo, H. J., Mao, S., & White, S. D. M. 1998 *MNRAS* **295**, 319.

Moderski, R., Sikora, M., & Lasota, J.-P. 1998 *MNRAS* **301**, 142.

Natarajan, P. & Pringle, J. E. 1998 *ApJ* **506**, L97.

Nipoti, C., Blundell, K. M., & Binney, J. 2005 *MNRAS* **361**, 633.

Oh, S. P. & Haiman, Z. 2002 *ApJ* **569**, 558.

Papaloizou, J. C. B. & Pringle, J. E. 1983 *MNRAS* **202**, 1181.

Perets, H. B., Hopman, C., & Alexander, T. 2007 *ApJ* **656**, 709.

Quinlan, G. D. 1996 *New Astronomy* **1**, 35.

REES, M. J. 1978 In *Structure and Properties of Nearby Galaxies* (eds. E. M. Berkhuijsen & R. Wielebinski). IAU Symposium Ser., vol. 77, p. 237.

REES, M. J. 1988 *Nature* **333**, 523.

RICHSTONE, D., ET AL. 1998 *Nature* **395**, A14.

SAGE, L. J., WELCH, G. A., & YOUNG, L. M. 2007 *ApJ* **657**, 232.

SALVATERRA, R. 2007 *Memorie della Societa Astronomica Italiana* **78**, 796.

SCHEUER, P. A. G. & FEILER, R. 1996 *MNRAS* **282**, 291.

SCHNITTMAN, J. D. 2007 *ApJ* **667**, L133.

SESANA, A., HAARDT, F., & MADAU, P. 2007a *ApJ* **660**, 546.

SESANA, A., VOLONTERI, M., & HAARDT, F. 2007b *MNRAS* **377**, 1711.

SHAKURA, N. I. & SUNYAEV, R. A. 1973 *A&A* **24**, 337.

SHLOSMAN, I., FRANK, J., & BEGELMAN, M. C. 1989 *Nature* **338**, 45.

SIKORA, M., STAWARZ, L., & LASOTA, J.-P. 2007 *ApJ* **658**, 815.

SOLTAN, A. 1982 *MNRAS* **200**, 115.

SPERGEL, D. N., BEAN, R., DORÉ, O., NOLTA, M. R., BENNETT, C. L., DUNKLEY, J., HINSHAW, G., JAROSIK, N., KOMATSU, E., PAGE, L., PEIRIS, H. V., VERDE, L., HALPERN, M., HILL, R. S., KOGUT, A., LIMON, M., MEYER, S. S., ODEGARD, N., TUCKER, G. S., WEILAND, J. L., WOLLACK, E., & WRIGHT, E. L. 2007 *ApJS* **170**, 377.

STREBLYANSKA, A., HASINGER, G., FINOGUENOV, A., BARCONS, X., MATEOS, S., & FABIAN, A. C. 2005 *A&A* **432**, 395.

THORNE, K. S. 1974 *ApJ* **191**, 507.

TREMAINE, S., ET AL. 2002 *ApJ* **574**, 740.

ULVESTAD, J. S. & HO, L. C. 2001 *ApJ* **558**, 561.

VAN DEN BOSCH, F. C., ABEL, T., CROFT, R. A. C., HERNQUIST, L., & WHITE, S. D. M. 2002 *ApJ* **576**, 21.

VECCHIO, A. 2004 *Phys. Rev. D* **70** (4), 042001.

VOLONTERI, M. 2007 *ApJ* **663**, L5.

VOLONTERI, M., HAARDT, F., & MADAU, P. 2003 *ApJ* **582**, 559.

VOLONTERI, M., LODATO, G., & NATARAJAN, P. 2008 *MNRAS* **383**, 1079.

VOLONTERI, M., MADAU, P., QUATAERT, E., & REES, M. J. 2005 *ApJ* **620**, 69.

VOLONTERI, M. & REES, M. J. 2005 *ApJ* **633**, 624.

VOLONTERI, M. & REES, M. J. 2006 *ApJ* **650**, 669.

VOLONTERI, M., SIKORA, M., & LASOTA, J.-P. 2007 *ApJ* **667**, 704.

WANG, J.-M., CHEN, Y.-M., HO, L. C., & MCLURE, R. J. 2006 *ApJ* **642**, L111.

WILLOTT, C. J., DELFOSSE, X., FORVEILLE, T., DELORME, P., & GWYN, S. D. J. 2005 *ApJ* **633**, 630.

YU, Q. & TREMAINE, S. 2002 *MNRAS* **335**, 965.

Supermassive black holes in deep multiwavelength surveys

By C. MEGAN URRY[1] AND EZEQUIEL TREISTER[2]

[1]Department of Physics, Yale University, P.O. Box 208121, New Haven, CT 06520-8121, USA
[2]European Southern Observatory, Casilla 19001, Santiago 19 Chile

In recent years deep x-ray and infrared surveys have provided an efficient way to find accreting supermassive black holes, otherwise known as active galactic nuclei (AGN), in the young universe. Such surveys can, unlike optical surveys, find AGN obscured by high column densities of gas and dust. In those cases, deep optical data show only the host galaxy, which can then be studied in greater detail than in unobscured AGN. Some years ago the hard spectrum of the x-ray "background" suggested that most AGN were obscured. Now GOODS, MUSYC, COSMOS, and other surveys have confirmed this picture and given important quantitative constraints on AGN demographics. Specifically, we show that most AGN are obscured at all redshifts and the amount of obscuration depends on both luminosity and redshift, at least out to redshift $z \sim 2$, the epoch of substantial black hole and galaxy growth. Larger-area deep infrared and hard x-ray surveys will be needed to reach higher redshifts and to fully probe the co-evolution of galaxies and black holes.

1. Cosmic growth of black holes and galaxies

Abundant evidence indicates that the growth of a supermassive black hole is closely tied to the formation and evolution of the surrounding galaxy. The energy released from accretion onto the black hole affects star formation in the galaxy, probably limiting growth at the high- and low-mass ends and, of course, the distribution and angular momentum of matter in the galaxy governs the amount of matter accumulated by the black hole (Silk & Rees 1998; King 2005; Rovilos et al. 2007). Emergent energy from accretion is also a factor in understanding ionization and radiation backgrounds (Hasinger 2000; Lawrence 2001). Understanding the growth history of these black holes is therefore critical to understanding the global evolution of structure in the Universe.

Yet the demographics of supermassive black holes remain elusive. The largest samples of quasars and AGN, by which we mean supermassive black holes in a high accretion-rate phase,† have been found through optical selection (e.g., the Sloan Digital Sky Survey quasar sample; Schneider et al. 2002, 2007), but, at least locally, these are not representative of the larger AGN population. Instead we need surveys less biased against obscured AGN.

There are three reasons to suspect that most AGN are obscured by large column densities of gas and dust. First, a large body of evidence suggests that local AGN have geometries that are not spherically symmetric, and that different aspect angles present markedly different observed characteristics; this is referred to as AGN unification (Antonucci 1993; Urry & Padovani 1995). Second, AGN are more common at high redshift ($z \sim 2$–3), where the average star formation rate is higher and thus it is even more likely that gas and dust surround the galaxy nucleus than at $z \sim 0$. Third, and most important, obscured AGN are required to explain the shape of the x-ray "background" radiation.

---

† Some make a distinction between AGN and quasars, with the latter being above an arbitrary luminosity, typically $M_B = -23$ mag. Here we use the term AGN to refer to an actively accreting supermassive black hole above or below this luminosity.

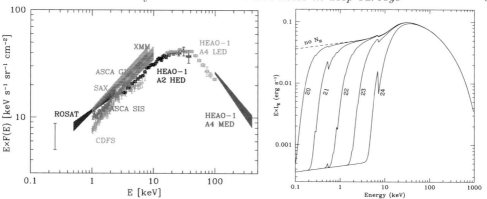

FIGURE 1. Left: The x-ray background spectrum (data as described by Urry & Treister 2005, shown in units of energy per logarithmic band) is very hard, peaking near 30 keV, in contrast to the spectra of unobscured AGN which are roughly horizontal in these units. (Figure from Gilli 2004.) Right: Large column densities of gas absorb low-energy photons via the photoelectric effect, such that emission from an obscured AGN peaks at an energy that increases with column density. Each line represents a factor of 10 increase in equivalent N_H (lines are marked with $\log N_H$ in atoms/cm^2; assumes solar abundances and the cross sections of Morrison & McCammon 1983).

The x-ray "background" is actually the superposition of individual AGN that were not resolved in early x-ray experiments (hence the designation "background"). As shown in Figure 1a, its spectrum peaks at $\sim$30 keV (i.e., this is where most of the energy is produced), much harder than the typical spectrum of an unobscured AGN (roughly flat in these units). In obscured AGN, however, the softest x-ray photons have been absorbed via the photoelectric effect, and Compton-thick AGN (those with $N_H \gtrsim 10^{24}$ cm^{-2}) actually peak at roughly 30 keV (Figure 1b). Thus x-ray observations have long indicated there is a large population of heavily obscured AGN (Setti & Woltjer 1989; Comastri et al. 1995; Gilli et al. 2001). While much of the x-ray background has been resolved at low energies, recent work on x-ray deep fields suggests the hardest (most obscured) x-ray sources have yet to be detected (e.g., Worsley et al. 2005).

Relatively unbiased AGN samples require joint hard x-ray and infrared surveys. Hard x-rays ($E > 2$ keV) penetrate all but the thickest column densities and efficiently locate black hole accretion. The absorbed optical through soft x-ray emission heats the surrounding dust and is re-radiated in the infrared, at wavelengths that depend on the dust temperature. Meanwhile, high-resolution optical surveys (e.g., with *HST*) allow separation of nuclear (AGN) and host galaxy light. Today, NASA's three operating Great Observatories—the *Chandra X-ray Observatory*, the *Spitzer Space Telescope*, and the *Hubble Space Telescope*—enable matched x-ray, infrared, and optical surveys at unprecedented depth and resolution.

2. Deep multiwavelength surveys

The announcement in spring 2000 of the first Spitzer Legacy opportunity led to the Great Observatories Origins Deep Survey (GOODS), to date the deepest wide-area multiwavelength survey carried out with *Spitzer*, *HST*, and *Chandra* (Dickinson et al. 2003; Giavalisco et al. 2004). By targeting the Spitzer Legacy, and later the Hubble Treasury, observations on the pre-existing *Chandra* deep fields, we leveraged substantial investments of observing time, probed AGN demographics at the peak of quasar activity

($z \sim$ 1–2), and enabled a wide range of other science, described in the *Astrophysical Journal Letters* special edition of January 10, 2004.[†]

The GOODS data are deep enough to detect AGN to very high redshift ($z \gtrsim 6$), and the volume sampled ensures sizable AGN samples out to $z \sim 3$. To sample larger volumes and to search effectively for rare objects like AGN or massive galaxies at high redshift, in 2002 we designed the Multiwavelength Survey by Yale and Chile (MUSYC) survey, covering one square degree in two equatorial and two southern fields. Three of the four regions had already been observed extensively, including the Extended Chandra Deep Field South (ECDF-S) and the Extended Hubble Deep Field South (EHDF-S), thus leveraging substantial investments with *Chandra*, *XMM-Newton*, *HST*, and the largest ground-based telescopes.[‡] Soon after starting MUSYC we helped start the COSMOS survey (Scoville et al. 2007b), a 2-square degree field centered at 1000+02 that has now been imaged extensively with *HST*, *Spitzer*, *XMM*, and *Chandra*.[¶] The unprecedented combination of area and depth allow a wide variety of science, described in the September 2007 special edition of the *Astrophysical Journal Supplement*. Other relevant multiwavelength surveys include the Lockman Hole (Hasinger et al. 2001), CLASXS (Yang et al. 2004), the Extended Groth Strip (Davis et al. 2007), SWIRE (Lonsdale et al. 2003), XBOOTES (Hickox et al. 2008), HELLAS2XMM (Baldi et al. 2002), SEXSI (Harrison et al. 2003), CYDER (Treister et al. 2005), CHAMP (Kim et al. 2004), and AMSS (Akiyama et al. 2003).

3. Finding obscured AGN

The question we set out to answer with GOODS, MUSYC, and COSMOS was, "Is there a substantial population of obscured AGN that is missed by traditional optical surveys?" To answer this question we need to sample the AGN population at $z \sim$ 1–2, at the peak of the number density. The GOODS survey, and the Chandra Deep Field South (CDF-S) x-ray survey on which GOODS piggy-backed (Giacconi et al. 2002), were designed to sample the AGN population at $z \sim$ 0.5–2, which includes the AGN that make up the x-ray background. Luminous AGN at higher redshifts could certainly be detected but the volume surveyed is too small to expect to see a reasonable number of them.

Early results in the CDFS and other deep x-ray fields showed there was a population of optically faint hard x-ray sources (Alexander et al. 2001; Franceschini et al. 2002a,b; Mainieri et al. 2005), which collectively comprised a large fraction of the integrated x-ray background intensity. However, as optical counterparts were identified and spectra obtained, several apparent problems emerged. First, the redshift distribution peaked at relatively low values, $z \lesssim 1$, lower than expected from the early population synthesis models for the x-ray background. Second, the fraction of x-ray sources that were identified

---

[†] For more details see http://www.stsci.edu/science/goods.

[‡] The four MUSYC fields include the ECDF-S, for which substantial additional *Chandra* observations were later acquired by N. Brandt (Lehmer et al. 2005; Virani et al. 2006); the EHDF-S, for which there not yet any x-ray data; a field centered on a $z = 6.3$ quasar at 1030+05, which has relatively deep XMM data; and a new field centered at 1256+01 ("Castander's Window"), an equatorial region with low 100-micron dust emission. The first two have now been imaged with *Spitzer* (IRAC and MIPS) and *HST* (ACS and NICMOS), and all fields are accessible with ALMA. For more details see http://www.astro.yale.edu/MUSYC and Gawiser et al. (2006b); Quadri et al. (2007); Gawiser et al. (2006a); van Dokkum et al. (2006).

[¶] COSMOS is now one of the largest and best covered deep multiwavelength-survey fields, representing the investment of thousands of hours of major telescope time (Scoville et al. 2007a; Sanders et al. 2007; Hasinger et al. 2007; Capak et al. 2007; Lilly et al. 2007). For more details, see http://cosmos.astro.caltech.edu.

as obscured, either because of high N_H or absence of broad lines, was less than the canonical 3/4 seen at low redshift, and appeared to decline with redshift rather than increase, as specified by the best population synthesis model at that time (Gilli et al. 2001). A number of authors pointed out these problems (e.g., Mainieri et al. 2005; Rosati et al. 2002; Brandt & Hasinger 2005), and Franceschini et al. (2002a) suggested that "the unification scheme based on a simple orientation effect fails at high redshifts" and that the production of the x-ray background by a collection of obscured AGN "requires major revision."

At the same time, these faint red x-ray AGN were copious emitters of infrared radiation (Treister et al. 2004, 2006), fully consistent with the unification paradigm. It also became apparent early on that, as for optically selected quasars, the evolution of x-ray–selected AGN is luminosity dependent, with high-luminosity AGN evolving earlier and more rapidly than low-luminosity AGN (Ueda et al. 2003; Cowie et al. 2003); this luminosity-dependent density evolution had not been incorporated in earlier population synthesis models for the x-ray background. Finally, we suspected that selection effects could play a significant role in affecting the redshifts and optical identifications of survey sources.

Accordingly, we developed a comprehensive quantitative approach, based on the unification scenario and incorporating the most recent, best luminosity function and evolution, to predict the number counts and redshift distributions at any wavelength (for the moment, infrared through x-ray, though it could be generalized) for surveys of arbitrary area, depth, and wavelength. Here we describe the quantitative interpretation of the multiwavelength data from GOODS, MUSYC, COSMOS and other deep multiwavelength surveys, including the dependence of the obscured fraction of AGN on luminosity and redshift. Specifically, we show that most AGN are obscured at all redshifts; that the fraction of obscured AGN decreases with luminosity; and that it increases with redshift, at least out to redshift $z \sim 2.5$, an epoch of substantial black hole and galaxy growth. Deep infrared and hard x-ray surveys over larger areas will be needed to reach higher redshifts and to probe fully the co-evolution of galaxies and black holes.

3.1. *Connecting x-ray, optical, and infrared surveys*

Our approach was to connect surveys at different wavelengths by assuming something sensible about AGN spectral energy distributions (SEDs), then combining those with well-measured luminosity functions and evolution to understand the source counts and redshift distributions of AGN selected at a given flux limit at any wavelength (Treister et al. 2004, 2006; Treister & Urry 2006; analogous approaches have been taken by Ballantyne et al. 2006 and Dwelly & Page 2006). Simultaneously, we constrain the same AGN population to fit the x-ray background (Treister & Urry 2005). An alternative approach is to model only the x-ray spectra of AGN and to fit the x-ray background alone with a mixture of obscured and unobscured AGN (Comastri et al. 1995; Gilli et al. 2001, 2007); this constrains the demographics but does not connect the x-ray sources to those detected at optical and infrared wavelengths—and thus does not use those additional constraints on the AGN demographics, nor does it allow a quantitative estimate of the important effect of optical or infrared flux limits on the survey content or spectroscopic identifications.

Briefly, our procedure was as follows: We started with the underlying AGN demographics, described by an AGN luminosity function that incorporates dependence on absorbing column density, and the luminosity-dependent evolution of this function (Ueda et al. 2003). Because this luminosity function is based on hard x-ray observations, it is relatively free of bias against obscured AGN.

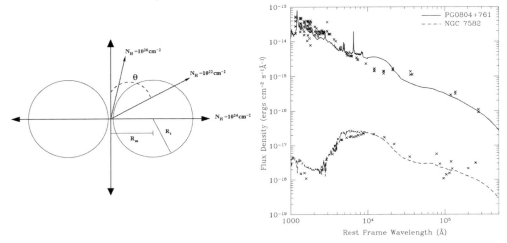

FIGURE 2. The infrared spectra of AGN were modeled by dust emission (Nenkova et al. 2002) from a clumpy torus geometry (left) that, assuming random orientations and adjusting the torus geometry to give a $3:1$ ratio of column densities above:below $N_H = 10^{22}$ cm^{-2}, yields an N_H distribution consistent with various observational estimates (see text for details). Right: Two examples of model SEDs (*lines*), which are fully determined from L_X and N_H. These fit very well the observed SEDs of local unobscured (*data points, top*) and obscured (*bottom*) AGN, with no free parameters. The composite model SEDs include infrared dust emission, reddened quasar spectra (keyed to N_H value), and an L_* host galaxy, linked to the x-rays by the known optical-to-x-ray ratio (which depends on L_X).

We then developed a set of SEDs, based on the unification paradigm, that represent AGN with a wide range in intrinsic luminosity and absorbing column density (parameterized in terms of the neutral hydrogen column density, N_H, along the line of sight). Specifically, at optical wavelengths ($\lambda = 0.1$–1 microns), we used a Sloan Digital Sky Survey composite quasar spectrum (Vanden Berk et al. 2001) plus Milky Way-type reddening laws and a standard dust-to-gas ratio to convert N_H to A_V; we also added an L_* elliptical host galaxy, which is the dominant component for heavily obscured AGN. In the x-ray ($E > 0.5$ keV), we assumed a power-law spectrum with photon index $\Gamma = 1.9$, typical of unobscured AGN, absorbed by column densities in the range $\log N_H = 20$–24 cm^{-2}. To describe the infrared part of the SEDs ($\lambda > 1$ micron), which in the unification paradigm includes radiation from dust heated by absorbed ultraviolet through soft x-ray photons, we used dust emission models by Nenkova et al. (2002) in a clumpy torus geometry (Figure 2a), converting to N_H from viewing angle assuming random orientations. The resulting N_H distribution is completely consistent with various observational estimates (Ueda et al. 2003; Dwelly & Page 2006; Risaliti et al. 1999; Comastri et al. 1995; Tozzi et al. 2006; Gilli et al. 2007). AGN models with the same intrinsic x-ray luminosities were normalized at 100 microns. To connect the ultraviolet and x-ray parts of the SED, we used the standard dependence of x-ray to optical luminosity ratio on luminosity (e.g., Steffen et al. 2006).

The SED models, parameterized in terms of L_X (as a proxy for intrinsic luminosity) and N_H (which depends only on viewing angle and torus geometry), describe extremely well the local population of AGN. There is some freedom in the choice of torus geometry, of course; we selected an aspect ratio that would produce three times as many AGN with N_H greater than 10^{22} cm^{-2} compared to smaller column densities (Figure 2a), which is roughly the observed local ratio (Risaliti et al. 1999). Figure 2b shows the observed SEDs of two local AGN, one unobscured (top) and one heavily obscured (bottom), compared

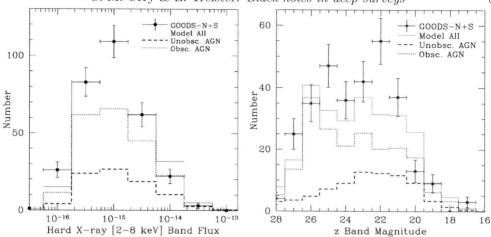

FIGURE 3. X-ray (left) and optical (right) counts observed in the GOODS fields (*data points*) agree well with a simple unification model (*solid line*), even with no dependence of torus geometry on luminosity or redshift (Treister et al. 2004). Inclusion of the luminosity dependence (found later in larger AGN samples) gives essentially the same result for the GOODS survey. Unobscured AGN ($N_H < 10^{22}$ cm^{-2}) dominate at bright optical magnitudes (*dashed line*) and obscured AGN ($N_H > 10^{22}$ cm^{-2}) at faint z-band magnitudes (*dotted line*). For $R > 24$ mag, the approximate limit for obtaining decent optical spectral identifications with 8-meter class telescopes, the vast majority of sources are obscured AGN.

to the model SED corresponding to the observed L_X and N_H. With no free parameters, the fit is remarkably good overall, meaning that this representation accurately reflects the distribution of photons across a very wide wavelength range.

Combining these model SEDs with the AGN luminosity function and evolution yields a distribution of sources as a function of spatial location and wavelength. For a given wavelength and flux limit, then, our calculation produces the expected number counts and redshift distribution of AGN. This population can be filtered by multiple flux limits (for example, one could look at the x-ray counts and redshift distribution of sources above arbitrary flux limits in the x-ray, optical, and infrared bands) and/or by other properties (e.g., N_H value). Our procedure was to generate the expected content of real surveys, and to compare to observations, in order to constrain the underlying AGN demographics, specifically, the AGN luminosity function, its evolution, its dependence on N_H, and the family of AGN SEDs (which includes the torus properties).

3.2. Quantitative results from the Population Synthesis Model: Number counts, redshift distributions, and the x-ray background

In a series of papers we showed that this straightforward model matches very well the observed properties of AGN samples. The first paper, Treister et al. (2004), assumed for the sake of simplicity that the torus geometry (and hence the N_H distribution) did not depend on luminosity or redshift. Even this simplest population synthesis model fits beautifully the observed x-ray and optical counts in the deep GOODS survey (Figure 3). The source population at faint optical magnitudes, particularly the AGN too faint for identification with optical spectroscopy, is dominated by obscured AGN.

The unification-inspired model described above, like other population synthesis models (e.g., Gilli et al. 2001), predicts a large population of obscured AGN out to high redshift. The predicted peak of the redshift distribution (*dashed histogram*, Figure 4a) is near $z \sim 1$, not much higher than that observed in the GOODS data (*shaded/open histogram*,

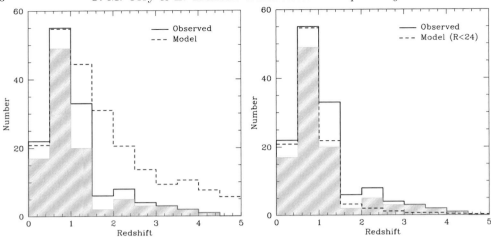

FIGURE 4. Left: Observed spectroscopic redshift distribution (*hatched histogram*) and photometric redshift distribution (*open histogram*) for the GOODS-North field, compared to that expected from the unification model (*dashed line*), applying only an x-ray flux limit. There is poor agreement at high redshifts. Right: Imposing an optical flux limit as well on the model leads to an expected distribution (*dashed line*) that agrees very well with the observed distribution.

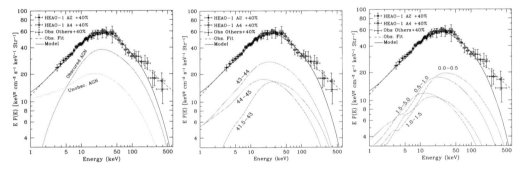

FIGURE 5. The predicted x-ray background spectrum (*heavy line*; as in Treister & Urry 2005, including luminosity-dependent density evolution from Ueda et al. 2003 and luminosity dependence of the obscured fraction) from the unification-inspired population-synthesis model agrees very well with the observed data (*data points*). Left: Obscured AGN (*thin, dark gray, peaked curve*) dominate at high energies ($E > 10$ keV), while unobscured AGN (*thin, light gray, broad curve*) dominate in the soft x-ray. There is little freedom in the fit parameters, especially at low energies, so this agreement is remarkably good. (Emission at $E > 100$ keV is dominated by blazars, which are not modeled here.) The x-ray background is dominated by moderate luminosity AGN, in the range 10^{43-44} ergs s^{-1} (middle panel) and by intermediate redshifts, $z \sim 0.5$–1.5 (right panel).

Figure 4a), but there discrepancy at $z > 1$. Many of these high-redshift obscured AGN, because they are optically faint, are not identified in follow-up optical spectra of x-ray sources. This can be clearly seen imposing the spectroscopic limit, $R \sim 24$ mag, on the expected redshift distribution (*dashed histogram*, Figure 4b); this shows the observed and expected distributions agree well because the higher-redshift AGN remain largely unidentified. Equivalently, AGN without spectroscopic identifications will preferentially be faint, obscured sources.

The advent of multiple large AGN samples with high spectroscopic completeness allowed Barger et al. (2005) to deduce a dependence of obscured fraction on optical

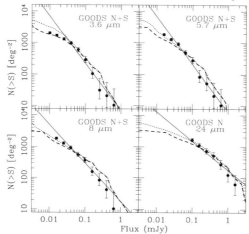

FIGURE 6. *Filled circles*: observed cumulative infrared flux distributions for x-ray detected AGN in the GOODS-North and South fields in three IRAC bands, and for the North field only in the MIPS 24-micron band. Error bars correspond to the 84% confidence level in the number of sources in that bin (Gehrels 1986). Because this is a cumulative plot, the bins are not independent. *Solid line*: Power-law fit to the observed number counts. *Dashed line*: expected infrared flux distribution (Treister et al. 2006) based on our population synthesis model, taking into account the x-ray flux limit in the GOODS fields and the luminosity dependence of the obscured AGN fraction, and correcting for the effects of undersampling at the bright end caused by the small volume of the GOODS fields. *Dotted line*: expected infrared flux distribution not considering the x-ray flux limit (i.e., all sources expected from extrapolating the AGN luminosity function). In general, the model explains the normalization and shape of the counts quite well, especially considering the poor statistics at the bright end. At the faint end, where the host galaxy light dominates, the model diverges more significantly, since the model assumptions (no distribution in host galaxy magnitude or spectrum) are clearly far too simple.

luminosity. We incorporated a simple functional form of that dependence in our model—namely, a linear transition between 100% of AGN with $L_X = 10^{42}$ ergs s^{-1} being obscured and no AGN with $L_X = 10^{46}$ ergs s^{-1} being obscured. Taking into account the selection effects due to flux limits, the model then matches the observed dependence very well (Figure 11). The resulting number counts and redshift distributions for GOODS do not change, since only a restricted luminosity range (primarily 10^{43-44} ergs s^{-1}) is probed by that survey. In all work subsequent to (Treister et al. 2004)—on infrared counts, x-ray background, and evolution of obscuration (see below)—we incorporated this luminosity dependence.

Our population synthesis model predicts, with very little freedom, the spectrum of the x-ray background. We simply sum the x-ray emission from all the AGN. There is some freedom in that we (like everyone else who does this kind of calculation) have to model the x-ray spectrum. We used an absorbed power law plus an iron line and a Compton-reflection hump, consistent with the AGN spectra that are well measured locally; we also assumed solar abundances and used the N_H distribution described above for log $N_H = 20$–24 cm^{-2}. The N_H distribution for higher column densities ($N_H > 10^{24}$ cm^{-2}) is poorly constrained at present, so we extrapolated from $N_H < 10^{24}$ cm^{-2} with a flat slope (again, much like what others have assumed). Beyond this, one can adjust the metallicity (this increases the hard x-rays, holding everything else constant) or the include a small range of power-law slopes (ditto; Gilli et al. 2007), but we did not feel the data could constrain those parameters and so left them fixed. In any case, the integral constraint of the x-ray

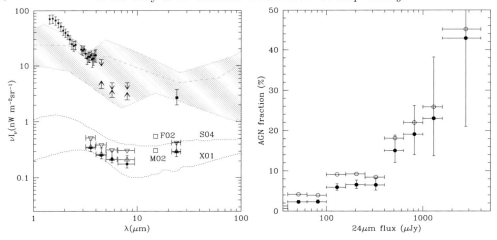

FIGURE 7. Left: Extragalactic infrared background intensity as a function of wavelength. *Shaded region*: allowed values for the total intensity compiled by Hauser & Dwek (2001). 1–4 µm: Measurements reported by Matsumoto et al. (2005). 4–8 µm: lower limits from Fazio et al. (2004), upper limits from Kashlinsky et al. (1996) at 4.5 µm and Stanev & Franceschini (1998) at 5.8 and 8 µm. Measurement at 24 µm from Papovich et al. (2004). *Solid circles*: integrated infrared emission from x-ray-detected AGN in the GOODS fields, with 1σ error bars from source number statistics (which dominate over measured flux errors) calculated as in Gehrels (1986). *Triangles*: Integrated AGN emission from the models of Treister & Urry (2005) if only x-ray-detected AGN are considered (*lower*) or including all AGN (*upper*). *Dotted lines*: Expected AGN integrated emission from the AGN models of Silva et al. (2005; *upper line*) and Xu et al. (2001; *lower line*). *Open squares*: Expected AGN integrated emission at 15 µm from the extrapolation to fainter fluxes of *IRAS* and *ISO* (Matute et al. 2002 [M02]; corrected by a factor of 5 to include the contribution from obscured AGN) and *ISO* only (Fadda et al. 2002 [F02]) observations. Right: The fraction of sources that are AGN rises sharply with 24 µm infrared flux (*filled circles*). Vertical error bars show the 1σ Poissonian errors on the number of sources (Gehrels 1986). Also shown is the contribution corrected by the AGN expected to be missed by x-ray selection (*open circles*), as estimated using our AGN population synthesis model; this shows the same strong dependence on infrared flux, indicating that particular dependence is not a selection effect induced by the x-ray flux limit.

background is an excellent check on whether the demographics of AGN assumed here is realistic. Figure 5 shows that, with almost no free parameters, the data are very well fit indeed. (The normalization of the background at $E > 10$ keV is discussed below in §3.3.)

Another strong prediction of the Treister model concerns the infrared data, which at the time the model was developed had not yet been taken. We calculated the expected Spitzer counts in IRAC and MIPS 24-micron bands; Figure 6 shows the excellent agreement with observations (Treister et al. 2006). Small discrepancies at the faint end are due to the overly simple assumption of a single host-galaxy magnitude (which dominates at low fluxes), rather than a distribution of magnitudes.

From the *Spitzer* observations, we calculate the minimum AGN contribution to the extragalactic infrared background, obtaining a lower value than previously estimated, ranging from 2% to 10% in the 3–24 micron range (Treister et al. 2006). Accounting for heavily obscured AGN that, according to our population synthesis model, are not detected in x-rays, the AGN contribution to the infrared background increases by ∼45%, to ∼3–15%. Figure 7a shows the AGN contributions deduced from GOODS data (Treister et al. 2006) compared to other estimates and to the (uncertain) total extragalactic background light, as a function of infrared wavelength. The GOODS measurements place the strongest constraints to date on the AGN contribution to the extragalactic background light, indicating that stars dominate completely over AGN at infrared wavelengths.

The fraction of sources that are AGN rises sharply with 24 μm infrared flux (Figure 7b). Thus in deep surveys like GOODS or COSMOS, AGN are a small fraction of the total infrared source population, while in high flux-limit surveys like SWIRE (Lonsdale et al. 2003), they constitute a much higher fraction. AGN detected in large-area, shallow surveys are on average closer and/or more luminous than AGN found in deep pencil-beam surveys. We show in Figure 7b that this very strong dependence of AGN fraction on infrared flux is not an artifact of the x-ray flux limit. Specifically, using our AGN population synthesis model, we plot the number of AGN as a function of infrared flux including those too faint to be detected in x-rays in the Chandra Deep Fields (*open circles*). Clearly the trend is independent of x-ray flux limit.

3.3. *INTEGRAL and Swift hard x-ray surveys*

The population synthesis model presented here implies ∼50% of AGN are currently missed even in deep x-ray surveys with *Chandra* or *XMM*. These are very obscured AGN, many of them Compton thick (i.e., $N_H > 10^{24}$ cm^{-2}), so all but the hardest x-ray surveys are biased against them. Fortunately, hard x-ray instruments on the *INTEGRAL* and *Swift* satellites can now reach fluxes below ∼10^{-11} ergs cm^{-2} s^{-1} at energies above 20 keV. Serendipitous AGN surveys at these energies covering almost the full sky have been carried out using *Swift*/BAT (Markwardt et al. 2005) and *INTEGRAL*/IBIS (Beckmann et al. 2006; Sazonov et al. 2007), yielding samples of ∼100 AGN each. These surveys provide an unbiased view of the AGN population, independent of the amount of obscuration, although at present the sensitivity is sufficient to reach only local populations.

Figure 8a shows the logN-logS distribution from the *INTEGRAL* catalog (Sazonov et al. 2007) of AGN confirmed as Compton thick with N_H measurements from x-ray data. Clearly the original population synthesis model of Treister & Urry (2005) overpredicted the density of Compton-thick AGN by a factor of ∼4. This is not too surprising, as the model included assumptions that, at the time of the publication of that work, were unconstrained—for example, the reflection fraction† was assumed to be unity, and the number of Compton-thick AGN was simply extrapolated with a flat slope from the N_H distribution at lower column densities.

Now we explore the constraints on these quantities that come from the new *Swift* and *INTEGRAL* surveys. First, we define a "Compton-thick AGN correction factor," which simply multiplies the original flat-extrapolation assumption to match the observed number density of AGN with $N_H > 10^{24}$ cm^{-2}. According to a χ^2 minimization, the best-fit value for the Compton-thick AGN factor is 0.25; this produces a good agreement between model and observations, with a reduced χ^2 of 0.3.

Second, we consider the ratio of direct and reflected x-rays. The spectrum of Compton-thick AGN at high energies is dominated by the Compton reflection component (e.g., Matt et al. 2000), which has a strong peak at $E \sim 30$ keV (Magdziarz & Zdziarski 1995). The observed spectrum of the x-ray background, which we now understand as the integrated emission from previously unresolved AGN, also has peak at about the same energy (Gruber et al. 1999). The normalization of the reflection component relative to the direct emission is known only for a few nearby AGN, mostly from *BeppoSAX* observations (e.g., Perola et al. 2002), and therefore this parameter is usually taken as an assumption

---

† The reflection fraction is the geometrical factor describing the solid angle, relative to 2π, subtended by cold reflecting material as seen from the primary x-ray source.

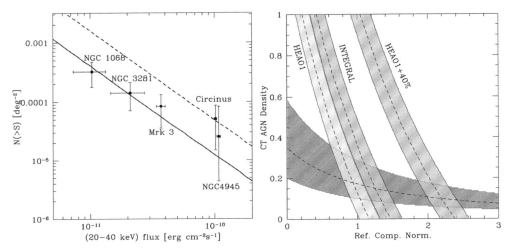

FIGURE 8. Left: Log N–logS relation for Compton-thick AGN only. The *dashed line* shows the original prediction of Treister & Urry (2005), which assumed a reflection component normalization of 1 and a Compton-thick AGN factor of 1 (i.e., as many Compton-thick AGN as Compton-thin AGN in the next lowest decade of the N_H distribution). Data points are from the *INTEGRAL* survey of Sazonov et al. (2007). The population synthesis model can be brought to agreement with the observations if there are a factor of 4 fewer Compton-thick AGN (i.e., the Compton-thick AGN factor is 0.25). Right: Compton-thick AGN factor versus normalization of the Compton reflection component. The *dark gray region* shows the values of these parameters that produce a space density of Compton-thick AGN consistent with the observed value in the Sazonov et al. (2007) sample, considering 1-σ statistical fluctuations. The allowed values of the parameters needed to fit the intensity of the x-ray background at 20–40 keV, assuming a 5% uncertainty in each case, are shown by the *light gray region* for the original Gruber et al. (1999) measurements, the *medium gray regions* for the *INTEGRAL* values and for the Gruber et al. (1999) points increased by 40% (which was the assumption in Treister & Urry 2005). A large value of the reflection component is required to obtain values consistent with the latter x-ray background intensity, inconsistent with observations of the reflection component in individual AGN. Both the reflection value and the renormalization of the Gruber et al. (1999) x-ray background intensity must be lower.

in AGN population synthesis models that can explain the spectral shape and intensity of the x-ray background. The resulting peak x-ray background intensity depends on both the assumed space density of Compton-thick AGN and the normalization of the reflection component, R; while satisfying the overall intensity constraint, one can trade increased reflection for fewer Compton-thick AGN, or vice versa. Figure 8b shows the allowed regions in terms of the Compton-thick AGN factor versus reflection parameter. That is, the shaded regions show the values of these two parameters that produce an integrated x-ray background intensity in the 20–40 keV range, the region most affected by both the number of Compton-thick AGN and R, consistent with: (*medium gray regions*) the Marshall et al. 1983 data increased by 40% (the maximum suggested renormalization to match higher estimates with imaging instruments at lower energies; Treister & Urry 2005) and the recent *INTEGRAL* x-ray background intensity measured using Earth occultation by Churazov et al. (2007); (*light gray region*) the original HEAO-1 measurements, not renormalized, which are about 10% below Churazov et al. In each case uncertainties in the x-ray background intensity were assumed to be 5%. The dark gray-shaded region shows the parameter space that produces a density of Compton-thick AGN consistent with the observations in Figure 8a, roughly one-quarter what everyone has assumed

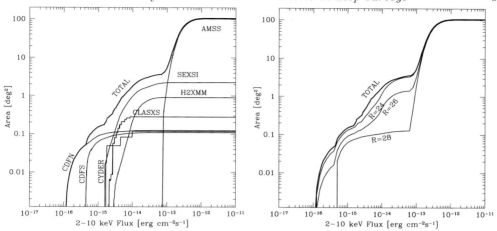

FIGURE 9. Left: Area versus hard x-ray flux for each of the seven surveys combined in this work and for the total sample. Right: Total effective area as a function of x-ray flux and optical magnitude (for $R = 24$, 26 and 28 mag), taking into account the spectroscopic incompleteness of each survey (see text for details). These curves are used to compute the expected fraction of identified obscured AGN for an intrinsically non-evolving ratio.

to date (including Gilli et al. 2007), and consistent with a reflection component with normalization near 1.

Previously, Treister & Urry (2005) assumed an x-ray background intensity consistent with the Gruber et al. (1999) value increased by 40%, which was well fit by a Compton-thick AGN factor of 1 (flat extrapolation of the N_H distribution; Figure 5). However, such a high value of the Compton-thick AGN factor is clearly inconsistent with the new observational constraints on the density of Compton-thick AGN. So, either the intensity of the x-ray background is lower, as suggested by recent *INTEGRAL* measurements (Churazov et al. 2007), or the average value of the reflection parameter is high, $R \sim 2$, or some combination of the two. Observations of individual sources seem to indicate that such a high value for the reflection component is unlikely. From a sample of 22 Seyfert galaxies, excluding Compton-thick sources, Malizia et al. (2003) concluded that both obscured and unobscured sources have similar reflections component with normalization values in the 0.6–1 range. A similar value of $R \simeq 1$ was reported by Perola et al. (2002) based on *BeppoSAX* observations of a sample of nine Seyfert 1 galaxies. Although with large scatter, normalizations for the average reflection component of 0.9 for Seyfert 1 and 1.5 for Seyfert 2 were measured from *BeppoSAX* observations of a sample of 36 sources (Deluit & Courvoisier 2003). Therefore, a value of $R \sim 1$ for the normalization of the reflection component, required by both the observed Compton-thick AGN space density and the x-ray background intensity reported by *INTEGRAL* and *HEAO-1*, is in good agreement with the observed values in nearby Seyfert galaxies.

4. Evolution of the obscured fraction of AGN

The x-ray background, being an integral constraint, is not a strong probe of the fraction of AGN that is obscured (as we showed in the previous section), much less of the evolution of that fraction with cosmic epoch. As shown in Figure 5, a simple population synthesis model in which obscured and unobscured AGN have the same evolution (or equivalently, the fraction of AGN that is obscured does not evolve) is fully consistent with the spectrum of the x-ray background. Now, however, the large x-ray samples

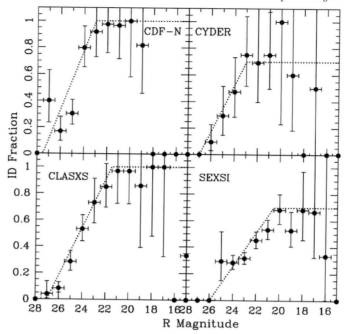

FIGURE 10. Fraction of optically identified AGN as a function of optical magnitude (*data points*), for four of the seven surveys comprising our super-sample. All seven are well described by a simple linear increase to a constant fraction at bright magnitudes (*lines*). This allows us to derive the effective survey area as a function of both x-ray flux and optical magnitude.

that have become available in the past few years, spanning a range of depths and with high spectroscopic completeness, allow us to determine whether and how the fraction of obscured AGN depends on redshift (Treister & Urry 2006).

To study the evolution of the obscured AGN fraction, one needs to distinguish between the effects of redshift and luminosity, which are correlated in any flux-limited sample. Wide-area, shallow x-ray surveys (e.g., XBOOTES; Hickox et al. 2008) sample moderate luminosity AGN at low redshifts and high luminosity sources at high redshifts, while deep pencil-beam surveys (e.g., CDFS; Giacconi et al. 2002) find moderate luminosity AGN at high redshifts but lack rare, high-luminosity sources because of the small volume sampled. Combining the two extremes covers the luminosity-redshift plane effectively.

We generated an AGN super-sample comprising seven large surveys with high identification fractions: AMSS (Akiyama et al. 2003), SEXSI (Harrison et al. 2003), HELLAS2XMM (Baldi et al. 2002), CLASXS (Yang et al. 2004), CYDER (Treister et al. 2005), and the two *Chandra* deep fields (Giacconi et al. 2002; Alexander et al. 2003). This super-sample contains a total of 2341 hard x-ray-selected AGN (x-ray sources with $L_X > 10^{42}$ ergs s^{-1}), the largest such sample to date by a factor of $\sim$4 (Treister & Urry 2006). It spans a range of luminosities at each redshift, over a broad redshift range. The total area of this super-sample as a function of x-ray flux is shown in Figure 9a.

More than half the super-sample is optically identified. We define an AGN as unobscured when there is evidence for broad emission lines in their spectra, and as obscured otherwise. The "obscured fraction" is then the number of obscured AGN divided by the total number of AGN. For obvious reasons, the identified fraction of any survey depends on the brightness of the optical counterparts. We parameterized the identified fraction of each of the seven surveys with a simple function that is constant at bright magnitudes

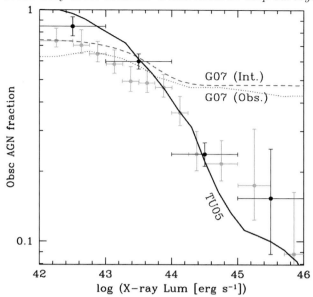

FIGURE 11. Fraction of obscured AGN in the super-sample as a function of hard x-ray luminosity. Black data points: Observed distribution for the super-sample described in this work. Light data points: Compilation of Hasinger et al. (2007) used in the work reported by Gilli et al. (2007). Black solid line: Expected luminosity dependence for the Treister population synthesis models, taking into account the x-ray flux limit and spectroscopic completeness of the super-sample described here (Treister & Urry 2006). Dashed and dotted lines: Intrinsic and bias-corrected (i.e., expected) luminosity dependences for the population synthesis model of Gilli et al. (2007). This assumed luminosity dependence is not a good description of AGN demographics at higher luminosities; the Gilli et al. 2007 model is constrained primarily by the x-ray background intensity, which is dominated by moderate luminosity AGN.

and falls linearly to faint magnitudes (Figure 10; see Treister & Urry 2006 for details). We then weight the area versus x-ray flux curve (Figure 9a) by the completeness at each optical magnitude, for each survey, and sum those to get the effective area of the super-sample as a function of both x-ray flux and optical magnitude (Figure 9b). This allows us to calculate the expected numbers of optically identified AGN for the super-sample, i.e., we can now correct for both x-ray and optical spectroscopic limits.

As discussed earlier (Section 3.2), the fraction of obscured AGN depends on luminosity (Barger et al. 2005). Our population synthesis model assumed a linear transition between 100% obscured fraction at $L_X = 10^{42}$ ergs s^{-1} and 0 obscured fraction at $L_X = 10^{46}$ ergs s^{-1}. Taking into account the selection effects due to flux limits (Figure 9b), this assumption (*black line*, Figure 11) matches the observed dependence very well in the present super-sample (*black points*). The somewhat shallower luminosity dependence adopted by (Gilli et al. 2007; *dashed line* is their assumed dependence, *dotted line* incorporates flux limits), in contrast, does not describe the obscured fraction well for high-luminosity AGN. Their model still gives an adequate fit to the x-ray background, because the dominant contribution is from moderate luminosity AGN.

Our population synthesis model fits the x-ray, optical, and infrared counts of AGN; the x-ray background (modulo the trade-off between the number of Compton-thick AGN and the reflection fraction of each); the hard x-ray counts measured with *Swift* and *INTEGRAL* (ditto); and the observed luminosity dependence of the obscured fraction of

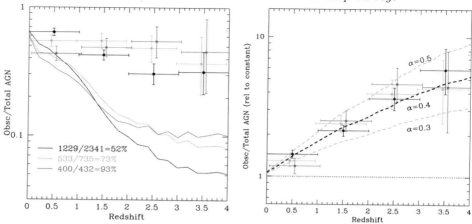

FIGURE 12. Left: Data points indicate the observed fraction of obscured AGN as a function of redshift. Measured fraction for our super-sample of 1229 optically-classified AGN (*black points*), and for sub-samples cut at brighter optical magnitudes to increase the spectroscopic complete-ness to 73% (*light-gray points*) and 93% (*dark-gray points*). The solid lines represent the expected fraction as a function of redshift for an intrinsically non-evolving ratio, taking into account the effects of spectroscopic incompleteness and flux limits for the super-sample and two sub-samples. Right: fraction of obscured AGN relative to the expectations for a non-evolving obscured AGN ratio, incorporating the effects of spectroscopic incompleteness. A significant increase with red-shift is clearly seen, independent of the spectroscopic completeness of the sample. For an intrinsic evolution of the form $(1+z)^{\alpha}$, the *thick-dashed line* shows $\alpha = 0.4$ and the *thin-gray-dashed lines* show $\alpha = 0.5$ (upper) and 0.3 (lower).

AGN in the largest AGN sample to date. What can it tell us about the evolution of this obscured fraction?

Figure 12a shows the observed fraction as a function of redshift (*black data points*). As many have noted previously, the observed fraction declines with redshift, from roughly 3/4 locally to ∼1/3 at $z \sim 4$. This has been interpreted to mean that obscured AGN are rare at high redshift. However, the lines in Figure 12a show the expected decline for an underlying population whose obscured fraction is actually constant with redshift, calculated for our super-sample using the appropriate corrections for x-ray and optical flux limits and spectroscopic completeness. That is, an even steeper decline is expected in the observed samples even when the underlying population does not change at all with redshift.

Three significant selection effects cause this strong decline. First, obscured AGN have smaller x-ray fluxes, so there is a bias against their detection in the x-ray sample in the first place. This is a small effect, particularly because it becomes relatively less important with increasing redshift (for which rest-frame emission is in an increasingly harder x-ray band affected only by higher and higher column densities). Second, obscured AGN are fainter in the optical, so there is a bias against spectroscopic identifications for $z \gtrsim 1$ (Treister et al. 2004); this is not important for samples with highly complete spectroscopic identifications. Third and quite important, the obscured fraction depends strongly on luminosity (Figure 11). Given the luminosity-redshift correlation inherent in flux-limited samples, the mean AGN luminosity increases with redshift and therefore the obscured fraction that is observed decreases—even if the same population of obscured lower-luminosity AGN is present. Our analysis shows that this is the dominant selection effect. It is an important effect even in AGN samples with 100% complete spectroscopic identifications, as indicated by the blue line in Figure 12a. Simply put, high redshift AGN samples are biased to higher luminosity, and thus contain lower obscured fractions, even if there is no underlying evolution of the ratio between obscured and unobscured AGN at all.

The observed more gradual decline in obscured fraction of AGN actually implies an increase in obscured fraction with redshift. Figure 12b shows the data relative to (i.e., divided by) the expectation for a non-evolving population, for the super-sample and the two sub-samples with higher identification fractions. The effect is largely independent of the completeness of the optical identifications. Fitting the increasing fraction with a simple power-law dependence on redshift, $\propto (1 + z)^\alpha$, gives a good fit for $\alpha = 0.4 \pm 0.1$. This means that the fraction of AGN at redshift $z \sim 4$ that are obscured is observed to be twice as high as would be the case were the intrinsic fraction constant. AGN obscuration is substantially greater in the young Universe.

5. Summary

GOODS, MUSYC, COSMOS and other deep multiwavelength surveys provide overwhelming evidence for a large population of obscured AGN that dominate AGN demographics out to high redshifts. Optical surveys are biased against detecting these objects, and even hard x-ray surveys, which are considerably less biased, suffer strong selection effects, primarily due to the luminosity dependence of obscuration.

Taking these effects into account quantitatively, with a realistic, well-constrained population synthesis model that uses AGN spectral energy distributions based on a unification paradigm, we deduce that the ratio of obscured ($N_H > 10^{22}$ cm^{-2}) to unobscured ($N_H < 10^{22}$ cm^{-2}) AGN is roughly 3:1 locally (integrated over all luminosities), and increases with redshift. Low-luminosity AGN (10^{42} ergs s$^{-1} < L_X < 10^{44}$ ergs s^{-1}) are much more likely to be obscured than high-luminosity AGN ($L_X > 10^{44}$ ergs s^{-1}).

To the extent that our assumed infrared through x-ray spectral energy distributions are reasonable, and our assumed N_H distribution is reasonable (it is essentially the same as that used by others in the field; see Gilli et al. 2007 for a comparison of the different distributions), these results are completely robust.

How might one avoid the biases inherent in optical and x-ray surveys? In principle, far-infrared surveys are unbiased because the absorbed energy is re-radiated thermally; however, infrared surveys are very inefficient for AGN selection since the infrared sky is strongly dominated by starlight (Figure 7a). In addition, AGN signatures (such as broad emission lines, strong power-law continua, and rapid variability) may well be hidden in obscured objects. Thus, identifying complete samples of AGN from far-infrared surveys will never be a simple matter, although it can potentially put useful limits on the fraction of galaxies with buried AGN.

Selection effects are very important to take into account for any survey, even those that are 100% identified. For example, consider a deep hard x-ray survey for which all sources have optical and/or infrared counterparts with known redshifts and classification. Many would assume the survey itself is not strongly biased and that the identifications yield the underlying demographics, since no x-ray sources remain unidentified. However, missing from the x-ray sample are the most heavily obscured AGN; even more important, unobscured AGN are over-represented (relative to their fraction of the underlying population), especially at high redshift, because of the dependence of obscuration on luminosity. This in fact is the dominant effect for existing surveys.

Therefore, to understand the distribution of black holes in the universe and to estimate cosmic accretion rates, the selection effects must be modeled using reasonable assumptions about the underlying population. That in turns yields a picture of the universe that matches well our picture of nascent accreting black holes at the centers of dusty, star-forming, young galaxies in the early Universe.

This work would not have been possible without the exquisite data made possible by NASA's Great Observatories and by great observatories on the ground. We gratefully acknowledge support from NSF grant AST-0407295 and NASA grants NAG5-10301, HST-AR-10689.01-A, HST-GO-09822.09-A, HST-GO-09425.13-A, and NNG05GM79G.

REFERENCES

AKIYAMA, M., UEDA, Y., OHTA, K., TAKAHASHI, T., & YAMADA, T. 2003 *ApJ* **148**, 275.

ALEXANDER, D. M., BAUER, F. E., BRANDT, W. N., SCHNEIDER, D. P., HORNSCHEMEIER, A. E., VIGNALI, C., BARGER, A. J., BROOS, P. S., COWIE, L. L., GARMIRE, G. P., TOWNSLEY, L. K., BAUTZ, M. W., CHARTAS, G., & SARGENT, W. L. W. 2003 *AJ* **126**, 539.

ALEXANDER, D. M., BRANDT, W. N., HORNSCHEMEIER, A. E., GARMIRE, G. P., SCHNEIDER, D. P., BAUER, F. E., & GRIFFITHS, R. E. 2001 *AJ* **122**, 2156.

ANTONUCCI, R. 1993 *ARA&A* **31**, 473.

BALDI, A., MOLENDI, S., COMASTRI, A., FIORE, F., MATT, G., & VIGNALI, C. 2002 *ApJ* **564**, 190.

BALLANTYNE, D. R., EVERETT, J. E., & MURRAY, N. 2006 *ApJ* **639**, 740.

BARGER, A. J., COWIE, L. L., MUSHOTZKY, R. F., YANG, Y., WANG, W.-H., STEFFEN, A. T., & CAPAK, P. 2005 *AJ* **129**, 578.

BECKMANN, V., GEHRELS, N., SHRADER, C. R., & SOLDI, S. 2006 *ApJ* **638**, 642.

BRANDT, W. N. & HASINGER, G. 2005 *ARA&A* **43**, 827.

CAPAK, P., AUSSEL, H., AJIKI, M., MCCRACKEN, H. J., MOBASHER, B., SCOVILLE, N., SHOPBELL, P., TANIGUCHI, Y., THOMPSON, D., TRIBIANO, S., SASAKI, S., BLAIN, A. W., BRUSA, M., CARILLI, C., COMASTRI, A., CAROLLO, C. M., CASSATA, P., COLBERT, J., ELLIS, R. S., ELVIS, M., GIAVALISCO, M., GREEN, W., GUZZO, L., HASINGER, G., ILBERT, O., IMPEY, C., JAHNKE, K., KARTALTEPE, J., KNEIB, J.-P., KODA, J., KOEKEMOER, A., KOMIYAMA, Y., LEAUTHAUD, A., LEFEVRE, O., LILLY, S., LIU, C., MASSEY, R., MIYAZAKI, S., MURAYAMA, T., NAGAO, T., PEACOCK, J. A., PICKLES, A., PORCIANI, C., RENZINI, A., RHODES, J., RICH, M., SALVATO, M., SANDERS, D. B., SCARLATA, C., SCHIMINOVICH, D., SCHINNERER, E., SCODEGGIO, M., SHETH, K., SHIOYA, Y., TASCA, L. A. M., TAYLOR, J. E., YAN, L., & ZAMORANI, G. 2007 *ApJ* **172**, 99.

CHURAZOV, E., SUNYAEV, R., REVNIVTSEV, M., SAZONOV, S., MOLKOV, S., GREBENEV, S., WINKLER, C., PARMAR, A., BAZZANO, A., FALANGA, M., GROS, A., LEBRUN, F., NATALUCCI, L., UBERTINI, P., ROQUES, J., BOUCHET, L., JOURDAIN, E., KNOEDLSEDER, J., DIEHL, R., BUDTZ-JORGENSEN, C., BRANDT, S., LUND, N., WESTERGAARD, N. J., NERONOV, A., TURLER, M., CHERNYAKOVA, M., WALTER, R., PRODUIT, N., MOWLAVI, N., MAS-HESSE, J. M., DOMINGO, A., GEHRELS, N., KUULKERS, E., KRETSCHMAR, P., & SCHMIDT, M. 2007 *A&A* **467**, 529.

COMASTRI, A., SETTI, G., ZAMORANI, G., & HASINGER, G. 1995 *A&A* **296**, 1.

COWIE, L. L., BARGER, A. J., BAUTZ, M. W., BRANDT, W. N., & GARMIRE, G. P. 2003 *ApJ* **584**, L57.

DAVIS, M., ET AL. 2007 *ApJ* **660**, L1.

DELUIT, S. & COURVOISIER, T. J.-L. 2003 *A&A* **399**, 77.

DICKINSON, M., GIAVALISCO, M., & THE GOODS TEAM 2003. In *The Mass of Galaxies at Low and High Redshift* (eds. R. Bender & A. Renzini). p. 324. Springer-Verlag.

DWELLY, T. & PAGE, M. J. 2006 *MNRAS* **372**, 1755.

FADDA, D., FLORES, H., HASINGER, G., FRANCESCHINI, A., ALTIERI, B., CESARSKY, C. J., ELBAZ, D., & FERRANDO, P. 2002 *A&A* **383**, 838.

FAZIO, G. G., ASHBY, M. L. N., BARMBY, P., HORA, J. L., HUANG, J.-S., PAHRE, M. A., WANG, Z., WILLNER, S. P., ARENDT, R. G., MOSELEY, S. H., BRODWIN, M., EISENHARDT, P., STERN, D., TOLLESTRUP, E. V., & WRIGHT, E. L. 2004 *ApJ* **154**, 39.

FRANCESCHINI, A., BRAITO, V., & FADDA, D. 2002a *MNRAS* **335**, L51.

FRANCESCHINI, A., FADDA, D., CESARSKY, C. J., ELBAZ, D., FLORES, H., & GRANATO, G. L. 2002b *ApJ* **568**, 470.

GAWISER, E., VAN DOKKUM, P. G., GRONWALL, C., CIARDULLO, R., BLANC, G. A., CASTANDER, F. J., FELDMEIER, J., FRANCKE, H., FRANX, M., HABERZETTL, L., HERRERA, D., HICKEY, T., INFANTE, L., LIRA, P., MAZA, J., QUADRI, R., RICHARDSON, A., SCHAWINSKI, K., SCHIRMER, M., TAYLOR, E. N., TREISTER, E., URRY, C. M., & VIRANI, S. N. 2006a *ApJ* **642**, L13.

GAWISER, E., VAN DOKKUM, P. G., HERRERA, D., MAZA, J., CASTANDER, F. J., INFANTE, L., LIRA, P., QUADRI, R., TONER, R., TREISTER, E., URRY, C. M., ALTMANN, M., ASSEF, R., CHRISTLEIN, D., COPPI, P. S., DURÁN, M. F., FRANX, M., GALAZ, G., HUERTA, L., LIU, C., LÓPEZ, S., MÉNDEZ, R., MOORE, D. C., RUBIO, M., RUIZ, M. T., TOFT, S., & YI, S. K. 2006b *ApJ* **162**, 1.

GEHRELS, N. 1986 *ApJ* **303**, 336.

GIACCONI, R., ZIRM, A., WANG, J., ROSATI, P., NONINO, M., TOZZI, P., GILLI, R., MAINIERI, V., HASINGER, G., KEWLEY, L., BERGERON, J., BORGANI, S., GILMOZZI, R., GROGIN, N., KOEKEMOER, A., SCHREIER, E., ZHENG, W., & NORMAN, C. 2002 *ApJ* **139**, 369.

GIAVALISCO, M., ET AL. 2004 *ApJ* **600**, L93.

GILLI, R. 2004 *Advances in Space Research* **34**, 2470.

GILLI, R., COMASTRI, A., & HASINGER, G. 2007 *A&A* **463**, 79.

GILLI, R., SALVATI, M. & HASINGER, G. 2001 *A&A* **366**, 407.

GRUBER, D. E., MATTESON, J. L., PETERSON, L. E., & JUNG, G. V. 1999 *ApJ* **520**, 124.

HARRISON, F. A., ECKART, M. E., MAO, P. H., HELFAND, D. J., & STERN, D. 2003 *ApJ* **596**, 944.

HASINGER, G. 2000. In *ISO Survey of a Dusty Universe* (eds. D. Lemke, M. Stickel & K. Wilke), *Lecture Notes in Physics*, vol. 548, p. 423. Springer Verlag.

HASINGER, G., CAPPELLUTI, N., BRUNNER, H., BRUSA, M., COMASTRI, A., ELVIS, M., FINOGUENOV, A., FIORE, F., FRANCESCHINI, A., GILLI, R., GRIFFITHS, R. E., LEHMANN, I., MAINIERI, V., MATT, G., MATUTE, I., MIYAJI, T., MOLENDI, S., PALTANI, S., SANDERS, D. B., SCOVILLE, N., TRESSE, L., URRY, C. M., VETTOLANI, P., & ZAMORANI, G. 2007 *ApJ* **172**, 29.

HASINGER, G., ET AL. 2001 *A&A* **365**, L45.

HAUSER, M. G. & DWEK, E. 2001 *ARA&A* **39**, 249.

HICKOX, R. C., JONES, C., FORMAN, W. R., MURRAY, S. S., BRODWIN, M., XBOOTES, T. C., IRAC SHALLOW SURVEY, S., AGES, & NOAO DWFS TEAMS 2008. In *Infrared Diagnostics of Galaxy Evolution* (eds. R. R. Chary, H. I. Teplitz, & K. Sheth). Conf. Ser. 381, p. 418. Astronomical Society of the Pacific.

KASHLINSKY, A., MATHER, J. C., & ODENWALD, S. 1996 *ApJ* **473**, L9.

KIM, D.-W., ET AL. 2004 *ApJ* **150**, 19.

KING, A. 2005 *ApJ* **635**, L121.

LAWRENCE, A. 2001. In *The Promise of the Herschel Space Observatory* (eds. G. L. Pilbratt, J. Cernicharo, A. M. Heras, T. Prusti, & R. Harris) ESA Special Publication, vol. 460, p. 95. ESA.

LEHMER, B. D., BRANDT, W. N., ALEXANDER, D. M., BAUER, F. E., SCHNEIDER, D. P., TOZZI, P., BERGERON, J., GARMIRE, G. P., GIACCONI, R., GILLI, R., HASINGER, G., HORNSCHEMEIER, A. E., KOEKEMOER, A. M., MAINIERI, V., MIYAJI, T., NONINO, M., ROSATI, P., SILVERMAN, J. D., SZOKOLY, G., & VIGNALI, C. 2005 *ApJ* **161**, 21.

LILLY, S. J., FÈVRE, O. L., RENZINI, A., ZAMORANI, G., SCODEGGIO, M., CONTINI, T., CAROLLO, C. M., HASINGER, G., KNEIB, J.-P., IOVINO, A., LE BRUN, V., MAIER, C., MAINIERI, V., MIGNOLI, M., SILVERMAN, J., TASCA, L. A. M., BOLZONELLA, M., BONGIORNO, A., BOTTINI, D., CAPAK, P., CAPUTI, K., CIMATTI, A., CUCCIATI, O., DADDI, E., FELDMANN, R., FRANZETTI, P., GARILLI, B., GUZZO, L., ILBERT, O., KAMPCZYK, P., KOVAC, K., LAMAREILLE, F., LEAUTHAUD, A., BORGNE, J.-F. L., MCCRACKEN, H. J., MARINONI, C., PELLO, R., RICCIARDELLI, E., SCARLATA, C., VERGANI, D., SANDERS, D. B., SCHINNERER, E., SCOVILLE, N., TANIGUCHI, Y., ARNOUTS, S., AUSSEL, H., BARDELLI, S., BRUSA, M., CAPPI, A., CILIEGI, P.,

FINOGUENOV, A., FOUCAUD, S., FRANCESCHINI, R., HALLIDAY, C., IMPEY, C., KNO-BEL, C., KOEKEMOER, A., KURK, J., MACCAGNI, D., MADDOX, S., MARANO, B., MAR-CONI, G., MENEUX, B., MOBASHER, B., MOREAU, C., PEACOCK, J. A., PORCIANI, C., POZZETTI, L., SCARAMELLA, R., SCHIMINOVICH, D., SHOPBELL, P., SMAIL, I., THOMP-SON, D., TRESSE, L., VETTOLANI, G., ZANICHELLI, A., & ZUCCA, E. 2007 *ApJ* **172**, 70.

LONSDALE, C. J., ET AL. 2003 *PASP* **115**, 897.

MAGDZIARZ, P. & ZDZIARSKI, A. A. 1995 *MNRAS* **273**, 837.

MAINIERI, V., ROSATI, P., TOZZI, P., BERGERON, J., GILLI, R., HASINGER, G., NONINO, M., LEHMANN, I., ALEXANDER, D. M., IDZI, R., KOEKEMOER, A. M., NORMAN, C., SZO-KOLY, G., & ZHENG, W. 2005 *A&A* **437**, 805.

MALIZIA, A., BASSANI, L., STEPHEN, J. B., DI COCCO, G., FIORE, F., & DEAN, A. J. 2003 *ApJ* **589**, L17.

MARKWARDT, C. B., TUELLER, J., SKINNER, G. K., GEHRELS, N., BARTHELMY, S. D., & MUSHOTZKY, R. F. 2005 *ApJ* **633**, L77.

MATSUMOTO, T., MATSUURA, S., MURAKAMI, H., TANAKA, M., FREUND, M., LIM, M., CO-HEN, M., KAWADA, M., & NODA, M. 2005 *ApJ* **626**, 31.

MATT, G., FABIAN, A. C., GUAINAZZI, M., IWASAWA, K., BASSANI, L., & MALAGUTI, G. 2000 *MNRAS* **318**, 173.

MATUTE, I., LA FRANCA, F., POZZI, F., GRUPPIONI, C., LARI, C., ZAMORANI, G., ALEXAN-DER, D. M., DANESE, L., OLIVER, S., SERJEANT, S., & ROWAN-ROBINSON, M. 2002 *MNRAS* **332**, L11.

MORRISON, R. & MCCAMMON, D. 1983 *ApJ* **270**, 119.

NENKOVA, M., IVEZIĆ, Ž. &, ELITZUR, M. 2002 *ApJ* **570**, L9.

PAPOVICH, C., DOLE, H., EGAMI, E., LE FLOC'H, E., PÉREZ-GONZÁLEZ, P. G., ALONSO-HERRERO, A., BAI, L., BEICHMAN, C. A., BLAYLOCK, M., ENGELBRACHT, C. W., GOR-DON, K. D., HINES, D. C., MISSELT, K. A., MORRISON, J. E., MOULD, J., MUZEROLLE, J., NEUGEBAUER, G., RICHARDS, P. L., RIEKE, G. H., RIEKE, M. J., RIGBY, J. R., SU, K. Y. L., & YOUNG, E. T. 2004 *ApJ* **154**, 70.

PEROLA, G. C., MATT, G., CAPPI, M., FIORE, F., GUAINAZZI, M., MARASCHI, L., PETRUCCI, P. O., & PIRO, L. 2002 *A&A* **389**, 802.

QUADRI, R., MARCHESINI, D., VAN DOKKUM, P., GAWISER, E., FRANX, M., LIRA, P., RUD-NICK, G., URRY, C. M., MAZA, J., KRIEK, M., BARRIENTOS, L. F., BLANC, G. A., CAS-TANDER, F. J., CHRISTLEIN, D., COPPI, P. S., HALL, P. B., HERRERA, D., INFANTE, L., TAYLOR, E. N., TREISTER, E., & WILLIS, J. P. 2007 *AJ* **134**, 1103.

RISALITI, G., MAIOLINO, R., & SALVATI, M. 1999 *ApJ* **522**, 157.

ROSATI, P., TOZZI, P., GIACCONI, R., GILLI, R., HASINGER, G., KEWLEY, L., MAINIERI, V., NONINO, M., NORMAN, C., SZOKOLY, G., WANG, J. X., ZIRM, A., BERGERON, J., BOR-GANI, S., GILMOZZI, R., GROGIN, N., KOEKEMOER, A., SCHREIER, E., & ZHENG, W. 2002 *ApJ* **566**, 667.

ROVILOS, E., GEORGAKAKIS, A., GEORGANTOPOULOS, I., AFONSO, J., KOEKEMOER, A. M., MOBASHER, B., & GOUDIS, C. 2007 *A&A* **466**, 119.

SANDERS, D. B., SALVATO, M., AUSSEL, H., ILBERT, O., SCOVILLE, N., SURACE, J. A., FRAYER, D. T., SHETH, K., HELOU, G., BROOKE, T., BHATTACHARYA, B., YAN, L., KARTALTEPE, J. S., BARNES, J. E., BLAIN, A. W., CALZETTI, D., CAPAK, P., CAR-ILLI, C., CAROLLO, C. M., COMASTRI, A., DADDI, E., ELLIS, R. S., ELVIS, M., FALL, S. M., FRANCESCHINI, A., GIAVALISCO, M., HASINGER, G., IMPEY, C., KOEKEMOER, A., LE FÈVRE, O., LILLY, S., LIU, M. C., MCCRACKEN, H. J., MOBASHER, B., RENZINI, A., RICH, M., SCHINNERER, E., SHOPBELL, P. L., TANIGUCHI, Y., THOMPSON, D. J., URRY, C. M., & WILLIAMS, J. P. 2007 *ApJ* **172**, 86.

SAZONOV, S., REVNIVTSEV, M., KRIVONOS, R., CHURAZOV, E., & SUNYAEV, R. 2007 *A&A* **462**, 57.

SCHNEIDER, D. P., HALL, P. B., RICHARDS, G. T., STRAUSS, M. A., VANDEN BERK, D. E., ANDERSON, S. F., BRANDT, W. N., FAN, X., JESTER, S., GRAY, J., GUNN, J. E., SUBBARAO, M. U., THAKAR, A. R., STOUGHTON, C., SZALAY, A. S., YANNY, B., YORK, D. G., BAHCALL, N. A., BARENTINE, J., BLANTON, M. R., BREWINGTON, H., BRINKMANN, J., BRUNNER, R. J., CASTANDER, F. J., CSABAI, I., FRIEMAN, J. A.,

FUKUGITA, M., HARVANEK, M., HOGG, D. W., IVEZIĆ, Ž., KENT, S. M., KLEINMAN, S. J., KNAPP, G. R., KRON, R. G., KRZESIŃSKI, J., LONG, D. C., LUPTON, R. H., NITTA, A., PIER, J. R., SAXE, D. H., SHEN, Y., SNEDDEN, S. A., WEINBERG, D. H., & WU, J. 2007 *AJ* **134**, 102.

SCHNEIDER, D. P., ET AL. 2002 *AJ* **123**, 567.

SCOVILLE, N., ABRAHAM, R. G., AUSSEL, H., BARNES, J. E., BENSON, A., BLAIN, A. W., CALZETTI, D., COMASTRI, A., CAPAK, P., CARILLI, C., CARLSTROM, J. E., CAROLLO, C. M., COLBERT, J., DADDI, E., ELLIS, R. S., ELVIS, M., EWALD, S. P., FALL, M., FRANCESCHINI, A., GIAVALISCO, M., GREEN, W., GRIFFITHS, R. E., GUZZO, L., HASINGER, G., IMPEY, C., KNEIB, J.-P., KODA, J., KOEKEMOER, A., LEFEVRE, O., LILLY, S., LIU, C. T., McCRACKEN, H. J., MASSEY, R., MELLIER, Y., MIYAZAKI, S., MOBASHER, B., MOULD, J., NORMAN, C., REFREGIER, A., RENZINI, A., RHODES, J., RICH, M., SANDERS, D. B., SCHIMINOVICH, D., SCHINNERER, E., SCODEGGIO, M., SHETH, K., SHOPBELL, P. L., TANIGUCHI, Y., TYSON, N. D., URRY, C. M., VAN WAER-BEKE, L., VETTOLANI, P., WHITE, S. D. M., & YAN, L. 2007a *ApJ* **172**, 38.

SCOVILLE, N., AUSSEL, H., BRUSA, M., CAPAK, P., CAROLLO, C. M., ELVIS, M., GIAVALIS-CO, M., GUZZO, L., HASINGER, G., IMPEY, C., KNEIB, J.-P., LeFEVRE, O., LILLY, S. J., MOBASHER, B., RENZINI, A., RICH, R. M., SANDERS, D. B., SCHINNERER, E., SCHMI-NOVICH, D., SHOPBELL, P., TANIGUCHI, Y., & TYSON, N. D. 2007b *ApJ* **172**, 1.

SETTI, G. & WOLTJER, L. 1989 *A&A* **224**, L21.

SILK, J. & REES, M. J. 1998 *A&A* **331**, L1.

SILVA, L., DE ZOTTI, G., GRANATO, G. L., MAIOLINO, R., & DANESE, L. 2005 *MNRAS* **357**, 1295.

STANEV, T. & FRANCESCHINI, A. 1998 *ApJ* **494**, L159.

STEFFEN, A. T., STRATEVA, I., BRANDT, W. N., ALEXANDER, D. M., KOEKEMOER, A. M., LEHMER, B. D., SCHNEIDER, D. P., & VIGNALI, C. 2006 *AJ* **131**, 2826.

TOZZI, P., GILLI, R., MAINIERI, V., NORMAN, C., RISALITI, G., ROSATI, P., BERG-ERON, J., BORGANI, S., GIACCONI, R., HASINGER, G., NONINO, M., STREBLYANSKA, A., SZOKOLY, G., WANG, J. X., & ZHENG, W. 2006 *A&A* **451**, 457.

TREISTER, E., CASTANDER, F. J., MACCARONE, T. J., GAWISER, E., COPPI, P. S., URRY, C. M., MAZA, J., HERRERA, D., GONZALEZ, V., MONTOYA, C., & PINEDA, P. 2005 *ApJ* **621**, 104.

TREISTER, E. & URRY, C. M. 2005 *ApJ* **630**, 115.

TREISTER, E. & URRY, C. M. 2006 *ApJ* **652**, L79.

TREISTER, E., URRY, C. M., VAN DUYNE, J., DICKINSON, M., CHARY, R.-R., ALEXANDER, D. M., BAUER, F., NATARAJAN, P., LIRA, P., & GROGIN, N. A. 2006 *ApJ* **640**, 603.

TREISTER, E., ET AL. 2004 *ApJ* **616**, 123.

UEDA, Y., AKIYAMA, M., OHTA, K., & MIYAJI, T. 2003 *ApJ* **598**, 886.

URRY, C. M. & PADOVANI, P. 1995 *PASP* **107**, 803.

URRY, C. M. & TREISTER, E. 2005. In *Growing Black Holes: Accretion in a Cosmological Context* (eds. A. Merloni, S. Nayakshin, & R. A. Sunyaev). ESO Astrophysics Symposia, p. 432. Springer-Verlag.

VAN DOKKUM, P. G., QUADRI, R., MARCHESINI, D., RUDNICK, G., FRANX, M., GAWISER, E., HERRERA, D., WUYTS, S., LIRA, P., LABBÉ, I., MAZA, J., ILLINGWORTH, G. D., FÖRSTER SCHREIBER, N. M., KRIEK, M., RIX, H.-W., TAYLOR, E. N., TOFT, S., WEBB, T., & YI, S. K. 2006 *ApJ* **638**, L59.

VANDEN BERK, D. E., ET AL. 2001 *AJ* **122**, 549.

VIRANI, S. N., TREISTER, E., URRY, C. M., & GAWISER, E. 2006 *AJ* **131**, 2373.

WORSLEY, M. A., FABIAN, A. C., BAUER, F. E., ALEXANDER, D. M., HASINGER, G., MA-TEOS, S., BRUNNER, H., BRANDT, W. N., & SCHNEIDER, D. P. 2005 *MNRAS* **357**, 1281.

XU, D.-W., WEI, J.-Y., & HU, J.-Y. 2001 *Chinese Journal of Astronomy & Astrophysics* **1**, 46.

YANG, Y., MUSHOTZKY, R. F., STEFFEN, A. T., BARGER, A. J., & COWIE, L. L. 2004 *AJ* **128**, 1501.

Black hole masses from reverberation mapping

By BRADLEY M. PETERSON[1] AND MISTY C. BENTZ[1,2]

[1]Department of Astronomy, The Ohio State University, 140 West 18th Avenue,
Columbus, OH 43210, USA

[2]Current address: Department of Physics and Astronomy, University of California at Irvine,
4129 Frederick Reines Hall, Irvine, CA 92697, USA

Emission-line reverberation mapping, which explores the dynamics of gas at distances of order 1000 gravitational radii from the supermassive black holes in Type 1 (broad emission-line) active galactic nuclei, can be used to measure the masses of the black holes that power these sources. Evidence that these masses are reliable is based on the relationships between black hole mass and (a) host-galaxy bulge velocity (the M_{BH}-σ_* relationship) and (b) host-galaxy bulge luminosity (or mass). The scatter around the M_{BH}-σ_* relationship suggests that reverberation-based masses are accurate to about a factor of three. We revisit the relationship between BLR radius R and luminosity L and find consistency with $R \propto L^{1/2}$ once the luminosity measurement is corrected for contamination by starlight from the host galaxy.

1. Introduction

The existence of supermassive black holes at the centers of galaxies was suspected long before it was possible to accurately measure the masses of such objects. In the case of active galactic nuclei (AGNs), Zeldovich & Novikov (1964), Salpeter (1964), and Lynden-Bell (1969) concluded early that AGNs are plausibly powered by gravitational infall onto supermassive collapsed objects. In the case of quiescent galaxies, it was suspected that M 87† harbors a supermassive black hole based on both stellar dynamics (Sargent et al. 1978) and its strongly peaked central surface brightness (Young et al. 1978). However, it was only in the *HST* era that sufficient angular resolution has been attained to resolve the black hole radius of influence,

$$ r_* = \frac{GM_{BH}}{\sigma_*^2} \quad , \tag{1.1} $$

where M_{BH} is the black hole mass and σ_* is the velocity dispersion of the stellar bulge; inside r_*, stellar dynamics are dominated by the presence of the black hole, and dynamical measurements on this scale or smaller make it possible to measure the black hole mass with some confidence. Of course, this is not without some small ironies:

(*a*) While most of the black hole masses measured by stellar or gas dynamics have required use of *HST*, the two most accurately measured black hole masses—that at our own Galactic Center (Genzel et al. 2003; Ghez et al. 2005, and references therein) and the megamaser source in NGC 4258 (Miyoshi et al. 1995)—have relied solely on ground-based observations.

(*b*) Measurement of black hole masses by modeling of stellar and gas dynamics has proven to be much more difficult for AGNs, the sources in which supermassive black holes were first suspected to exist, than for quiescent galaxies. In AGNs, the glare from the

---

† Strictly speaking, M 87 is not exactly a quiescent galaxy, as evidenced by its well-known jet. However, it is certainly not a quasar-like source. In modern terms, the nucleus of M 87 could be described as a supermassive black hole accreting at a low Eddington ratio and/or in a radiatively inefficient mode.

active nucleus makes detection of weak stellar absorption features difficult, particularly on the smallest angular scales.

For AGNs, however, there is another method available to measure the black hole mass, namely emission-line reverberation mapping (Blandford & McKee 1982; Peterson 1993). The broad emission lines that characterize the ultraviolet through infrared spectra of Seyfert 1 galaxies and quasars are observed to vary in flux in response to the changing flux of the continuum source, but with a time delay due to light travel-time effects within the broad-line region (BLR). With careful monitoring of the flux variability of the continuum and emission lines, we can measure this time delay, or lag, τ, between continuum and emission-line variations, and thus determine the size of the line-emitting region, $R = c\tau$. By combining the size of the line-emitting region with the velocity dispersion of the line-emitting gas (as given by the Doppler width of the emission line ΔV), the mass of the central black hole is

$$M_{\rm BH} = f \, \frac{c\tau \, \Delta V^2}{G} \quad , \tag{1.2}$$

where f is a dimensionless factor of order unity that depends on the geometry, kinematics, and inclination of the BLR, as we will discuss in more detail below.

Relative to other methods of black hole mass measurement, a particular advantage of reverberation mapping is the proximity of the BLR gas to the central black hole (which is, ultimately, precisely why the broad lines are broad, as the line emission arises in the deep gravitational potential of the black hole). Reverberation shows that the line emission typically arises on scales of a few hundred to a few thousand gravitational radii ($R_g = GM_{\rm BH}/c^2$); this is near enough to ensure that the enclosed mass measured is dominated by the black hole, but is far enough away that relativistic effects are weak (with the possible exception of a gravitational redshift, as mentioned in Section 7). This is also a much smaller scale ($\sim$10 mas for the nearest AGNs) than is resolvable for stellar and gas dynamical measurements, which are typically on scales of $\sim$10$^4 R_g$ or greater. Of course, a corresponding disadvantage is that the BLR gas is subject to forces other than gravity; radiation pressure, in particular, is expected to play a significant role in the BLR dynamics. It is therefore very important to demonstrate that reverberation is actually measuring the black hole mass: consistency checks and comparisons with independent methods are needed. These are the subject of this contribution.

2. The virial relationship

The first well-sampled reverberation data on NGC 5548 (Clavel et al. 1991; Peterson et al. 1991) showed evidence for ionization stratification of the BLR: lines more characteristic of highly ionized gas (e.g., He II λ1640) had shorter lags than those that are strong in less highly ionized gas (e.g., the Balmer lines). It had been known for years (e.g., Osterbrock & Shuder 1982) that there were differences in the widths of different lines in AGN spectra, and it made sense to look for a relationship between line width and lag, with the expectation that if things are simple one might find a "virial relationship" [cf. Eq. (1.2)] of the form $\Delta V \propto \tau^{-1/2}$. Krolik et al. (1991) looked for such a relationship (their Figure 4), but the results were not convincing. Indeed, the absence of a virial relationship cast doubt on whether one could use reverberation to measure black hole masses (e.g., Richstone et al. 1998; Ho 1999), since it was not clear that gravity dominated the BLR dynamics or, if it did, how the mass could be deduced since each line would give a different virial product $c\tau\Delta V^2/G$. The problem was reconsidered by Peterson & Wandel (1999) who, in contrast to Krolik et al. (1991), found strong evidence for a virial

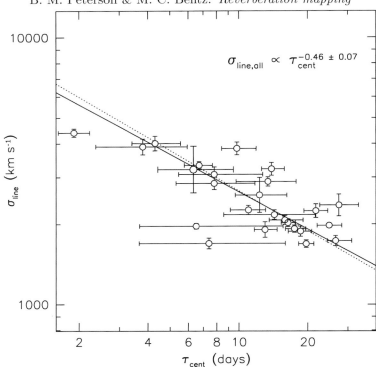

FIGURE 1. Line width σ_{line} versus time lag τ for emission lines in the spectrum of the well-studied AGN NGC 5548. The solid line shows the best-fit slope to the data and the dotted line shows the best fit for a fixed slope of $-1/2$. The slope of the relationship is consistent with that expected if the BLR dynamics are dominated by the gravitational acceleration by the central mass, i.e., Equation (1.2). From Bentz et al. (2007).

relationship based on the strong emission lines in NGC 5548. Why did these two studies get such different results? The simple answer is that a lot of experience in dealing with real reverberation data was obtained during the 1990s. More specifically, the things that Peterson & Wandel did differently included, in order of importance, (a) rejection of badly blended (e.g., Lyα+N v λ1240), very weak or weakly variable (e.g., O i λ1304), and clearly aliased (e.g., Mg ii λ2800) emission lines from the analysis, (b) use of a cross-correlation method with superior time resolution,† (c) inclusion of six individual measurements of the Hβ lag and an additional set of lags for UV lines measured with FOS on *HST* that were not available to Krolik et al., and (d) use of a more reliable methodology for determining the lag uncertainties (Peterson et al. 1998). Since then, in every case where it can be reasonably tested, consistency with a similar virial relationship between lag and line width has been demonstrated (Peterson & Wandel 2000; Onken & Peterson 2002; Kollatschny 2003a; Metzroth, Onken, & Peterson 2006).

Subsequently, based on a suggestion from Fromerth & Melia (2000), the virial relationship was improved by using the line dispersion (second moment of the line profile) rather

† Krolik et al. (1991) employed the Discrete Correlation Function (DCF) method (Edelson & Krolik 1988), which does not require any assumptions about the smoothness of continuum behavior, but unfortunately introduced in this case discretization effects: all lags they used were integral multiples of the four-day bin width they chose to match the mean sampling interval of the light curves.

than FWHM as the line width measure ΔV (Peterson et al. 2004), as shown in Figure 1. This is actually quite an important point: the ratio of the two line width measures, FWHM/σ_{line}, is a crude characterization of the line profile.‡ The observed values of this ratio for reverberation mapped AGNs are in the range $0.71 \leqslant \text{FWHM}/\sigma_{\text{line}} \leqslant 3.45$, with a mean value for the sample FWHM/$\sigma_{\text{line}} \approx 2$. The lower values of the line-width ratio are found in objects that are often identified as "narrow-line Seyfert 1 (NLS1) galaxies," which have "peaky" profiles and are often supposed to be high Eddington rate accretors. At the other extreme are AGNs like Akn 120, which have "boxy" profiles, often double-peaked, as expected for a Keplerian disk. Since the black hole mass inferred is proportional to the square of the line width, we see that the ratio of the masses in our two most extreme cases differs by a factor of $(3.45/0.71)^2 \approx 24$, depending on which line-width measure we use! Collin et al. (2006) demonstrate that the choice of line-width measure is not arbitrary: σ_{line} is a less-biased estimator of the mass than FWHM.

3. The M_{BH}-σ_* relationship

As is now well known, there is a remarkably tight correlation between the masses of the central black holes in galaxies and the velocity dispersion σ_* of the stars in the bulge component (Ferrarese & Merritt 2000; Gebhardt et al. 2000a), the "M_{BH}-σ_* relationship." This correlation was initially established on the basis of quiescent galaxies, although the megamaser source NGC 4258, a *bona fide* AGN (Wilkes et al. 1995), was also included. Gebhardt et al. (2000b) subsequently showed that the relatively small number of AGNs with both reverberation-based mass estimates and published stellar velocity dispersions showed consistency with the same M_{BH}-σ_* relationship, although the uncertainties in the σ_* values in particular were rather large. New efforts to obtain better velocity dispersions were undertaken (Ferrarese et al. 2001; Onken et al. 2004; Nelson et al. 2004) that solidified the earlier conclusion. Onken et al. (2004) went a step further and assumed that the zero point of the M_{BH}-σ_* relationship is the same for active and quiescent galaxies, and thus established statistically a mean value for the AGN mass scaling factor $\langle f \rangle \approx 5.5 \pm 1.8$ [Eq. (1.2)], providing the first empirical calibration of the AGN black hole mass scale. The AGN M_{BH}-σ_* relationship based on this value of the scaling factor is shown in Figure 2.

While the value $f = 5.5$ might seem rather high, it should be remembered that in AGN unified schemes, Type 1 AGNs are viewed preferentially at low inclinations so projected disk velocities are significantly lower than the disk rotation speed. Again, not knowing the details of the BLR geometry or velocity field there is little else that we can say at this time (though see Section 7).

Inspection of Figure 2 shows that most of the reverberation-mapped AGNs for which stellar velocity dispersions have been measured are at the low-mass end of the distribution. This is because measurements of the stellar velocity dispersion are generally made using the Ca II triplet at $\sim$8500 Å, and at only very modest redshift ($z \approx 0.06$) this feature is shifted into the telluric water vapor bands and becomes unobservable from the ground. The AGN mass function is steep enough that there are thus very few high-mass black holes in galaxies at $z < 0.06$. A way around this is to instead use the CO bandhead in the IR H-band for the stellar velocity dispersion measurements and some pioneering efforts in this direction have been made, as shown in Figure 2 (Dasyra et al. 2007). While

---

‡ For example, it is well known that for a Gaussian profile, FWHM/$\sigma_{\text{line}} = 2(2\ln 2)^{1/2} \approx 2.35$. For a triangular profile, this ratio has a value of $\sqrt{6} \approx 2.45$, and for a rectangular profile, it has a value $2\sqrt{3} \approx 3.46$.

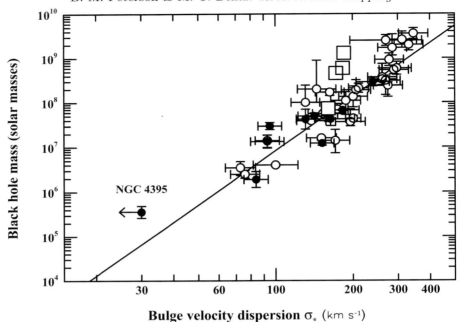

FIGURE 2. The relationship between black hole mass and stellar bulge velocity dispersion. The open circles represent (mostly) quiescent galaxies, from Tremaine et al. (2002). The filled circles are AGNs whose black hole masses are based on reverberation mapping, and whose bulge velocity dispersions are measured from the Ca II triplet (Onken et al. 2004; Nelson et al. 2004, and references therein). The large open squares represent reverberation-mapped PG quasars whose bulge velocity dispersions were measured from CO in the *H*-band (Dasyra et al. 2007).

it is impossible to draw any firm conclusion based on only four sources, Figure 2 suggests that the CO velocity dispersions are perhaps somewhat smaller than expected. Clearly additional higher-quality observations are needed to clarify this issue.

4. Direct comparison with other methods

Obviously the definitive test of reverberation-based masses has to be direct comparison with other methods, which, practically speaking, means comparison with masses based on stellar dynamics.† This is extremely difficult. First, to do this reliably, we must be able to resolve the black hole radius of influence [Eq. (1.1)]. At the present time, there are only two good candidates whose radius of influence might be resolvable with *HST*, NGC 3227 and NGC 4151.

NGC 3227 was observed in *K*-band with an adaptive-optics integral field unit on the VLT that resulted in spatial resolution of 0.″085 (Davies et al. 2006). Stellar dynamical modeling yields a black hole mass in the range 7–20 million solar masses, which is broadly consistent with the reverberation-based mass of 42 ± 21 million solar masses (Peterson et al. 2004). In the case of NGC 4151, an attempt to make use of the high angular resolution of *HST* was frustrated by the brightness of the nuclear source, even though the observation was made when this AGN was close to its historical minimum brightness during the STIS era (Onken et al. 2007). Nevertheless, an attempt was made

† Megamasers are also observed in AGNs, but only in Type 2 Seyferts, which are objects in which the BLR cannot be observed directly.

to determine the black hole mass based on two ground-based long-slit spectra, yielding only an upper limit of 70 million solar masses, which is consistent with the reverberation measurement of 46 ± 5 million solar masses (Metzroth, Onken, & Peterson 2006; Bentz et al. 2006b). In both NGC 3227 and NGC 4151, the stellar dynamical mass estimates are somewhat dubious since the existing data are insufficient to deal appropriately with what are clearly non-axisymmetric systems. The reverberation mass for NGC 3227 is also poorly determined, so there is clearly a lot more work to be done before any meaningful cross-checking of stellar dynamical and reverberation masses can be done. But at this time, there is at least no demonstrated gross inconsistency between the two methods.

5. The BLR radius-luminosity relationship

The general similarity of AGN spectra over several orders of magnitude in luminosity† (Vanden Berk et al. 2004) suggests that the physical conditions in the line-emitting regions must be approximately the same for all AGNs. This requires that the incident flux from the central source must be about the same for all AGNs, which in turn requires that the size of the BLR scales to geometrically dilute the ionizing radiation. The simple prediction is thus that the radius of the BLR, as measured by any particular emission line, should scale as $R \propto L^{1/2}$. This long-anticipated relationship was not clearly demonstrated until the first reverberation results on PG quasars were published (Kaspi et al. 2000), and even then the slope of the relationship was steeper than expected, $R \propto L^{0.7}$. It is often stated that it was the expansion of the luminosity range that allowed this relationship to emerge from the scatter due to uncertainties in the measurement of R. This is actually somewhat misleading: it was not so much a matter of extending the luminosity range as it was going to high enough luminosity that the host-galaxy starlight was not a major contributor to the measured optical luminosity. Precision spectrophotometry is a key requirement for successful reverberation mapping and to attain this with a seeing-dependent point-spread function, a large spectrograph entrance aperture needs to be employed. A consequence of this is that a significant amount of host-galaxy starlight contaminates the observations.

Since the BLR radius should depend only on the luminosity of the active nucleus proper, we need to correct the observed luminosities for starlight. To enable this, we obtained images of the reverberation-mapped AGNs with the High Resolution Channel (HRC) of the Advanced Camera for Surveys on *HST*. The HRC has an image scale of $0\rlap{.}''025$/pixel and makes full use of the nearly diffraction-limited capability of *HST*. For each galaxy, we obtained multiple graduated exposures to ensure that at least one image would not be saturated. High-resolution unsaturated images allowed us to cleanly separate the point-like AGN from the host galaxy bulge and thus produce a high-quality map of the host-galaxy surface-brightness distribution. We could then perform simulated aperture photometry by integrating the host-galaxy image over the projected spectrograph aperture that was used in the reverberation program.

By this process, we were able to obtain the luminosity of the AGN proper by subtracting the starlight contribution measured by simulated aperture photometry. The results of this are shown in Figure 3. The scatter among the lower-luminosity sources is markedly decreased from what it was before the starlight correction was effected. Also, as anticipated, the slope of the relationship is much flatter and is in fact consistent with $R \propto L^{1/2}$.

---

† The one strong exception to this statement remains the Baldwin Effect, i.e., that the equivalent width of the C IV $\lambda 1549$ decreases with increasing luminosity. The origin of this effect is still unknown.

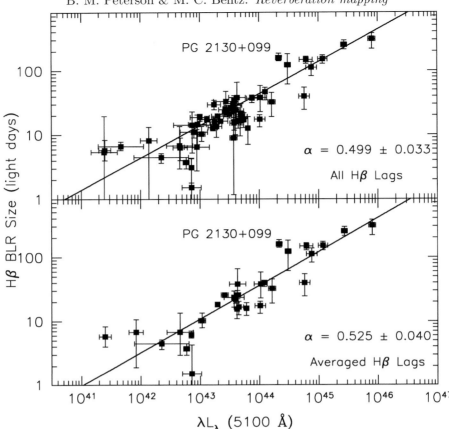

FIGURE 3. The BLR radius (as measured from the Hβ lag) versus luminosity relationship, after correction of the luminosity for host-galaxy starlight contamination. The measured lag of the outlier PG 2130+099 is somewhat suspect, but the slope of the relationship is insensitive to its inclusion. The upper panel shows all Hβ measurements for all reverberation-mapped AGNs. In the lower panel, all measurements for individual AGNs have been averaged (based on Bentz et al. 2006a, with additional unpublished data).

At the present time, the BLR radius-luminosity relationship is only well established for the Balmer lines. Until relatively recently, there have been measurements of C IV λ1549 lags for a handful of AGNs, but these were all too close in luminosity to provide a reasonable test. Two recent results have extended the luminosity range over which C IV lags have been measured. First, Peterson et al. (2005) used *HST*'s STIS to measure a lag of about one hour in the well-known dwarf Seyfert galaxy NGC 4395. Second, Kaspi et al. (2007) reaped the first fruits of a six-year campaign on the Hobby-Eberly telescope and measured a rest-frame lag of 188 days for a $\sim 10^{47}$ erg s^{-1} quasar. Based on these few data, but extending over more than 7 orders of magnitude in luminosity, the slope of the C IV BLR radius-luminosity relationship seems consistent with that for Hβ.

The importance of the radius-luminosity relationship extends far beyond what it tells us about the physics of the BLR. If the relationship is sufficiently reliable, one can bypass the resource-intensive process of reverberation mapping and simply measure the luminosity of the AGN and infer the BLR size from it, as Vestergaard describes in this volume.

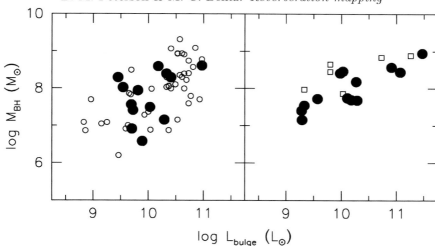

FIGURE 4. The relationship between black hole mass and bulge luminosity for AGNs. The left panel shows data from Wandel (2002), and the right panel shows a revised version based on improved reverberation-based masses and our ACS images that have led to much improved bulge luminosities. The AGNs common to both data sets are shown as filled circles.

6. The M_{BH}-L_{bulge} Relationship

Another remarkable correlation between nuclear and larger-scale properties of galaxies is that of the central black hole mass with the luminosity (or mass) of the bulge component of the galaxy, the M_{BH}-L_{bulge} relationship (Kormendy & Richstone 1995; Magorrian et al. 1998).

We are in the early stages of revisiting this relationship for AGNs based on the unsaturated ACS images of reverberation-mapped AGNs obtained to study the BLR radius-luminosity relationship described in Section 5. Wandel (2002) presented a compilation of AGN bulge luminosities, largely based on disk+bulge fits to WFPC2 images, mostly from McLure & Dunlop (2001). Unfortunately, many of the images from which the bulge luminosities were derived were archival images meant for other studies, and not particularly suited to bulge+disk decompositions. Most are badly saturated, with blooming that affects tens and even hundreds of pixels. Additional images were from low-resolution ground-based CCDs and photographic plates. All of these effects make it difficult to accurately separate the images into their constituent parts—bulge, disk, and AGN.

Our new results show a much tighter correlation between black hole mass and bulge luminosity (or mass) than previously determined for AGNs, as shown in Figure 4. The correlations are marginally consistent with those determined for quiescent galaxies. There is still much work to be done in comparing the results for active and quiescent galaxies, as different methods were utilized in the study of each set of objects. Further conclusions would be premature.

7. Uncertainties in reverberation-based mass measurements

The unknown geometry, kinematics, and inclination of the BLR ultimately limit the accuracy to which black hole masses can be determined. While the details remain uncertain, a paradigm that has emerged (e.g., Elvis 2000) is that the BLR is comprised of two components, a rotating disk-like structure and a disk-wind that is perhaps launched and accelerated hydromagnetically, but with radiation pressure possibly playing a role.

The disk component is higher density and probably produces much of the Balmer-line emission. The disk-wind is thought to be a lower density, more highly ionized gas that produces the strong UV lines.

Inclination effects are certain to be important in such a system, especially as unification models (e.g., Urry 2004) indicate that Type 1 objects, which afford a direct view of the BLR and continuum source, are seen at relatively low angles of inclination. In such a case, the observed line width underestimates the disk rotation speed, and thus yields an underestimate of the mass. For a flat Keplerian disk, projection effects will lead to an order of magnitude underestimate for an inclination of only $i = 20°$ since the projected velocity is proportional to $\sin i$ and the mass is proportional to the velocity dispersion squared. Of course, the absence of Type 1 Seyferts with extremely narrow lines argues that there must be some significant velocity dispersion in the polar direction in such a disk system, whether it is due to the vertical velocity of the disk-wind or warping or some other structure associated with the disk. We noted earlier that the empirically determined scale factor $f = 5.5$ [Eq. (1.2)] is rather high, suggesting that the typical projection correction to the line width is of the order of a factor of two or so.

There are several lines of evidence indicating that inclination effects are important. Wills and Browne (1986) consider the distribution of Hβ line widths in radio-loud sources and find that very broad Hβ lines are missing among sources with high radio core-to-lobe flux ratios. This is expected if the BLR has a disk-like configuration, since core-dominated radio sources are seen closer to the axis of the system, i.e., at lower inclination. Vestergaard, Wilkes, and Barthel (2000) show that there is a similar effect in the base of the C IV $\lambda1549$ emission line. Similarly, Jarvis & McLure (2006) find that flat-spectrum radios sources (again, those which are observed closer to the axis of the system) have, on average, smaller line widths than steep-spectrum sources. Moreover, the distribution of line widths for radio-quiet sources is more like those of the flat-spectrum sources.

So *statistically*, inclination seems to be important. But can we determine inclination in individual sources? It has been suggested in the literature (Wu & Han 2001; McLure & Dunlop 2001; Zhang & Wu 2002) that it might be possible to infer the inclination of the BLR in a particular AGN by comparing the reverberation-based mass M_{reverb} to the mass predicted by the M_{BH}-σ_* relationship $M_{\sigma*}$, since the M_{BH}-σ_* relationship seems to have little intrinsic scatter. The expectation is that σ_* is invariant with inclination. Again, Collin et al. (2006) show that *statistically*, this is apparently plausible and consistent in particular with the data on the subclass of objects related to NLS1 galaxies (their Figure 8), but that this hypothesis fails in individual cases. For example, the inclination of the AGN system in 3C 120 is constrained to $i < 20°$ by its superluminal jet (Marscher et al. 2002). As noted earlier, at an inclination this low, the projection factor should be huge, about an order of magnitude for a Keplerian disk. But in this case, the reverberation-based mass falls precisely on the M_{BH}-σ_* relationship, not an order of magnitude below it. In the case of the NLS1 Mrk 110 (noting again that among NLS1s are likely to be objects whose lines are narrow due to inclination effects), the mass predicted by the M_{BH}-σ_* relationship is *lower* than the reverberation mass ($M_{\sigma*} = 4.8$ million solar masses versus $M_{\mathrm{reverb}} = 25 \pm 6$ million solar masses). Moreover, in this particular source, there is an independent mass estimate available from the gravitational redshift of the broad emission lines (Kollatschny 2003b), $M_{\mathrm{grav}} = 14 \pm 3$ million solar masses, closer to the reverberation result than to the M_{BH}-σ_* prediction. It thus appears that M_{BH}-σ_* relationship is not an accurate enough predictor of black hole mass to use to infer inclination of individual sources. The AGN M_{BH}-σ_* relationship may have greater scatter than that for quiescent galaxies, though whether that is due to more uncertain mass measurements or the greater difficulty in measuring bulge velocity dispersions accurately (bearing in mind that errors in σ_* have a huge effect as the mass prediction is based on the fourth power of this quantity) is not entirely clear.

Inclination seems to be important, but can we actually measure it? There are a few ways that this can be done. As alluded to above in the case of 3C 120, the presence of a superluminal jet can at least place an upper limit on the inclination. Conversely, spectropolarimetry of the broad lines sometimes show evidence of "equatorial scattering" (Smith et al. 2005), which indicates a relatively high inclination.†

Another possibility is reverberation mapping—while at the present time, reverberation results have been largely confined to measuring lags or mean time delays, in principle it is possible to make use of the fact that the emission lines are resolved in Doppler velocity and to form a "velocity-delay map" that would allow us to determine the geometry and kinematics of the BLR, particularly if maps of multiple lines could be obtained. To date, attempts to recover velocity-delay maps from real data (Ulrich & Horne 1996; Kollatschny 2003a) have failed to show much structure. However, extensive realistic simulations (e.g., Horne et al. 2004) show that the minimum data requirements are not far beyond what has been done already and modest improvements (in terms of time-sampling interval and duration, signal-to-noise ratio, and spectral resolution) should enable recovery of a velocity-delay map. A complication that is sometimes glossed over is that reverberation mapping requires high-precision spectrophotometry: systematic errors need to be beaten down to the percent level in order for reverberation mapping to succeed.

8. Conclusions

Good progress has been made in using reverberation mapping to measure the masses of the supermassive black holes in the nuclei of Type 1 active galaxies. Multiple lines of evidence suggest that these masses can in fact be trusted, particularly:

(*a*) In every case where it has been testable, the data have been consistent with a virial relationship between emission-line width and lag, as expected in a Keplerian system where the central mass dominates.

(*b*) As in quiescent galaxies with black hole masses measured by modeling of stellar or gas dynamics, reverberation-mapped AGNs follow an M_{BH}-σ_* relationship.

(*c*) Like quiescent galaxies, AGNs show a relationship between central black hole mass and bulge luminosity (or mass).

(*d*) The two cases where attempts have been made to measure AGN black hole masses from stellar dynamics have not shown any inconsistency, although the uncertainties are still too large to conclude anything definitively.

Based on the scatter around the AGN M_{BH}-σ_* relationship, reverberation masses seem to be accurate to about a factor of three, with the largest systematic source of uncertainty probably being the inclination of the BLR. It is also clear that the full potential of reverberation mapping has not yet been realized—once it is possible to obtain full velocity-delay maps for multiple emission lines, we will be able to explore the structure and kinematics of the BLR in more detail, which at the present time, we can only indirectly discern.

We are grateful for support of this work by the National Science Foundation through grant AST-0604066 and by NASA through grants HST-GO-10516, HST-GO-10833, and

† A scattering medium roughly in the disk plane that is close to the BLR polarizes the light differently from opposite sides of the disk, resulting in a change of position angle across the broad emission line. This can only happen when the disk is seen at relatively high inclination.

HST-AR-10691 from the Space Telescope Science Institute. M.C.B. has been supported by an NSF Graduate Fellowship.

REFERENCES

Bentz, M. C., et al. 2006a *ApJ* **644**, 133.

Bentz, M. C., et al. 2006b *ApJ* **651**, 775.

Bentz, M. C., et al. 2007 *ApJ* **662**, 205.

Blandford, R. D. & McKee, C. F. 1982 *ApJ* **255**, 419.

Clavel, J., et al. 1991 *ApJ* **366**, 64.

Collin, S., Kawaguchi, T., Peterson, B. M., & Vestergaard, M. 2006 *A&A* **456**, 75.

Dasyra, K., et al. 2006 *ApJ* **657**, 102.

Davies, R., et al. 2006 *ApJ* **646**, 754.

Edelson, R. & Krolik, J. H. 1988 *ApJ* **333**, 646.

Elvis, M. 2000 *ApJ* **545**, 63.

Ferrarese, L. & Merritt, D. 2000 *ApJ* **539**, L9.

Ferrarese, L., et al. 2001 *ApJ* **555**, L79.

Fromerth, M. J. & Melia, F. 2000 *ApJ* **533**, 172.

Gebhardt, K., et al. 2000a *ApJ* **539**, L13.

Gebhardt, K., et al. 2000b *ApJ* **543**, L5.

Genzel, R., et al. 2003 *ApJ* **594**, 812.

Ghez, A. M., et al. 2005 *ApJ* **620**, 744.

Ho, L. C. 1999. In *Astrophysics and Space Science Library* (ed. S. K. Chakrabarti). Vol. 234, p. 157.

Horne, K., Peterson, B. M., Collier, S., & Netzer, H. 2004 *PASP* **116**, 465.

Jarvis, M. J. & McLure, R. J. 2006 *MNRAS* **369**, 182.

Kaspi, S., et al. 2000 *ApJ* **533**, 631.

Kaspi, S., et al. 2007 *ApJ* **659**, 997.

Kollatschny, W. 2003a *A&A* **407**, 461.

Kollatschny, W. 2003b *A&A* **412**, 61.

Kormendy, J. & Richstone, D. 1995 *ARA&A* **33**, 581.

Krolik, J. H., et al. 1991 *ApJ* **371**, 541.

Lynden-Bell, D. 1969 *Nature* **223**, 690.

Magorrian, J., et al. 1998 *AJ* **115**, 2285.

Marscher, A. P., et al. 2002 *Nature* **417**, 625.

McLure, R. J. & Dunlop, J. S. 2001 *MNRAS* **327**, 199.

Metzroth, K. G., Onken, C. A., & Peterson, B. M. 2006 *ApJ* **647**, 901.

Miyoshi, M., et al. 1995 *Nature* **373**, 127.

Nelson, C. H., et al. 2004 *ApJ* **615**, 652.

Onken, C. A. & Peterson, B. M. 2002 *ApJ* **572**, 746.

Onken, C. A., et al. 2004 *ApJ* **615**, 645.

Onken, C. A., et al. 2007 *ApJ* **670**, 105.

Osterbrock, D. E. & Shuder, J. M. 1982 *ApJS* **49**, 149.

Peterson, B. M. 1993 *PASP* **105**, 247.

Peterson, B. M. & Wandel, A. 1999 *ApJ* **521**, L95.

Peterson, B. M. & Wandel, A. 2000 *ApJ* **540**, L13.

Peterson, B. M., et al. 1991 *ApJ* **368**, 119.

Peterson, B. M., et al. 1998 *PASP* **110**, 660.

Peterson, B. M., et al. 2004 *ApJ* **613**, 682.

Peterson, B. M., et al. 2005 *ApJ* **633**, 799.

Richstone, D., et al. 1998 *Nature* **395**, A14.

Salpeter, E. E. 1964 *ApJ* **140**, 796.

Sargent, W. L. W., et al. 1978 *ApJ* **221**, 731.

Smith, J. E., et al. 2002 *MNRAS* **359**, 846.

TREMAINE, S., ET AL. 2002 *ApJ* **74**, 740.

ULRICH, M.-H. & HORNE, K. 1996 *MNRAS* **283**, 748.

URRY, C. M. 2004. In *AGN Physics with the Sloan Digital Sky Survey* (eds. G. T. Richards & P. B. Hall). ASP Conference Series, Vol. 311, p. 49.

VANDEN BERK, D., ET AL. 2004. In *AGN Physics with the Sloan Digital Sky Survey* (eds. G. T. Richards & P. B. Hall). ASP Conference Series, Vol. 311, p. 21.

VESTERGAARD, M., WILKES, B. J., & BARTHEL, P. D. 2000 *ApJ* **538**, L103.

WANDEL, A. 2002 *ApJ* **565**, 762.

WILKES, B. J., ET AL. 1995 *ApJ* **455**, 13.

WILLS, B. J. & BROWNE, I. W. A. 1986 *ApJ* **302**, 56.

WU, X.-B. & HAN, J. L. 2001 *A&A* **561**, 59.

YOUNG, P., ET AL. 1978 *ApJ* **221**, 721.

ZELDOVICH, YA. B. & NOVIKOV, I. D. 1964 *Sov. Phys. Dokl.* **158**, 811.

ZHANG, T.-Z., & WU, X.-B. 2002 *Chinese J. of Astron. & Astrophys.* **2**, 487.

Black-hole masses from gas dynamics

By F. DUCCIO MACCHETTO

Space Telescope Science Institute and European Space Agency, 3700 San Martin Drive,
Baltimore, MD 21218, USA

Since the advent of the *Hubble Space Telescope* (*HST*), the progress in studying and understanding black holes has been impressive. Early questions regarding the very existence of black holes have been replaced by questions regarding the role that they play in the formation and evolution of galaxies, particularly at early epochs in the universe. However, the apparently well-established relationship between the mass of the black hole and the mass or luminosity of the galactic bulge rests on a relatively small number of direct observations, and while very few doubt that this relationship exists, it is essential to actually measure the properties of a number of black holes over a range of masses and host galaxies. The direct methods adopted to measure black holes in the nearby universe use gas or stellar kinematics to gather information on the gravitational potential in the nuclear region of the galaxy. The stellar-kinematical method has the advantage that stars are present in all galactic nuclei and their motion is always gravitational. The drawbacks are that it requires relatively long observation times in order to obtain high-quality observations, and that stellar-dynamical models are very complex and potentially plagued by indeterminacy. Conversely, the gas-kinematical method is relatively simple; it requires relatively short observation times for the brightest emission-line nuclei, even if not all galactic nuclei present detectable emission lines. However, an important drawback is that noncircular or non-gravitational motions can completely invalidate this method. Since the observed correlations are based on black-hole masses obtained with different methods, it is important to check whether these methods provide consistent and robust results. Over the last several years we have carried out *HST* observations of a variety of elliptical, Seyfert, and spiral galaxies. In particular, we have undertaken a major STIS survey of 54 late-type spiral galaxies to study the scaling relations between black holes and their host spheroids at the low-mass end. Our measurements of BH masses in late-type spiral galaxies have shown that they are very challenging and at the limit of the highest spatial resolution currently available. Nonetheless, our measurements generally support the scaling relations between black holes and their host spheroids, suggesting that (i) they are reliable and (ii) black holes in spiral galaxies follow the same scaling relations as those in more massive early-type galaxies. A crucial test for the gas-kinematical method, the correct recovery of the known BH mass in NGC 4258, has been successful.

1. Introduction

It is widely accepted that Active Galactic Nuclei (AGN) are powered by accretion of matter onto massive black holes (BHs). AGN activity peaked at $z \sim 1$–2 (e.g., Maloney & Petrosian 1999) and the high ($\gtrsim 10^{12}\ L_\odot$) luminosities of quasi-stellar objects (QSOs) are explained by accretion onto supermassive $\sim 10^8\ M_\odot$, $\sim 10^{10}\ M_\odot$ black holes at or close to the Eddington limit. The observed evolution of the space density of AGN (Chokshi & Turner 1992; Faber et al. 1997; Marconi & Salvati 2002) implies that a significant fraction of luminous galaxies must host black holes—relics of past activity. Indeed, it is now clear that a large fraction of hot spheroids contains a massive black hole (e.g., Magorrian et al. 1998; van der Marel 1999), and it appears that the BH mass is proportional to both the mass of the host spheroid (Kormendy & Richstone 1995) and its velocity dispersion (Ferrarese & Merritt 2000; Gebhardt et al. 2000; Merritt & Ferrarese 2001; Ferrarese & Ford 2005). Several radio galaxies, all associated with giant elliptical galaxies like M87 (Macchetto et al. 1997), M84 (Bower et al. 1998), NGC 7052 (van der Marel & van den Bosch 1998), and Centaurus A (Marconi et al. 2001, 2006), are now known to host supermassive black holes that are accreting at a low rate and/or low accretion efficiency (Chiaberge, Capetti, & Celotti 1999). They presumably sustained quasar activity in the past, but at the present epoch are emitting much below their Eddington limits ($L/L_{\rm Edd} \sim 10^5 \sim 10^7$). The study of the Seyfert BH-mass distribution provides a statistical method of investigating the interplay between accretion rate and BH growth. In order to achieve this, it is necessary to directly measure the BH masses in Seyfert galaxies and compare their Eddington and Bolometric luminosities using hard x-ray luminosities. Similarly important will be the comparison between the BH masses found in Seyfert galaxies with those of non-active galaxies. However, until recently, there were very few secure BH measurements or upper limits in spiral galaxies. It is important to directly establish how common BHs are in spiral galaxies, and whether they follow the same $M_{\rm BH}$–$M_{\rm sph}$, $M_{\rm BH}$–σ correlations as for elliptical galaxies. To detect and measure the masses of massive BHs requires spectral information at the highest possible angular resolution, the sphere of influence of massive BHs is typically $\leqslant 1''$ in radius even in the closest galaxies. Nuclear absorption-line spectra can be used to demonstrate the presence of a BH, but the interpretation of the data is complex because it involves stellar-dynamical models that have many degrees of freedom. In Seyfert galaxies the problems are compounded by the copious light from the AGN. Radio-frequency measurements of masers in disks around BHs provide some of the most spectacular evidence for BHs, but have the disadvantage that only a small fraction of the disks will be inclined such that their maser emission is directed toward us. In principle, studies at *HST* resolution of ordinary optical emission lines from gas disks provide a more widely applicable and readily interpreted way of detecting BHs (cf. M87; Macchetto et al. 1997; Barth et al. 2001) provided that the gas velocity fields are not dominated by non-gravitational motions. In early-type galaxies there are still worrying issues about the dynamical configuration of nuclear gas (e.g., misalignment with the major axis, irregular structure etc.). By contrast, nuclear gas in relatively quiescent spirals is believed to be organized into well-defined rotating disks seen in optical line images (e.g., M81; Devereux, Ford, & Jacoby 1997). Ho et al. (2002), found that the majority of spiral galaxies in their survey have irregular velocity fields in the nuclear gas, and are not well suited for kinematical analysis. Still, 25% of the galaxies where $H\alpha$ emission was detected all the way to the center have velocity curves consistent with circular rotation, and the galaxies with more complicated velocity curves can also be useful for BH mass measurement after detailed analysis of the spectra. Indeed, even in the most powerful Seyfert nuclei, such as NGC 4151—where the gas is known to be interacting with radio ejecta—it is possible to derive the mass of the BH from spatially resolved *HST* spectroscopy using careful analysis of the velocity field to separate the underlying quiescently rotating disk gas from that disturbed by the jets (Winge et al. 1999).

2. The sample, observational strategy and modeling

Following our earlier studies, we have undertaken a spectroscopic survey of 54 spirals using STIS on the *HST*. Our sample was extracted from a comprehensive ground-based study by Axon et al., who obtained $H\alpha$ and N II rotation curves of 128 Sb, SBb, Sc, and SBc spiral galaxies from RC3 at a seeing-limited resolution of $1''$. By restricting ourselves to galaxies with recession velocities $\sim V < 2000$ km s^{-1}, we obtained a volume-limited sample of 54 spirals that are known to have nuclear gas disks and to span wide ranges in bulge mass and concentration. The systemic velocity cutoff was chosen so that we can probe close to the nuclei of these galaxies to detect even lower-mass black holes. The frequency of AGN in our sample is typical of that found in other surveys of nearby spirals, with comparable numbers of weak nuclear radio sources and LINERS. The sample is described in detail by Hughes et al. (2003) and a photometric analysis of the STIS acquisition images was presented by Scarlata et al. (2004). All methods used to directly detect BHs and measure their masses require high spatial resolution to resolve the BH sphere of influence, i.e., the region where the gravitational influence of the BH dominates over that of the host galaxy. The radius of the BH sphere of influence is traditionally defined as $r_{\mathrm{BH}} = GM_{\mathrm{BH}}/\sigma^2$, where G is the gravitational constant and σ is the stellar velocity dispersion, and is usually $<1''$. *HST*'s resolution is approximately ~ 0.07 arcseconds (FWHM) at $\lambda 6500$ Å. For a spiral galaxy at a distance of 20 Mpc which hosts a BH with $M_{\mathrm{BH}} = 10^7 \, M_\odot$ and follows the M_{BH}–σ correlation (Tremaine et al. 2002), the expected value of σ is $\sim 10^5$ km s^{-1}. The radius of the black-hole sphere of influence expected for this object is thus $r_{\mathrm{BH}} \sim 0.04$ arcseconds, very close to the *HST* angular resolution, making these observations possible, but extremely challenging.

The observational strategy used for all the galaxies in the sample consisted in obtaining spectra at three parallel positions, with the central slit centered on the nucleus and the flanking ones at a distance of $0.2''$. These different slit positions are labeled NUC, OFF1 and OFF2. Each spectrum was obtained with the G750M grating and the $0.2''$ slit which provided a dispersion of $\delta\lambda = 1.108$ Å pix^{-1} and a spectral resolution of $R = \lambda/(2\delta\lambda) \sim 3000$. Where available, our observations were complemented with both imaging and spectroscopic archival data from other programs. This ensured that all the available data were used to derive the rotation curves and further constrain the modelling.

The raw spectra were first reprocessed through the *calstis* pipeline and standard pipeline tasks were used to obtain flat-field corrected images. The two exposures taken at a given slit position were then realigned with a shift along the slit direction (by an integer number of pixels), and the pipeline task *ocrreject* was used to reject cosmic rays and hot pixels. Subsequent calibration procedures followed the standard pipeline reduction; i.e., the spectra were wavelength calibrated and corrected for two-dimensional distortions. The expected accuracy of the wavelength calibration is 0.1–0.3 pixels within a single exposure and 0.2–0.5 pixels among different exposures, which corresponds to ~ 3–8 km s^{-1} (relative) and ~ 5–13 km s^{-1} (absolute).

The procedure used to model the kinematical data to fit the observed rotation curves was first described by Macchetto et al. (1997) and then refined by several authors (van der Marel & van der Bosch 1998; Barth et al. 2001; Marconi et al. 2003). This procedure is described in detail in Marconi et al. (2003). Briefly, the code computes the rotation curves of the gas, assuming that the gas is rotating in circular orbits within a thin disk in the galaxy potential; any hydrodynamical effects, such as gas pressure, are assumed to be negligible. The gravitational potential has two components: the stellar potential, determined from WFPC or NICMOS observations and characterized by its mass-to-light ratio, and a dark mass concentration (the black hole), spatially unresolved at *HST*+STIS resolution and characterized by its total mass M_{BH}. In computing the rotation curves (Figure 2), we take into account the finite spatial resolution of the observations, the intrinsic surface-brightness distribution, and we integrate over the slit and pixel area. The free parameters characterizing the best-fitting model are found by standard χ^2 minimization; they are:

- the black-hole mass M_{BH};
- the stellar mass-to-light ratio M/L;
- the coordinates of the BH position (X, Y) on the plane of the sky where the nuclear slit is centered at $(0,0)$;
- the galaxy systemic velocity V_{sys}; and
- the position angle of the disk line of nodes θ.

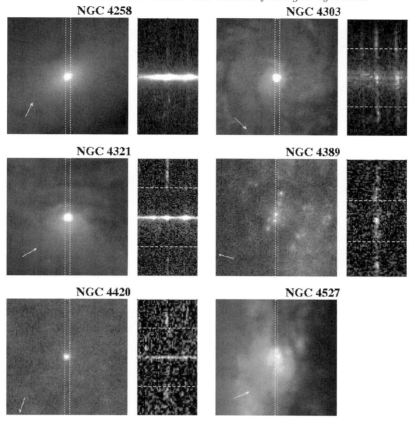

FIGURE 1. STIS images and spectra for some of the galaxies in the sample.

I will now describe some of the more interesting examples of the determination of black holes' masses using emission lines.

3. M87

Following the seminal paper of Sargent et al. (1978) discussing stellar kinematics in the nucleus of M87, Ford et al. (1994) and Harms et al. (1994) used *HST* to obtain narrow-band $H\alpha +$ [N II] images of M87, which showed the presence of a disk-like structure. Spectra, taken at five different locations with the Faint Object Spectrograph (FOS), showed that the gas velocity reached 500 km on either side of the nucleus. If interpreted as being due to Keplerian motion, this implied the presence of a massive black hole. Macchetto et al. (1997) followed these observations with the first long-slit, very high-spatial resolution observations by using the spectrograph on *HST*'s FOC. We observed the ionized gas disk in the emission line of [O II] $\lambda3727$ at three different positions, each separated by 0.2 arcsec, with a spatial sampling of 0.03 arcsec (or ~ 2 pc at the distance of M87), and measured the rotation curve of the inner ~ 1 arcsec of the ionized gas disk to a distance as close as 0.07 arcsec ($\simeq 5$ pc) to the dynamical center. We modeled the kinematics of the gas under the assumption of the existence of both a central black hole and an extended central mass distribution, taking into account the effects of the instrumental PSF, the intrinsic luminosity distribution of the line, and the finite size of

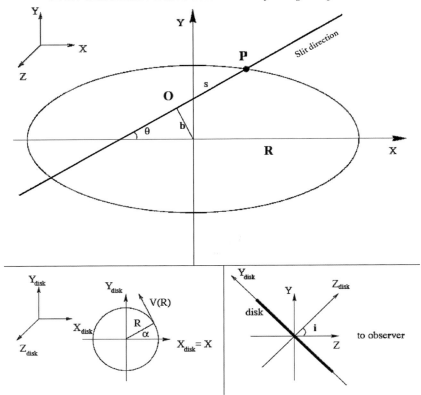

FIGURE 2. Schematic representation of the reference frames used in the determination of the Keplerian rotation curve. XY is the plane of the sky, X is directed along the major axis of the disk, and Z is directed toward the observer.

the slit. We found that the central mass must be concentrated within a sphere whose maximum radius is $\simeq 3.5$ pc, and showed that both the observed rotation curve and the line profiles are best explained by a thin disk in Keplerian motion (Figure 3). Finally, we proved that the observed motions are due to the presence of a supermassive black hole and derived a value of $M_{\mathrm{BH}} = (3.2 \pm 0.9) \times 10^9 \ M_\odot$ for its mass.

4. The black-hole mass of a Seyfert galaxy

In early studies, Winge et al. (1999) analyzed both ground-based and *HST*/FOC long-slit spectroscopy at subarcsecond spatial resolution of the narrow-line region (NLR) of NGC 4151. We found that the extended emission gas $(R > 4'')$ is, in a normal rotation in the galactic plane, a behavior that we were able to trace even across the nuclear region, where the gas is strongly disturbed by the interaction with the radio jet and connects smoothly with the large-scale rotation defined by the neutral gas emission. The *HST* data, at 0.03'' spatial resolution, allow us for the first time to truly isolate the kinematic behavior of the individual clouds in the inner narrow-line region. We found that underlying the perturbations introduced by the radio ejecta, the general velocity field can still be well represented by planar rotation down to a radius of $\sim 0.5''$ (30 pc), the distance at which the rotation curve has its turnover. The most striking result that emerges from the analysis is that the galaxy potential derived by fitting the rotation

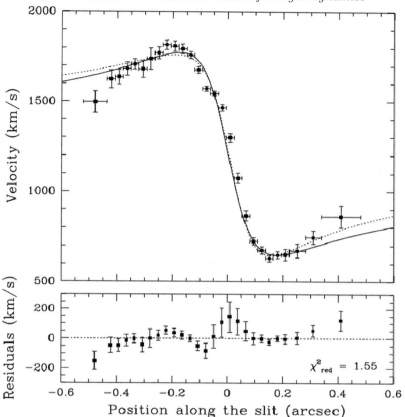

FIGURE 3. Best fits of the observed M87 rotation curve with the Keplerian thin-disk model. We find $M_{BH} = (3.2 \pm 0.9) \times 10^9 \ M_{\odot}$.

curve changes from a "dark halo" at the extended narrow-line region distances to being dominated by the central mass concentration in the NLR, with an almost Keplerian fall off in the $1'' < R < 4''$ interval. The observed velocity of the gas at $0.5''$ implies a mass of $M \sim 10^9 \ M_{\odot}$ within the inner 60 pc. The presence of a turnover in the rotation curve indicates that this central mass concentration is extended. The first measured velocity point (outside the region saturated by the nucleus) would imply an enclosed mass of $\sim 5 \times 10^7 \ M_{\odot}$ within $R \sim 0.15''$ (10 pc), which represents an upper limit to any nuclear point mass.

5. Cen A

Centaurus A (NGC 5128) is the closest (3.5 Mpc) giant elliptical galaxy hosting an AGN and a jet. A prominent dust lane, which obscures the inner half-kiloparsec of the galaxy with associated gas, young stars and H II regions, is interpreted as the result of a relatively recent merger event between a giant elliptical galaxy and a small, gas-rich disk galaxy. IR and CO observations of the dust lane have been modeled by a thin warped disk (Quillen et al. 1992, 1993) which dominates ground-based near-IR observations along with the extended galaxy emission. R-band imaging polarimetry from *HST* with WFPC (Schreier et al. 1996), are consistent with dichroic polarization from such a disk.

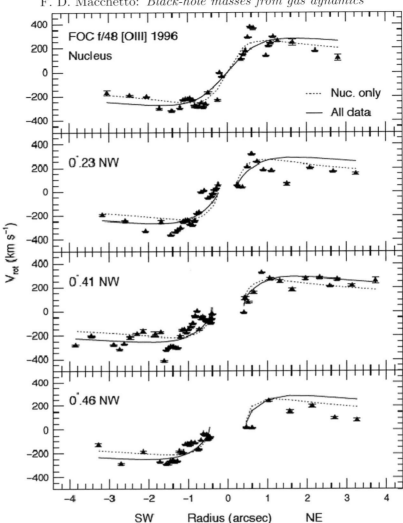

FIGURE 4. NGC 4151. Comparison between the 1996 FOC f/48 [O III] 5007 μm data set and the best-fitting circular rotation curve. The dotted line corresponds only to the fit to the nuclear slit position, and the solid line to the fit obtained simultaneously using all four slit positions.

Early *HST*/WFC2 and NICMOS observations of Cen A have shown that the 20 pc-scale nuclear disk previously detected by NICMOS in Paα (Schreier et al. 1996) has also been detected in the [Fe II] λ1.64 μm line, which shows a morphology similar to that observed in Paα with an [Fe II]/Paα ratio typical of low ionization Seyfert galaxies and LINERS. Marconi et al. (2000) derived a map of dust extinction $E(B-V)$ in a $20'' \times 20''$ circumnuclear region, and reveal a several-arcseconds-long dust feature near to, but just below the nucleus, oriented in a direction transverse to the large dust lane. This structure is related to the bar observed with ISO and SCUBA, as reported by Mirabel et al. (1999), who found rows of Paα emission knots along the top and bottom edges of the bar—which they interpret as star-formation regions, possibly caused by shocks driven into the gas. Gas and dust are supplied by a recent galaxy merger; a several-arcminute-scale bar allows the dissipation of angular momentum and infall of gas toward the center of the galaxy;

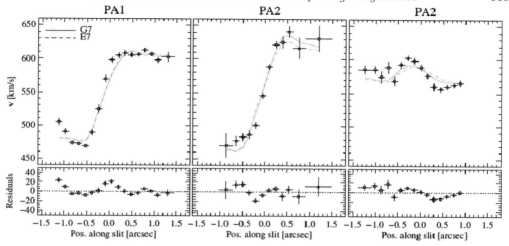

FIGURE 5. Cen A. Model fits of the velocity along the slit and residuals for the ISAAC data.

subsequent shocks trigger star formation; and the gas eventually accretes onto the AGN via the 20 pc disk.

By reconstructing the radial light profile of the galaxy to within $0.1''$ of the nucleus, Marconi et al. (2001) show that Cen A has a core profile. Using the models of van der Marel and van den Bosch (1998), they estimate a black-hole mass of $\sim 10^8$ $M_\odot$, consistent with ground-based kinematical measurements. In a later paper, Marconi et al. (2006), used STIS observations of the [S III] λ 9533 µm line to study the kinematics of the ionized gas in the nuclear region. The STIS data were analyzed in conjunction with the ground-based near-infrared Very Large Telescope ISAAC spectra used by Marconi et al. (2001) to infer the presence of a supermassive black hole and to measure its mass. The observed velocity dispersion in the ISAAC spectra is well matched with a circularly rotating disk; the observed line profiles and the higher-order moments in the Hermite expansion of the line profiles are consistent with emission from such a disk. The velocity dispersion in the STIS data is larger than expected from rotation, indicating that the more-ionized gas can be affected by non-circular and/or non-gravitational motions. Nonetheless, the STIS data still provide a consistent estimate of the BH mass, indicating that large velocity dispersion of the gas does not invalidate gas kinematical BH estimates.

Both sets of data, with spatial resolutions differing by an order of magnitude, provide independent and consistent measures of the BH mass, which are also in agreement with the previous estimate. The gas kinematical analysis (Figure 5) best fit is for a mass of $M_{\rm BH} = (1.1)^{+0.1}_{-0.1} \times 10^7$ $M_\odot$ for an assumed disk inclination of $i = 25°$ or $M_{\rm BH} = (5.5)^{+0.7}_{-0.6} \times 10^7$ $M_\odot$ for $i = 45°$.

The spatial resolution of *HST* allows them to constrain the size of any cluster of dark objects alternative to a BH to $r < 0.035$ arcseconds, i.e., 0.6 pc at the distance of Cen A. Thus, Cen A is among the best cases for supermassive BHs in galactic nuclei. Cen A was the first galaxy for which there were BH mass measurements—both with gas and stellar dynamics—and it is significant that the two estimates agree, thus strengthening their reliability.

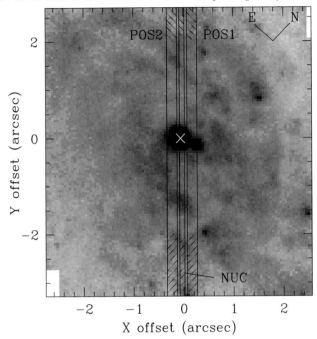

FIGURE 6. NGC 4041. Slit positions overlaid on the acquisition image. The 0,0 position of the target is derived from the STIS ACQ procedure. The white cross is the kinematic center derived from the fitting of the rotation curves.

6. NGC 4041

Marconi et al. (2003) carried out observations of NGC 4041 (Figure 6), which is classified as an Sbc spiral galaxy with no detected AGN activity. It is at a distance of $\simeq 19.5$ Mpc, which corresponds to a scale of 95 pc/arcsecond.

The *HST*/STIS spectra were used to map the velocity field of the gas in its nuclear region. They detected the presence of a compact ($r \simeq 0.4''$; $\simeq 40$ pc), high surface brightness, circularly rotating nuclear disk, cospatial with a nuclear star cluster. This disk is characterized by a rotation curve with a peak-to-peak amplitude of ~ 40 km s^{-1} and is systematically blueshifted by ~ 10–20 km s^{-1} with respect to the galaxy systemic velocity.

The standard approach followed in gas-kinematical analysis is to assume that (i) gas disks around black holes are not warped (i.e., they have the same line of nodes and inclinations as the more extended components), and (ii) the stellar population has a constant mass-to-light ratio with radius. In this case however, the blueshift of the inner disk indicates that the standard approach must be modified to allow for kinematical decoupling between the inner and large-scale disks.

Figure 7 shows the best-fit model (solid line) obtained with the mass density distribution derived from the K-band light profile with the assumption of spherical symmetry. The dotted line is the best-fit model without a black hole. The left panel of Figure 7 shows the fit of the NUC data from the model with the mass distribution derived from the I band. The best fit to the data is for $M_{\rm BH} = 10^7\ M_\odot$. If the standard assumptions are relaxed and stellar mass-to-light and disk inclination are allowed to vary, the kinematical data could be accounted for by the stellar mass without a black hole, provided that either the mass-to-light ratio is increased by a factor of ~ 2 or the inclination is allowed

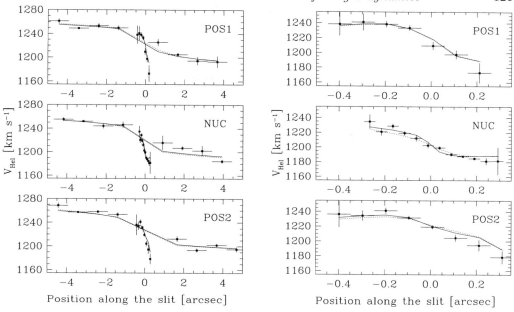

FIGURE 7. NGC 4041. Left: Best-fit standard model of the observed rotation curves (solid line) compared with the data. The dotted line is the best-fit model without a black hole. To guide the eye, the model values are connected by straight lines. Right: Zoom on the nuclear disk region.

to vary. This model resulted in a 3σ upper limit of $M_{\rm BH} < 6 \times 10^6 \ M_\odot$ on the mass of any nuclear black hole. Combining the results from the standard and alternative models, the present data only allow them to set an upper limit of $\sim \times 10^7 \ M_\odot$ to the mass of the nuclear BH. If this upper limit is taken in conjunction with an estimated bulge B magnitude of -17.7 and with a central stellar velocity dispersion of $\simeq 95$ km s^{-1}, the black hole in NGC 4041 is consistent with both the $M_{\rm BH}$–$L_{\rm sph}$ and the $M_{\rm BH}$–σ correlations.

7. NGC 5252

NGC 5252 is an early-type (S0) Seyfert 2 galaxy at redshift $z = 0.023$, (92 Mpc; $1'' =$ 450 pc), whose line emission shows a biconical morphology (Tadhunter & Tsvetanov 1989) extending out to 20 kpc from the nucleus along PA = 15°. Ground-based measurements of the large-scale velocity field of the gas were obtained by Morse et al. (1998) with a resolution of 1.4″. The inner part of the velocity field corresponds to the dusty spiral structure on a scale of $\sim$1 kpc, which Morse et al. (1998) argued to be part of a rotating disk, significantly inclined with respect to the galaxy plane. On a sub-arcsec scale, three emission line knots form a linear structure oriented at PA $\sim$ 35, close to the bulge major axis, suggestive of a small-scale gas disk. Capetti et al. (2005) carried out STIS observations of this galaxy (Figure 8).

The results for the slit centered along the nucleus (Figure 9) show the line central velocity, flux, and FWHM at each location along the slit. The line emission, detected out to a radius of $\sim$1.6″ ($\sim$720 pc) is strongly concentrated, showing a bright compact knot cospatial with the continuum peak. Two secondary emission-line maxima are also present at $\pm0.35''$ from the main peak. They correspond to the intersection of the slit with emission-line knots seen in the WFPC and STIS images. Two different gas systems

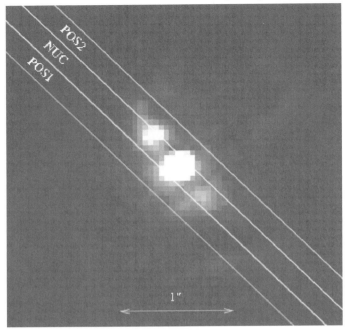

FIGURE 8. The nuclear region of NGC 5252 with superimposed STIS slit positions.

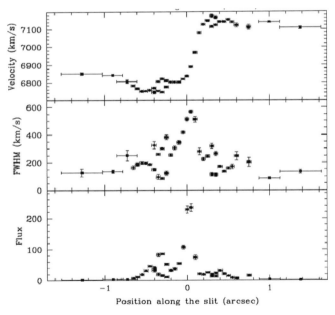

FIGURE 9. NGC 5252. The line's central velocity, flux, and FWHM at each location along the slit centered on the nucleus.

are present in the nuclear regions: the first shows a symmetric velocity field with decreasing line width, which can be interpreted as being produced by gas rotating around the nucleus; the second component, showing significant non-circular motions, is found to be exclusively associated with the off-nuclear blobs.

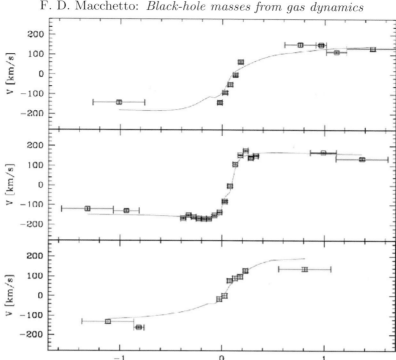

FIGURE 10. NGC 5252. Model fit to the three slit positions.

The velocity curve has a full amplitude of $\sim$400 km s^{-1} and it shows general reflection symmetry. Starting from the center of symmetry (at $r \sim 0.1$ and $v \sim 6950$ km s^{-1}), the velocity rapidly rises on both sides by $\sim$200 km s^{-1}, reaching a peak a $r \sim 0.2''$. In this central region, both the line flux and the line width rapidly decrease (from the maximum value of 600 km s^{-1} down to 100 km s^{-1}).

In correspondence with the off-nuclear knots, the situation is more complex as these are the regions where two velocity components are present. One component has a lower (and decreasing) velocity, width, and flux, and it appears to extend the trend seen at smaller radii. The second component is considerably broader and shows a higher velocity offset with respect to the center of symmetry. At radii larger than $r \sim 0.4''$ the narrow component intensity falls below the detection threshold and only the broad component is visible. Lines remain broad out to $\sim$0.8$''$ from the nucleus. At even larger distances, the velocity field flattens.

The best fit to the data, which is obtained for a black-hole mass $M_{\rm BH} = 0.95 \times 10^9 \ M_\odot$ (Figure 10), shows that the kinematics of the gas in the innermost regions can be successfully accounted for by circular motions in a thin disk when a supermassive black hole is added to the galaxy potential. The central velocity dispersion of NGC 5252 (Nelson & Whittle 1995) is 190 ± 27 km s^{-1}. The correlation between velocity dispersion and black-hole mass predicts a mass of $M = 1.0^{+1.0}_{-0.5} \times 10^8 \ M_\odot$, where the error is dominated by the uncertainty in σ_c.

Therefore, the black-hole mass derived for NGC 5252 is larger by a factor of $\sim$10 than the value expected from this correlation! (See Figure 11). This value, however, is in good agreement with the correlation between bulge and BH mass. As for its active nucleus, NGC 5252 is an outlier when compared to the available data for Seyfert galaxies, not

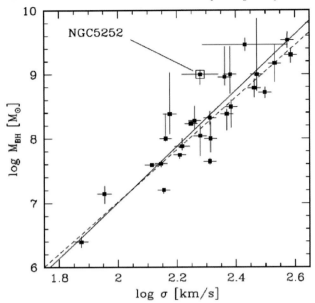

FIGURE 11. The M_{BH}–σ correlation showing that the BH mass for NGC 5252 is a factor of $\sim$10 above the expected value.

only because it harbors a black hole larger than is typical for these objects, but also because its host galaxy is substantially brighter than average for Seyfert galaxies. On the other hand, both the black hole and the bulge's mass are typical of the range for radio-quiet quasars. Combining the determined BH mass with the hard x-ray luminosity, NGC 5252 is estimated to emit at a fraction $\sim$0.005 of L_{Edd}. This active nucleus thus appears to be a quasar relic, now probably accreting at a relatively low rate, rather than a low black-hole mass counterpart of QSOs.

8. NGC 1300

NGC 1300 is a striking SBbc galaxy ($D = 18.8$ Mpc; 1 arcsec $= 91$ pc) with a strong, well-defined kpc-scale bar; it does not exhibit nuclear activity. The classical bar structure lends itself to dynamical modeling: the H I velocity field within the bar and spiral arms has been modeled with tilted rings (Lindblad et al. 1997); and the large-scale gravitational potential inferred from a mass model based on the decomposition of the galaxy into individual structural components—the bulge, bar, disk, and lens (Aguerri et al. 2001). The nuclear dust lanes connect to those of the leading edge of the bar, suggesting that the large- and small-scale gas flows are intimately linked. The position angle of the large-scale disk is $\simeq 106°$ and its inclination is $\simeq 49°$.

STIS observations of NGC 1300 and NGC 2748 were carried out by Atkinson et al. (2005; Figure 12). We investigated a number of models to fit the rotation curves of NGC 1300. The best base model was defined to be that which uses the spherical density model comprising the two components necessary to reproduce the complete surface brightness profile derived from the F160W image, and the $H\alpha$ flux map. The complete list of models investigated is:

1. a two-component spherical density model and the flux map derived from the $H\alpha$ flux profiles;

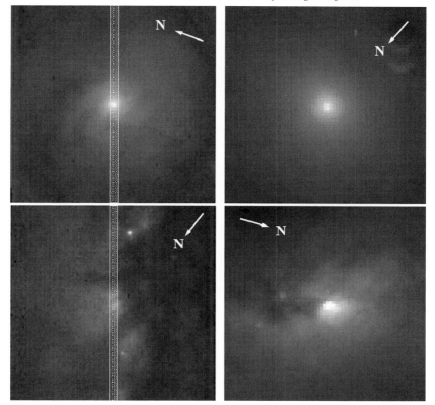

FIGURE 12. Images of NGC 1300 (top) and NGC 2748 (bottom). Left: STIS F28×50LP filter images. Right: NICMOS F160W filter images. The NUC slits are indicated in the two acquisition images: the OFF1/2 slits are the same size and are located immediately adjacent to the NUC slits.

 2. a two-component spherical density model and the flux map derived from the [N II] flux profiles;

 3. a one-component spherical density model and the flux map derived from the $H\alpha$ flux map; and

 4. a one-component 'disk-like' ($q = 0.1$) density model and the flux map derived from the $H\alpha$ flux map.

The base models produce similar geometric parameters and black-hole mass estimates to the best base model (Figure 13). The range of black-hole masses derived from all of the models is from $M_{\rm BH} = 4.3$ to $7.9 \times 10^7\ M_\odot$, however, the optimized model based on the best base model has a significantly lower value for χ^2. Our analysis shows that a black hole of mass $M_{\rm BH} = (6.6)^{+6.3}_{-3.2} \times 10^7\ M_\odot$ provides the best fit to the central rotation curves of NGC 1300.

9. NGC 2748

NGC 2748, (Figure 12), is an unbarred SAbc galaxy at a distance of 23.2 Mpc, (1 arcsec = 112 pc). The galaxy is highly inclined at $73°$. An edge-on disk provides the optimal orientation to maximize the amplitude of the observed rotation curve for an infinitesimally thin slit. For a finite slit width, or indeed a disk of non-negligible thickness, the velocities sampled at any point along the slit originate from a range of radii on the disk.

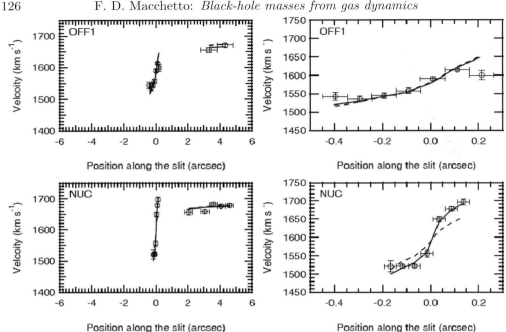

FIGURE 13. Model rotation curves derived for models with no black hole (dashed line) and a black hole of mass $M_{BH} = 6.6 \times 10^7 \, M_\odot$ (solid line) overlaid on the observed rotation curves of NGC 1300. Expanded views of the central regions are displayed on the right.

The corresponding average velocity can, for certain rotation curves, decrease. An obvious exception is that of a disk in solid-body rotation, where the radial velocity across the slit is constant. In this case, the edge-on disk yields the maximum amplitude in the rotation curve. Given the typical rotation curves we observe in the centers of spirals, we expect the inclination of NGC 2748 to yield a near-maximum amplitude rotation curve.

The best base model, (Atkinson et al. 2005), was defined as that which uses the spherical two-component density model based on the F160W surface brightness profile. The complete list of base models used in fitting NGC 2748 is, as for NGC 1300, based on the $H\alpha$ flux map, since there is little difference in shape between the $H\alpha$ and [N II] flux maps. The best fit to the observed central rotation curves of NGC 2748 (Figure 14), are given by a black hole of mass $M_{BH} = (4.4)^{+3.5}_{-3.6} \times 10^7 \, M_\odot$. Models without a black hole yield unsatisfactory fits and have a significantly larger χ^2. The black-hole masses based on the other base models all lie within the uncertainties of the optimized black-hole mass derived using the best base model.

10. NGC 3310

In a recent paper, Pastorini et al. (2007) have determined the black-hole mass for three galaxies: NGC 3310, NGC 4303, and NGC 4258. NGC 3310 (Arp217, UGC 5786) an Sbc galaxy in the RC3 catalogue, is at a distance of $D \sim 17.4$ Mpc (i.e., $1'' \sim 84$ pc), with an inclination of the galactic disk of about $i \sim 40°$, (Figure 15). It is a well-known starburst galaxy characterized by a disturbed morphology (see Elmegreen et al. 2002 for a recent review) which suggests that star formation has been triggered by a collision with a dwarf companion during the last $\sim 10^8$ yr, (Balick & Heckman 1981; Kregel & Sancisi 2001). Its central starburst region, as well as its young stellar clusters, have been studied in detail,

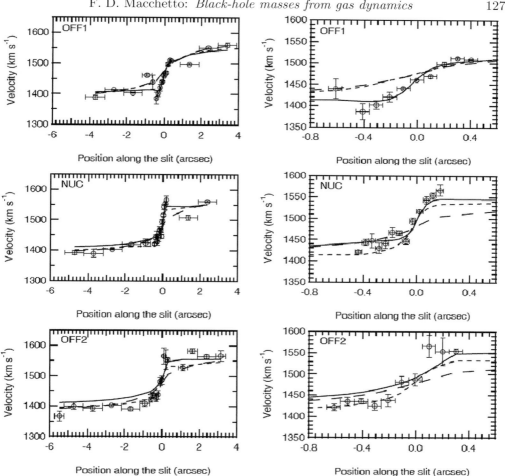

FIGURE 14. Model rotation curves derived for models with no black hole (long-dashed line), a black hole of mass $M_{BH} = 5 \times 10^7\ M_\odot$ with a coplanar disk (short-dashed line) and a black hole of mass $M_{BH} = 4 \times 10^7\ M_\odot$ with the angle between the slit and the line of nodes optimized at $-23°$ (solid line), overlaid on the observed rotation curves of NGC 2748. Expanded views of the central regions are displayed on the right.

revealing the presence of very young ($\sim 10^7$ to 10^8 yr) stars. The bright inner region is dominated by a two-armed open spiral pattern observed in Hα, and the inner part of this well-developed pattern connects to a $\sim$900-pc-diameter starburst ring, surrounding the compact blue nucleus. The gas kinematics in the galactic disk are disturbed, characterized by non-circular motions and strong streaming along the spiral arms—strengthening the idea of a recent merger event. In the nuclear region, the rotation center of the gas is offset with respect to the stellar continuum isophotes by $\sim 1.5'' \pm 0.3$ arcsec toward PA 142°. The STIS acquisition image of NGC 3310 shows the presence of a strong dust lane crossing the center of the galaxy (Scarlata et al. 2004). The spiral arms depart from a $\sim$6 arcsec ring, and beyond that radius motions might be significantly non-circular. Within the ring, the morphology of the emission-line region is characterized by a peaked nuclear blob about 1.2 arcsec in size, which appears unrelated with the starburst ring. In order

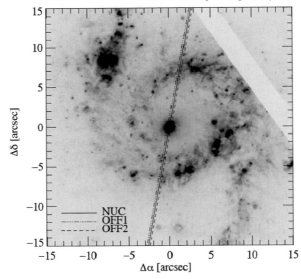

FIGURE 15. *HST*/ACS narrow-band image of NGC 3310 with slit positions overlaid. North is up and East is to the left.

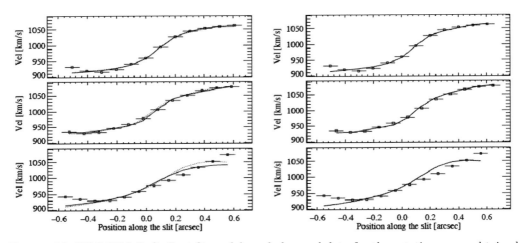

FIGURE 16. NGC 3310. Left: Best-fit models and observed data for the rotation curve obtained with A (solid line) and B (dotted line) flux distributions, assuming $i = 40°$. Right: Best-fit model of velocity obtained using ISBD A with the 'best' inclination ($i = 70°$) at the three different positions of the slit.

to avoid as much obvious contamination by non-circular motions as possible, the analysis was restricted to the nuclear blob, i.e., to the central 1.2-arcsec region.

To determine the best Intrinsic Surface Brightness Distribution (ISBD), a fundamental parameter in the modeling, different fits were made: A has two Gaussian components, an exponential plus a constant; and B has two exponential components, two Gaussians and a constant component. With both ISBD A and B, it is possible to reproduce the observed rotation curves.

The model with the lowest χ^2 has $i = 70°$; the comparison between observed and model rotation curves is shown in Figure 16. The inclination value of $i \sim 40°$, found by

Sanchez-Portal et al. (2000), refers to the large-scale disk, while the inner nuclear disk, which is clearly separated from the large-scale structure, is at a higher inclination. The observed gas kinematics are well matched by a circularly rotating disk model, but we are only able to set an upper limit to the BH mass which, taking into account all the allowed disk inclinations, varies in the range 5.0×10^6–$4.2 \times 10^7 \ M_\odot$ at the 95% confidence level.

11. NGC 4303

The second galaxy discussed by Pastorini et al. (2007) is the late-type Sbc galaxy NGC 4303 (M61), a double-barred galaxy in the Virgo Cluster at a distance of 16.1 Mpc ($1'' \sim 78$ pc). A large-scale bar of about 2.5 kpc ($\sim$35 arcsec) lies inside the outer spiral arm. In addition to the large-scale bar, this galaxy hosts a second inner bar of about 0.2 kpc ($\sim$2 arcsec) length that is surrounded by a circumnuclear starburst ring/spiral of $\sim$0.5 kpc ($\sim$6$''$) diameter. Based on the optical emission-line ratios, it is classified as a Seyfert 2/LINER borderline AGN. A young ($\sim$4 Myr) massive stellar cluster and a low-luminosity active nucleus coexist within the central 3 pc ($\sim$0.04$''$) of the galaxy, with the starburst contribution being dominant in the UV spectral region (Colina et al. 2002). The AGN presence is revealed through the luminous near-infrared point source (Colina et al. 2002) and through hard x-ray emission (Jimenez-Bailon et al. 2003). The gas kinematics of the galactic disk are affected by the presence of the large bar, but in the nuclear region the gas seems to show ordered rotation. From integral field observations, Colina & Arribas (1999) find that the Hβ velocity field at $\sim$1.5$''$ resolution is consistent with a circularly rotating massive disk of $\sim$300 pc radius ($\sim$4$''$). CO observations confirm this behavior and show that the gas within the central 0.7 kpc ($\sim$10$''$) follows a circular rotation with the dynamical center located within 1$''$ from the near-IR point source (Koda & Sofue 2006; Shinnerer et al. 2002).

The kinematics of NGC 4303 can be traced up to 4$''$ from the galaxy nucleus identified by the location of the near-IR point source. Although at 1–2$''$ spatial resolution the motion within the central region ($\sim$10$''$ of the nucleus) appears ordered and consistent with circular rotation, there is an inner bar with a size comparable to the spatial resolution ($\sim$2$''$ size). The curve of the extended component shows irregular behavior at about 3$''$ from the nucleus. Moreover, the Hα+[N II] surface brightness distribution along the slit clearly indicates the presence of a well-defined nuclear component as in the case for NGC 3310, NGC 4258 (Pastorini et al. 2007), and NGC 4041 (Marconi et al. 2003). Therefore the analysis was limited to the nuclear component, i.e., to the central $\sim$1$''$ from the nucleus.

As in the case of NGC 3310, different ISBDs were tested and only two can provide a reasonable fit of the observed data: D, composed of a constant and three exponential components; and E, composed of a constant and three Gaussian components. In NGC 4303, the best fit to the kinematical data, Figure 17, is for a BH with mass $M_{\rm BH} = (5.0)^{+0.87}_{-2.26} \times 10^6 \ M_\odot$ for a disk inclination $i = 70°$. However, the weak agreement between the data and the disk model makes this measurement somewhat unreliable. If all possible inclination values are allowed, then the $M_{\rm BH}$ varies in the range 6.0×10^5–$1.6 \times 10^7 \ M_\odot$ at the 95% confidence level.

12. NGC 4258

NGC 4258 (M106) is a nearby SABbc galaxy ($D = 7.2 \pm 0.3$ Mpc, $1'' \sim 35$ pc) spectroscopically classified as a 1.9 Seyfert galaxy. This galaxy is well known, since after the Milky Way, it represents the best case for a nuclear supermassive BH. Miyoshi et al.

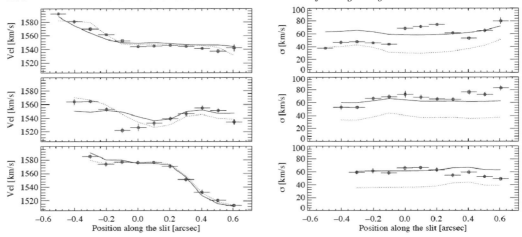

FIGURE 17. NGC 4303. Left: Best-fit model without a BH at $i = 70°$ and ISBD D. The dotted line represents the best fit model with a BH, obtained with the same inclination and ISBD. Right: Expected velocity dispersions for the best-fitting rotating-disk model with $i = 70°$ compared with observations (dots with error bars), assuming a constant intrinsic dispersion of 50 km s^{-1} (solid lines). The dotted line is the velocity dispersion expected only by unresolved rotation.

(1995) studied the H$_2$O maser line emission and their observations reveal individual masing knots revolving in a rotating gas disk at distances ranging from ∼0.13 pc to 0.25 pc around the central object. These data show a near-perfect Keplerian velocity distribution, implying that almost all the mass is located well within the inner radius, where the megamasers reside. Miyoshi et al. (1995) derived a central mass of ∼ 3.6×10^7 $M_\odot$ within the inner ∼0.13 pc. The available kinematical parameters for NGC 4258 are limited to the nuclear component, i.e., within 0.5″ of the continuum peak. The procedure followed to fit the data is as before, and six slit positions are fitted at the same time. Only one ISBD provides good agreement with the data. The left panel of Figure 18 shows the best-fit model. At the 68.3% confidence level, the derived black-hole mass is $M_{BH} = (7.9)^{+6.2}_{-3.5} \times 10^7$ $M_\odot$, while the 95% upper limit on M/L is $M/L < 1.8$ $M_\odot/L_{\odot,H}$. The non-perfect agreement between our model and the data suggests that rotation curves might be somehow affected by non-circular motions. Indeed, a jet-cloud interaction is responsible for the hot x-ray cocoons which give rise to the anomalous arms; their position angle, i.e., the region where the x-ray cocoons are interacting with the dense interstellar gas, is ∼ −33°, very close to the slit PA of the archival data (∼146° = −34°). The slit PA of our data is almost perpendicular to that of the archival data, and therefore less likely to be affected by non-circular motions, however we found no significant change in the black-hole mass estimate by excluding or including the archival data. This indicates that the non-circular motions do not affect the final black-hole mass estimate. The agreement between our mass estimate ($7.9^{+6.2}_{-3.5} \times 10^7$ $M_\odot$; $i = 60°$) and the H$_2$O-maser value (the difference is just slightly above 1σ) indicates that the gas-kinematical method is reliable even if the observed velocities present deviations from the circularly rotating-disk model.

The active nucleus has been detected through radio, x-ray, and infrared observations. The strong polarization of the relatively broad optical emission lines further supports the existence of an obscured active nucleus in NGC 4258. In particular, the AGN is characterized by a radio jet which propagates perpendicularly to the maser disk. Given the estimated BH mass of $7.9^{+6.2}_{-3.5} \times 10^7$ $M_\odot$ and the fact that NGC 4258 is a relatively low luminosity (∼10^{42} erg s^{-1}) object, the nuclear emission is sub-Eddington, with $L/L_E \sim$

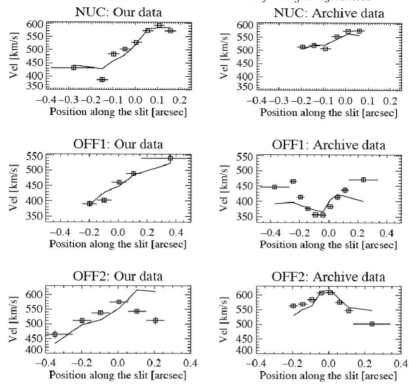

FIGURE 18. NGC 4258: Best-fit model of the observed velocity (solid line) compared with the observed data (dots with error bars). The model is computed for $i = 80°$.

3×10^{-4}. Such sub-Eddington sources are likely to have accretion-disk structures, where the accreting gas is optically thin and radiates inefficiently, and the accretion energy that is dissipated viscously, is advected with the accretion flow.

13. Conclusions

We can compare the BH-mass estimates in spiral galaxies with the known correlations between M_{BH} and host galaxy structural parameters. Since our work has more than doubled the number of gas-kinematical measurements in late-type spiral galaxies, it is worthwhile to consider the sample of all late-type spirals studied so far.

The results of these and other determinations of black-hole masses using emission-line kinematics are shown in Table 1, where for all galaxies studied in our project and those available from the literature, we show the measured BH masses or upper limits and the ratio with the expected values from the M_{BH}–L_{sph} (Marconi & Hunt 2003; Eq. 13.1) and M_{BH}/σ_* (Tremaine et al. 2002, Ferrarese & Ford 2005; Eqs. 13.2 and 13.3) correlations:

$$\log\left(M_{BH}/M_\odot\right) = 8.21 + 1.13\left(\log\left(L_{sph}/L_{\odot,K}\right) - 10.9\right) \tag{13.1}$$

$$\log\left(M_{BH}/M_\odot\right) = 8.13 + 4.02\log\left(\sigma_e/200\ \text{km s}^{-1}\right) \tag{13.2}$$

$$\log\left(M_{BH}/M_\odot\right) = 8.22 + 4.86\log\left(\sigma_c/200\ \text{km s}^{-1}\right) \tag{13.3}$$

Galaxy	Type	D†	$\log M_{BH}$‡	$\log L_K$¶	$\log \dfrac{M_{BH}}{M_{BH}(L_K)}$∥	σ_c††	$\log \dfrac{M_{BH}}{M_{BH}(\sigma_c)}$‡‡	σ_e††	$\log \dfrac{M_{BH}}{M_{BH}(\sigma_e)}$¶¶
NGC 1300	SB(rs)bc	18.8	$7.8^{+0.2}_{-0.2}$	10.0	0.6	90	1.3	87	1.1
NGC 2748	SAbc	23.2	$7.6^{+0.2}_{-0.4}$	9.8	0.6	79	1.4	83	1.0
NGC 3310	SAB(r)bc	17.4	<7.6	9.6	$\leqslant 0.9$	101	<0.8	84	<1.0
NGC 3516	S0	38.0	$\leqslant 7.3$	10.7	<0.00	144	<-0.18	132	<-0.06
NGC 4041	SA(rs)bc	19.5	<7.3	9.7	<0.4	92	<0.7	88	<0.6
NGC 4303	SAB(rs)bc	16.1	$6.6^{+0.5}_{-0.2}$	10.2	-0.8	108	-0.3	84	0.0
NGC 5252	S0	92.0	$8.98^{+0.66}_{-0.21}$	11.6	-0.02	192	0.7	190	1.0
NGC 4258	SAB(s)bc	7.2	$7.59^{+0.04}_{-0.04}$	10.3	0.09	120	0.45	148	-0.19
Milky Way	SbI–II	0.008	$6.60^{+0.03}_{-0.03}$	10.2	-0.9	100	-0.2	100	-0.3
M81	SA(s)ab	3.9	$7.84^{+0.1}_{-0.07}$	11.0	-0.4	174	-0.1	165	0.0

† Distance D is given in units of Mpc.
‡ black-hole masses are in units of $M_\odot$.
¶ Luminosity given in units of $L_{\odot,K}$.
∥ $M_{BH}(L_K)$ is the value expected from the Marconi & Hunt (2003) correlation.
†† σ has units of km s^{-1}.
‡‡ $M_{BH}(\sigma_c)$ is the value expected from the Ferrarese & Ford (2005) correlation.
¶¶ $M_{BH}(\sigma_e)$ is the value expected from the Tremaine et al. (2002) correlation.

TABLE 1. Comparison between M_{BH} estimates in late-type spirals with the expectations from the BH-spheroid scaling relations. The intrinsic scatter of the correlations is $\sim$0.3 in $\log M_{BH}$ and the uncertainty on the determination of the luminosity is 0.3 dex (see text).

BH masses are taken from Pastorini et al. (2007; NGC 3310, NGC 4303), from Atkinson et al. (2005; NGC 1300, NGC 2748), from Marconi et al. (2003; NGC 4041), and from the compilation by Ferrarese & Ford (2005; Milky Way, NGC 4258 and M81), to which we refer for the proper references for each galaxy. K-band luminosities of the spheroid are taken from Marconi & Hunt (2003; Milky Way, NGC 4258, M81), and from Dong & De Robertis (2006; NGC 2748 and NGC 4041). For the other galaxies (NGC 1300, NGC 3310 and NGC 4303), we took the total K-band magnitude from the 2MASS extended source catalogue (Jarrett et al. 2000) and we applied the near-IR bulge-total correction by Dong & De Robertis (2006). The correction factor depends on the galaxy morphological type T and for our galaxies ($T = 4.0$), it is $\Delta m = M_{\rm sph} - M_{\rm tot} = 2.0$. We then transformed the relative-to-absolute magnitudes using the distances shown in the table, and we obtained the luminosities using the solar K-band magnitude $M_{K\odot} = 3.28$. The uncertainty on the determination of K-band luminosity of each galaxy is 0.3 dex and is due to the observed dispersion on the bulge-total correction for a given morphological type (Dong & De Robertis 2006). The stellar velocity dispersions required by the correlations in Eqs. 13.2 and 13.3 are different: Tremaine et al. (2002) use the luminosity-weighted velocity dispersion σ_e measured within r_e, i.e., within the bulge half-light radius. On the other hand, Ferrarese & Ford (2005) use the central velocity dispersion σ_c, i.e., measured within an aperture of $r = 1/8 r_e$. For all galaxies except the Milky Way and M81, we used the values measured by Batchelor et al. (2005). Otherwise, we used the compilations by Tremaine et al. (2002) and Ferrarese & Ford (2005).

In Figure 19 we compare spiral galaxies (filled squares) with the known $M_{BH}/L_{\rm sph,K}$, M_{BH}/σ_e and M_{BH}/σ_c correlations from the compilations by Marconi & Hunt (2003), Tremaine et al. (2002), and Ferrarese & Ford (2005). The empty squares represent early-type galaxies and, for simplicity, we have not shown error bars.

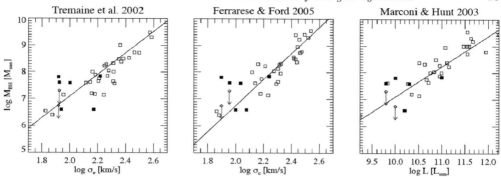

FIGURE 19. Correlations between M_{BH} and the host galaxy luminosity (right panel), the central stellar velocity dispersion, σ_c (middle panel) and the effective velocity dispersion σ_e (left panel). The empty squares denote early-type galaxies (E/S0) from the literature, while the filled ones refer to the spiral galaxies in Table 1.

When comparing the logarithms of the ratios between the measured BH masses and the expectations from the correlations, one has to bear in mind that the intrinsic dispersion of the correlations is ∼0.3 in log M_{BH}. The upper limit on the BH mass in NGC 3310 is consistent with all correlations, since it is larger than all expected M_{BH} values. The BH mass measurement in NGC 4303 is less reliable, as discussed previously. However, the measured M_{BH} value is in very good agreement with the M_{BH}–σ correlations, if their intrinsic dispersion is taken into account. M_{BH} is a factor of six smaller than expected from the M_{BH}–L_K correlation, however, it should be recalled that the bulge luminosity of NGC 4303 has been measured using the scaling relation by Dong & De Robertis (2006) and the observed dispersion for Sbc spiral galaxies is of the order of 0.3 in log L_K.

Considering that each correlation has an intrinsic dispersion of 0.3 in log M_{BH}, we can conclude that spiral galaxies do not show major deviations from the correlations followed by early-type galaxies, except that they appear to have a slightly larger scatter. From the above comparison, it is clear that a full statistical analysis on the behavior of spiral galaxies has to wait for a larger number of BH-mass detections.

Since the advent of *HST*, the progress in studying and understanding black holes has been impressive. Early questions regarding the very existence of black holes have been replaced by questions regarding the role that they play in the formation and evolution of galaxies, particularly at early epochs in the universe. The determination of black-hole masses using the gas-dynamics method has proven to yield solid results. Even when the observing limitations such as inclination, spatial resolution, or non-Keplerian motions complicate the analysis of the data, this method has the great advantge of providing believable results with well-defined errors, or at least solid upper limits to the black-hole masses. However, the apparently well-established relationship between the mass of the black hole and the mass or luminosity of the galactic bulge still rests on a relatively small number of direct observations, and it is imperative that many more high spatial resolution observations—using gas dynamics and other methods—be carried out with *HST*, since it is the only instrument that can provide the essential observational basis for much theoretical work on the characteristics, demographics, and evolution of black holes.

The work described here is the result of a long-standing collaboration with David Axon, Alessandro Capetti, Alessandro Marconi, Bill Sparks, and many others who have joined and supported our team through the years. They deserve the credit for the work and the results.

REFERENCES

AGUERRI, J. A. L., PRIETO, M., VARELA, A. M., MUÑOZ- TUÑÓN C. 2001 *Ap&SS* **276**, 611.

ATKINSON, J. W., ET AL. 2005 *MNRAS* **359**, 504.

BALICK, B. & HECKMAN, T. 1981 *A&A* **96**, 271.

BARTH, A. J., SARZI, M., RIX, H., HO, L. C., FILIPPENKO, A. V., & SARGENT, W. L. W. 2001 *ApJ* **555**, 685.

BATCHELDOR, D., ET AL. 2005 *ApJS* **160**, 78.

BOWER, G. A., ET AL. 1998 *ApJ* **492**, L111.

CAPETTI, A., MARCONI, A., MACCHETTO, D., & AXON, D. 2005 *A&A* **431**, 465.

CHIABERGE, M., CAPETTI, A., & CELOTTI, A. 1999 *A&A* **349**, 77.

CHOKSHI, A. & TURNER, E. L. 1992 *MNRAS* **259**, 421.

COLINA, L. & ARRIBAS, S. 1999 *ApJ* **514**, 637.

COLINA, L., ET AL. 2002 *ApJ* **579**, 545.

DEVEREUX, N., FORD, H., & JACOBY, G. 1997 *ApJ* **481**, L71.

DONG, X. Y., & DE ROBERTIS, M. M. 2006 *AJ* **131**, 1236.

ELMEGREEN, D. M., CHROMEY, F. R., McGRATH, E. J., & OSTENSON, J. M. 2002 *AJ* **123**, 1381.

FABER, S. M., ET AL. 1997 *AJ* **114**, 1771.

FERRARESE, L. & FORD, H. 2005 *Space Sci. Rev.* **116**, 523.

FERRARESE, L. & MERRITT, D. 2000 *ApJ* **539**, L9.

FORD, H. C., ET AL. 1994 *ApJ* **435**, L27.

GEBHARDT, K., ET AL. 2000 *ApJ* **539**, L13.

HARMS, R. J., ET AL. 1994 *ApJ* **435**, L35.

HO, L. C., SARZI, M., RIX, H., SHIELDS, J. C., RUDNICK, G., FILIPPENKO, A. V., & BARTH, A. J. 2002 *PASP* **114** 137.

HUGHES, M. A., ET AL. 2003 *AJ* **126**, 742.

JARRETT, T. H., ET AL. 2000 *AJ* **119**, 2498.

JIMENEZ-BAILON, E., ET AL. 2003 *AJ* **593**, 593.

KODA, J. & SOFUE, Y. 2006 *PASJ* **58**, 299.

KORMENDY, J., & RICHSTONE, D. 1995 *ARA&A* **33**, 581.

KREGEL, M. & SANCISI, R. 2001 *A&A* **376**, 59.

LINDBLAD, P. A. B., ET AL. 1997 *A&A* **317** 36.

MACCHETTO, F., MARCONI, A., AXON, D. J., CAPETTI, A., SPARKS, W., & CRANE, P. 1997 *ApJ* **489**, 579.

MAGORRIAN, J., ET AL. 1998 *AJ* **115**, 2285.

MALONEY, A. & PETROSIAN, V. 1999 *ApJ* **518**, 32.

MARCONI, A., AXON, D. J., CAPETTI, A., MACIEJEWSKI, W., ATKINSON, J., BATCHELDOR, D., BINNEY, J., CAROLLO, C. M., DRESSEL, L., FORD, H., GERSSEN, J., HUGHES, M. A., MACCHETTO, D., MERRIFIELD, M. R., SCARLATA, C., SPARKS, W., STIAVELLI, M., TSVE-TANOV, Z., & VAN DER MAREL, R. P. 2003 *ApJ* **586**, 868.

MARCONI, A., CAPETTI, A., AXON, D. J., KOEKEMOER, A., MACCHETTO, D., & SCHREIER, E. J. 2001 *ApJ* **549**, 915.

MARCONI, A. & HUNT, L. 2003 *ApJ* **589**, 21.

MARCONI, A., PASTORINI, G., PACINI, F., AXON, D. J.; CAPETTI, A., MACCHETTO, D., KOEKEMOER, A. M., & SCHREIER, E. J. 2006 *A&A* **448**, 921.

MARCONI, A. & SALVATI, M. 2002. In *Issues in Unification of Active Galactic Nuclei* (eds. R. Maiolino, A. Marconi, & N. Nagar). ASP Conference Proceedings, Vol. 258, p. 217. ASP.

MARCONI, A., SCHREIER, E. J., KOEKEMOER, A., CAPETTI, A., AXON, D., MACCHETTO, D., & CAON, N. 2000 *ApJ* **528**, 276.

MERRITT, D. & FERRARESE, L. 2001, *ApJ* **547**, 140.

MIRABEL, I. F., ET AL. 1999 *A&A* **341**, 667.

MIYOSHI, M., ET AL. 1995 *Nature* **373**, 127.

MORSE, J. A., CECIL, G., WILSON, A. S., & TSVETANOV, Z. I. 1998 *ApJ* **505**, 159.

NELSON, C. H. & WHITTLE, M. 1995 *ApJS* **99**, 67.

PASTORINI, G., ET AL. 2007 *A&A* **469**, 405.

QUILLEN, A. C., DE ZEEUW, P. T., PHILLEY, E. S., & PHILLIPS, T. G. 1992 *ApJ* **391**, 121.

QUILLEN, A. C., GRAHAM, J. R., & FROGEL, J. A. 1993 *ApJ* **412**, 550.

SANCHEZ-PORTAL, M., ET AL. 2000 *MNRAS* **312**, 2.

SARGENT, W. L. W., ET AL. 1978 *ApJ* **221**, 731.

SCARLATA, C., ET AL. 2004 *AJ* **128**, 1124.

SCHINNERER E., ET AL. 2002 *ApJ* **575**, 826.

SCHREIER, E. J., CAPETTI, A., MACCHETTO, F., SPARKS, W. B., & FORD, H. J. 1996 *ApJ* **459**, 535.

TADHUNTER, C. & TSVETANOV, Z. 1989 *Nature* **341**, 422.

TREMAINE, S., ET AL. 2002 *ApJ* **574**, 740.

VAN DER MAREL, R. P. 1999 *AJ* **117**, 744.

VAN DER MAREL, R. P. & VAN DEN BOSCH, F. C. 1998 *AJ* **116**, 2220.

WINGE, C., AXON, D. J., MACCHETTO, F. D., CAPETTI, A., & MARCONI, A. 1999 *ApJ* **519**, 134.

Evolution of supermassive black holes

By ANDREAS MÜLLER AND GÜNTHER HASINGER

Max-Planck-Institut für Extraterrestrische Physik, PO Box 1312, 85741 Garching, Germany

The cosmological evolution of supermassive black holes (SMBHs) seems to be intimately linked to their host galaxies. Active galactic nuclei (AGN) can be probed by deep x-ray surveys. We review results from selected large x-ray samples including the first results from the *XMM-Newton* COSMOS survey. A new picture arises from the fact that high-luminosity AGN grow earlier than low-luminosity AGN. In particular, the space density of low-luminosity AGN exhibits a significant decline for redshifts above $z = 1$. This "anti-hierarchical" growth scenario of SMBHs can be explained by two modes of accretion with different efficiency. The population of Compton-thick sources plays a key role in our understanding of the BH growth history. Their space density and redshift distribution is relevant to estimate the SMBH mass function. A comparison with the relic SMBH mass distribution in the local Universe constrains the average radiative efficiency and Eddington ratio of the accretion. We discuss a new synthesis model of Compton-thin and Compton-thick sources that is in concordance with deep x-ray observations, and in particular predicts the right level of contribution of the Compton-thick source population observed in the Chandra Deep Field South observations as well as the first INTEGRAL and Swift catalogs of AGN. Currently, one of the most important problems is the evolution of obscuration with redshift.

1. Introduction

1.1. *Deep surveys*

Deep field observations are a suitable observational technique to probe AGN physics. In multi-wavelength campaigns, astronomers select a field in the sky and produce images from several pointings. They use a combination of wide fields to collect many sources and so-called pencil beams to investigate sources in deep space. Today, there are many samples in different wavelength bands available, and large surveys are underway. Among the most important deep extragalactic surveys to date are the two fields related with the GOODS survey, the Hubble Deep Field North (HDF-N) and the Chandra Deep Field South (CDF-S), the latter field encompassing the Hubble Ultra Deep Field (HUDF) and embedded in the larger GEMS, COMBO-17 and ECDFS surveys. The COSMOS project is a multiwavelength survey centered around the largest *HST* observing project, covering two square degrees.

On the basis of multi-wavelength observations, and in comparison with galaxy and accretion models, it turns out that AGN are mainly determined by the following physical parameters: accretion rate, accretion efficiency, SMBH mass, SMBH spin, stellar formation rate, metallicity, dusty torus mass, and seed magnetic field. One geometric parameter is the inclination angle of the system. The challenge consists in determining these parameters from observations and to fit them into an AGN unification model.

The innermost part of AGN can be probed by means of x-ray observations. The hot inner accretion flows produce soft and hard x-rays. Hence, deep x-ray surveys provide an important chance to study AGN and their SMBHs within.

Astronomers are confronted with observational challenges as well. In deep x-ray surveys, the redshift of the sources is optically determined by using spectroscopic techniques (spectro-z) or by photometric techniques (photo-z). At high cosmological redshifts the I-band magnitudes are faint, and in certain redshift ranges spectroscopic features are

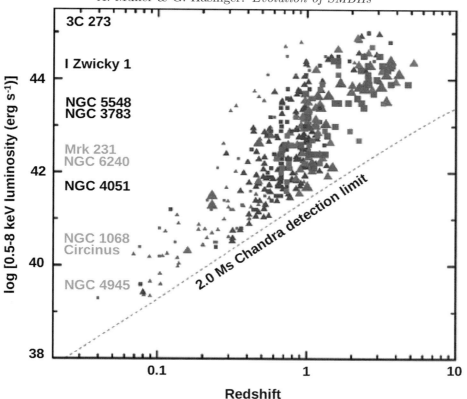

FIGURE 1. Rest-frame 0.5–8 keV luminosity as a function of redshift for sources form the CDF-N (triangles) and CDF-S (squares). The data points are grouped above the 2 Ms *Chandra* detection limit (dotted line). The 'spectroscopic desert' is visible on the right-hand side in the luminosity-redshift plane (taken from Brandt & Hasinger 2005).

weak or absent, therefore astronomers fail to determine spectro-z's. The problem of spectroscopic incompleteness emerges typically at $1 < z < 2$—a phenomenon that was coined 'spectroscopic desert' (see e.g., Brandt & Hasinger 2005). Figure 1 illustrates the sources in the luminosity-redshift plane. The I-band magnitudes successively grow from left to right until the emission drops below the spectroscopic sensitivity limit of 8–10 m-class telescopes, leaving a lack of spectroscopically identified sources on the right-hand side of the plot close to the sensitivity limit. In the CDF-S it is possible to fill this spectroscopic desert with photo-z thanks to a combination of the deep multiwavelength photometry available through GOODS, GEMS and COMBO-17 in this field (Szokoly et al. 2004; Zheng et al. 2004; Mainieri et al. 2005; Wolf et al. 2004; Grazian et al. 2006). As a result, it is possible to obtain 95% redshift completeness for the CDF-S. Redshift completeness of x-ray flux limited samples is vital to produce statistically and systematically reliable results. Therefore in the further analysis we typically cut all samples to a $>80\%$ redshift completeness, leading to an overall completeness of the sample of $\sim90\%$.

1.2. *AGN classification criteria*

The classical discrimination between type-1 (unobscured) and type-2 (obscured) AGN is done using a classification through optical spectroscopy. Optical type-1 AGN have broad permitted emission lines, while optical type-2 AGN do not show broad permitted lines, but still have high-excitation narrow emission lines. Therefore, many works in the field are primarily using the presence of broad lines as a discriminator and classify type-1 AGN only as Broad-line AGN (BLAGN). However, this AGN classification scheme breaks down when the optical spectrum is insufficient to discriminate the line widths. At high redshifts and low luminosities, there are several effects that compromise optical classification. First, in general high redshift objects are faint, so that they require very long observing times to obtain spectra of sufficient quality, and unambiguously discern broad emission line components. Secondly, there are redshift ranges where the classical broad lines shift out of the observed optical bands (see above). Thirdly, and probably most importantly, at high redshifts the spectroscopic slit is filled with the whole light of the AGN nucleus and the host galaxy. Depending on the ratio between nuclear and host luminosity, the host galaxy can easily outshine the AGN nucleus, rendering the AGN invisible in the optical light. This is the main reason why x-ray samples are much more efficient in picking up low-luminosity AGN at high redshift.

In principle, a classification purely based on x-ray properties would be possible. Large hydrogen column densities block soft x-rays. Therefore, a suitable AGN classification scheme in x-rays could involve the column density N_H that can be determined from the x-ray spectra. Usually, AGN x-ray spectra are fit by simple power law models. The effect of increasing column density is that the soft part of the spectrum is more and more suppressed, i.e., the spectrum becomes harder at high N_H. Correlations between optical and x-ray properties of obscured and unobscured AGN show a very good match ($\sim$90%) between optical obscuration and x-ray absorption. It is therefore suitable to define an x-ray type-1 AGN by $\log N_H < 22$, whereas AGN type-2 satisfy $\log N_H > 22$.† However, in practice is not possible to determine N_H values for samples with typically very faint x-ray sources, due to the small number of observed photons. Therefore, one has to resort to hardness ratios measured from coarse x-ray bands. The determination of N_H values from x-ray hardness ratios, in particular at high redshifts, is biased towards too high absorption values (see e.g., Tozzi et al. 2006), so that the x-ray classification alone also introduces significant systematic errors.

To overcome the difficulties of classification in either the optical or the x-ray bands, a combined optical/x-ray classification scheme has been introduced (Szokoly et al. 2004; Zheng et al. 2004). It involves a threshold x-ray luminosity L_X and the x-ray hardness ratio HR, defined as

$$HR = (H - S)/(H + S) \ , \qquad (1.1)$$

with the count rate in the hard band H and soft band S. The threshold AGN x-ray luminosity is set for an AGN $\log L_X > 42$, roughly at the upper luminosity range of vigorously star-forming galaxies. The practical combined optical/x-ray AGN classification is the following: An optical broad-line AGN (BLAGN) or a galaxy with $\log L_X > 42$ and hardness ratio HR < -0.2 is called an AGN type-1. An optical narrow-line AGN (NLAGN), or a galaxy with $\log L_X > 42$ and hardness ratio HR > -0.2 is called an AGN type-2.

This classification has the advantage that it can also be applied for sources, which have only photometric redshifts and no spectroscopic information at all. The other advantage is that it is based on a simple measurable quantity, the hardness ratio. There are also obvious systematic effects and difficulties. A small fraction of AGN are known not to follow the simple optical/x-ray obscuration/absorption correlation—either broad-line AGN have absorbed x-ray spectra (e.g., for BAL QSOs), or narrow-line AGN have unabsorbed

---

† Luminosities and column densities are given in the usual units erg s^{-1} and cm^{-2} respectively.

x-ray spectra (e.g., for narrow-line Seyfert-1 galaxies or Compton-thick, reflection dominated spectra). Also the hardness ratio threshold is a function of redshift. Finally, the dust obscuration versus gas absorption properties may be different, depending on environment and redshift. Nevertheless, we consider these systematic effects far smaller than those of any of the other classifications individually. All the systematics addressed here can, in principle, be quantified through appropriate theoretical population synthesis models.

2. Separate evolution of AGN classes

2.1. *Anti-hierarchical growth of SMBHs*

Deep surveys supply the x-ray luminosity functions (XLF) and bolometric luminosity functions of AGN. Using soft x-ray luminosity functions, Hasinger et al. (2005) could confirm that the density evolution depends significantly on AGN luminosity, i.e., the cosmological evolution of high-luminosity AGN (HLAGN) deviates strongly from that of low-luminosity AGN (LLAGN): the space density of HLAGN peaks at higher redshift $z \sim 2$, but the space density of LLAGN peaks at $z \sim 0.7$. This systematic trend of active SMBHs has been coined 'anti-hierarchical growth' or 'cosmic downsizing.' It is possible to understand this behavior in the framework of accretion theory (Merloni 2004). The observed local black hole mass function (BHMF) is taken as boundary condition to integrate the continuity equation backwards in time (mergers are neglected). It could be shown that the most massive SMBH with billions of solar masses were already in place at $z \simeq 3$. Indeed, this is observed for some SDSS quasars (Fan et al. 2003). Merloni (2004) has also demonstrated that radiatively inefficient accretion (e.g., ADAF solutions) dominate only at $z < 1$. Therefore, Seyfert galaxies and quasars evolve cosmologically in a radically different manner. The bulk of AGN has to wait much longer to grow or to be activated. The soft x-ray (0.5–2 keV) luminosity functions for AGN type-1 in six different redshift ranges are shown in Figure 2. Here, a successive evolution along redshift is visible. Only a luminosity-dependent density evolution (LDDE) can describe the data. Pure luminosity evolution (PLE) or pure density evolution (PDE) models are inadequate.

Now, there is firm evidence of a decline in space density of LLAGN towards higher redshift. For HLAGN, i.e., high-mass BHs deep x-ray surveys, so far provide rather poor constraints about their growth phase in the redshift range $3 \lesssim z \lesssim 6$ (Hasinger et al. 2005; Silverman et al. 2005), while in the optical and radio bands a clear decline of space density is observed for $z > 3$ (Fan et al. 2001; Wall et al. 2005).

The observed luminosity functions have also been compared to other models. Marconi et al. (2004) compared the mass function of black holes (BHs) in the local universe with that from AGN relics. The basic idea is that the relic BHs are grown from seed BHs by accretion in AGN phases. The local black hole mass function (BHMF) is estimated by using the prominent correlations between BH mass, stellar velocity dispersion and bulge luminosity. Again, merging is neglected in this model. Then, it is possible to derive the average radiative efficiency ϵ and the ratio between emitted and Eddington luminosity $\lambda = L/L_{\mathrm{Edd}}$ and the average lifetime of active BHs. Based on this model, typical efficiencies are found to be $\epsilon = 0.04$–0.16 and Eddington luminosities $\lambda = 0.1$–1.7. The BH lifetime strongly depends on BH mass and amounts to 450 million years for 100 solar-mass BHs and 150 million years for most massive BHs with more than billion solar masses.

The luminosity functions from SDSS (Richards et al. 2006), from hard x-rays (La Franca et al. 2005) and soft x-rays (Hasinger et al. 2005) are fairly consistent with the model

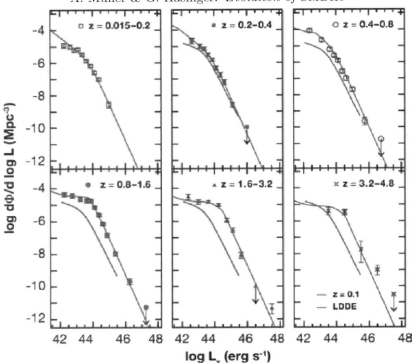

FIGURE 2. Observed 0.5–2 keV luminosity functions for AGN type-1 in six different redshift ranges. The solid curves through the data points are best luminosity-dependent density evolution (LDDE) fits. For comparison, the curve from the upper left panel ($z = 0.015$–0.2) is plotted in all other panels (adapted from Hasinger et al. 2005).

by Marconi et al. (2004).[†] Soltan's original argument (Soltan 1982) can be extended: one compares the accreted BHMF derived from the cosmic x-ray background with the MF of dormant relic BHs in local galaxies. Both can be reconciled if an energy conversion efficiency of $\epsilon = 0.1$ is assumed. From accretion theory it is known that such high efficiencies involve a Kerr BH, i.e., active BHs in AGN have to rotate. Other work on cosmological BH growth also indicates high BH spins (Volonteri et al. 2005).

Recently, bolometric quasar luminosity functions have been determined and they agree well with the x-ray data in each redshift slice. They demonstrate a perfect match between the soft x-ray, hard x-ray, optical, and even MIR wavebands, when the appropriate bolometric corrections and absorption distributions are taken into account (Hopkins et al. 2007). Because the optical and soft x-ray luminosity functions refer to type-1 AGN only, while the hard x-ray and MIR luminosity functions refer to all AGN, this agreement gives a kind of sanity-check on the type-1/type-2 classification discussed above. On the contrary, Barger et al. (2005), using a pure BLAGN classification, find a strong decline of the type-1 space density at the low-luminosity end.

Recent multi-scale simulations have shown that it is possible to explain the highest redshift, massive SDSS quasar SMBHs by gas accretion plus major mergers in a small group of protogalaxies within the framework of standard ΛCDM cosmology (Li et al. 2006). This picture immediately calls for deeper observations, because the known quasars with billion solar masses SMBHs at $z \simeq 6$ should possess progenitors, i.e., the so-called

[†] Even if bolometric corrections are not applied.

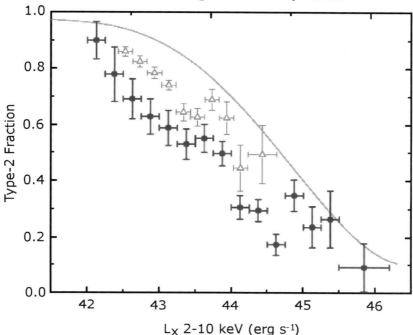

FIGURE 3. The luminosity dependence of the type-2 fraction compiled from different works. The x-ray data (filled circles) are from a sample of 1029 AGN selected in the 2–10 keV band, with >90% redshift completeness (Hasinger et al. 2007). The triangles are from a sample of [O III] selected AGN from the SDSS (Simpson 2005). The solid line is from silicate dust studies of a sample of AGN observed with *Spitzer* in the MIR (Maiolino et al. 2007). Given the known selection effects in the different bands, the data are consistent with each other and exhibit the same trend of decreasing absorption with luminosity.

mini-quasars with million solar masses at $z \simeq 10$ and stellar BHs with a few hundred solar masses at $z \simeq 20$. It is the task of future x-ray missions to probe these progenitors.

2.2. *Luminosity-dependence of AGN type-2 fraction*

A grouping into luminosity classes is one issue to prove the separate evolution of AGN classes. Here we summarize recent work on the AGN type-2 fraction. Ueda et al. (2003) studied the hard x-ray luminosity functions of an *ASCA* sample with ∼230 sources which was highly complete, ∼95%. They found that the fraction of type-2 AGN decreases with x-ray luminosity, $L_{2-10\,\mathrm{keV}}$. Similar results have been obtained with different systematics by the CLASXS team (Steffen et al. 2003) and by Hasinger (2004). Our preliminary analysis (Hasinger et al. 2007) confirms this trend in much more detail and better statistics using different data sets from *XMM-Newton*, *Chandra*, *ASCA* and *HEAO-1* (see Figure 3). Further, [O III]-selected Seyfert galaxies of the SDSS agree with this trend (Simpson 2005). It has been demonstrated in this work that for Seyfert galaxies selected in the SDSS, the fraction of BLAGN increases with the luminosity of the isotropically emitted [O III] narrow emission line. The [O III] luminosity can be converted into a 2–10 keV luminosity by empirical relations found for Seyfert galaxies (Mulchaey et al. 1994), i.e., $L_{\mathrm{[O\,III]}} = 0.015\,L_{2-10\,\mathrm{keV}}$.

Figure 3 illustrates all this work in direct comparison. Of course, both the x-ray and optical wavelength bands are subject to selection effects and biases: x-rays in the 2–10 keV band miss the population of Compton-thick Seyferts; optical surveys miss the AGN pop-

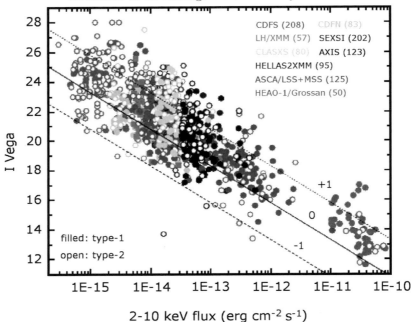

FIGURE 4. Hard x-ray (2–10 keV) sample of 1029 AGN in the I-flux plane. The individual samples have been cut a flux limit yielding >80% redshift completeness: CDF-S (Szokoly et al. 2004; Zheng et al. 2004; Grazian et al. 2006), CDF-N (Barger et al. 2003), LH/*XMM* (Mainieri et al. 2002; Szokoly et al. 2007), SEXSI (Eckart et al. 2006), CLASXS (Steffen et al. 2004), AXIS (Barcons et al. 2007), HELLAS2XMM (Fiore et al. 2003), *ASCA* LSS & MSS (Akiyama et al. 2000; Ueda et al. 2005), and *HEAO-1*/Grossan (Shinozaki et al. 2006).

ulation diluted by the host galaxy. Nevertheless, the agreement of better than 20% between the optical and x-ray selection is reassuring. Physically, this luminosity-dependent fraction might be interpreted as a 'leanout effect' because more luminous AGN can dissociate, ionize and finally blow away the dust in their environment. Recently, mid-infrared (MIR) *Spitzer* observations of 25 luminous and distant quasars additionally confirm this picture (Maiolino et al. 2007). They find a non-linear correlation between 6.7 μm MIR emission and 5100 Å optical emission. This is interpreted as a decreasing covering factor of circumnuclear dust as function of luminosity. Further, the silicate emission strength correlates with the luminosity, the accretion rate $L/L_{\rm Edd}$ and black hole mass $M_{\rm BH}$.

The evolution of this luminosity-dependent obscuration fraction with redshift is still a matter of intense debate. A redshift dependence of the obscured fraction was claimed by Fiore et al. (2003), using the HELLAS2XMM sample. On the other hand, other authors (Ueda et al. 2003; Gilli et al. 2007) did not find a redshift dependence. Recently, Treister and Urry (2006) performed a similar analysis, using an AGN meta-sample of 2300 AGN with ∼50% redshift completeness and find a shallow increase of the type-2 AGN fraction proportional to $(1 + z)^{0.3}$. The differences between the various results could be due to the way the redshift incompleteness, the different flux limits, and the AGN classification have been taken into account in the analysis.

Here, we briefly report on current work (Hasinger 2008) on a sample of ∼1000 AGN selected in the 2–10 keV band with ∼90% optical redshift completeness. In Figures 4 and 5 we show the 2–10 keV AGN samples utilized in this analysis in an x-ray flux/optical magnitude and an x-ray luminosity/redshift diagram. Type-1 and type-2 AGN of the

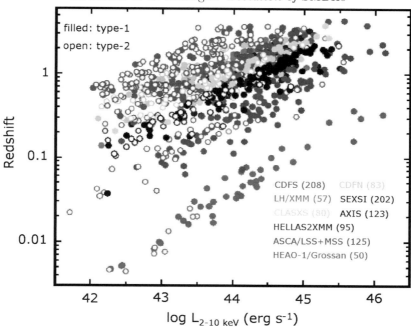

FIGURE 5. The same sample as in Figure 4, in the redshift-luminosity plane.

different samples are shown with filled and open circles, respectively. Just looking at the data, there is a noticeable change of the fraction of type-2 AGN as a function of luminosity. This trend becomes very obvious if the data is integrated in the redshift range $z = 0.2$–3.2 and the type-2 fraction is displayed as a function of x-ray luminosity (Figure 3.)

In order to address the possible evolution of the obscuration fraction with redshift, one can first simply determine the observed ratio of type-2 versus total AGN, integrated over all luminosities, in shells of increasing redshift. This is very similar to what was done by Treister & Urry (2006). Figure 6 shows the change of the observed fraction of absorbed objects with redshift. Despite the rather different selection effect and redshift incompleteness (90% vs. 50%), the observed data of Hasinger (2007) and Treister & Urry (2006) agree very well at redshifts above $z = 1.5$. At lower redshifts the Treister & Urry values are systematically above those of Hasinger, which is probably due to the fact that the former authors have only included BLAGN in their type-1 selection, and thus systematically overpredict the number of obscured sources.

The diagram also shows an attempt to estimate the possible effects of the small, but still significant, redshift incompleteness in the sample. For each redshift shell all the unidentified sources were placed at the center of the corresponding shell and included in the type-1 or type-2 classification, depending on their measured hardness ratios and assumed luminosities. This is clearly a very conservative limit, since it is highly unlikely that all the missing redshifts fall into one bin. Finally, the diagram also shows the type-2 fraction as a function of redshift predicted from the most recent background synthesis model (Gilli et al. 2007). This curve shows that the steep rise of the observed type-2 fraction at low redshift can be understood as an effect of the different flux limits for type-1 and type-2 sources in each survey (the latter ones are harder to find because of absorption). The Gilli et al. model predicts a smaller variation with redshift than what

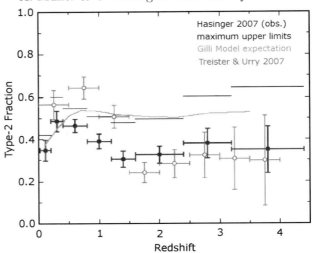

FIGURE 6. The observed fraction of type-2 AGN as a function of redshift. Filled Symbols and horizontal upper limits are from Hasinger (2008), open symbols from Treister & Urry (2006). The solid line shows the prediction of the Gilli et al. (2007) model, assuming no obscuration evolution (figure courtesy of Roberto Gilli).

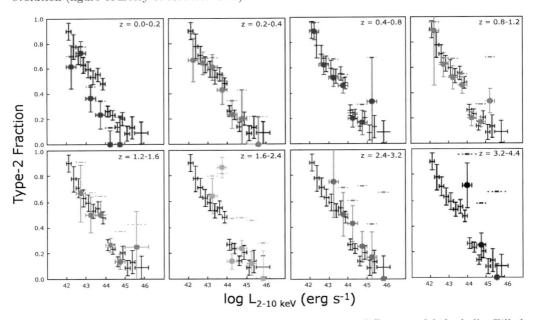

FIGURE 7. Type-2 fraction as a function of x-ray luminosity in different redshift shells. Filled circles show the data in the individual redshift shell indicated on each panel, while crosses give the average relation in the $z = 0.2$–3.2 range for comparison.

is actually observed. This is probably due to the fact that the model assumes flat type-2 fractions at high luminosities, while the observations in Figure 3 indicate a continuing decline.

Figure 6 is integrated over all luminosities in each redshift shell and therefore possibly hiding trends which are observable in a better resolved parameter space. In Figure 7 we therefore finally show the type-2 fraction as a function of luminosity, determined in the

FIGURE 8. X-ray deep field image taken in the Lockman Hole with *XMM-Newton* (Hasinger et al. 2001). The total exposure time amounts to ∼800 ks. This is a x-ray false-color composite image with red, green, and blue corresponding to 0.5–2, 2–4.5 and 4.5–10 keV photon energy. Point-like and extended sources are clearly visible that are associated with AGN and galaxy clusters respectively.

different redshift shells and compared to the mean relation in the $z = 0.2$–3.2 interval. A similar decreasing type-2 fraction as a function of luminosity is found in each redshift interval. At closer look, there is however a small trend of increasing obscuration fraction with redshift, even ignoring the lowest redshift shell, which is affected most by the flux limit incompleteness (see Figure 6). A more quantitative analysis is shown in Hasinger (2008).

3. The cosmic x-ray background

One major goal in x-ray astronomy is to explain the origin of the cosmic background emission in detail. The observed emission in x-rays originates mainly from AGN (point sources) and galaxy clusters (extended sources; Giacconi et al. 1962; Hasinger 2004; Brandt & Hasinger 2005). Figure 8 displays the typical appearance of an *XMM-Newton* x-ray deep field, located in the Lockman Hole in this case (Hasinger et al. 2001). The Lockman Hole is at an extraordinary location in the sky due to the fact that the column density is very low—$N_H = 5.7 \times 10^{19}$ cm^{-2} (Lockman et al. 1986).

Different groups developed detailed population synthesis models to reconstruct the observed cosmic x-ray background (XRB). Highly obscured AGN turn out to be crucial in this modeling. Even in the local universe, it has been shown that ≃40% of the Seyfert galaxies are Compton-thick (Risaliti et al. 1999). This means that these sources are missing from all current x-ray surveys because their column density is extremely large, $\log N_H \gtrsim 25$. As a consequence, only the reflected emission is seen in these x-ray spectra.

This hidden AGN population should show up in mid-infrared surveys. The flux criterion $F_{E>10}$ keV $\gg F_{SX}$ may serve as another good indicator for Compton-thick sources that could be investigated with *INTEGRAL* or *Swift*. Recently, Ueda et al. (2007) have presented *Suzaku* observations of two AGN out of a small sample detected in the *Swift* BAT survey, which show substantial x-ray absorption and may well be moderately obscured Compton-thick objects. Based on *Chandra* data from the CDF-S, Tozzi et al. (2006) have found ≃5% of AGN are Compton-thick and in the *XMM-Newton* observations of the COSMOS field, a similar fraction of candidate Compton-thick AGN is found (**?**). Beyond $z \simeq 0.1$ Compton-thick AGN are, however, very sparse (Martinez-Sansigre et al. 2007).

Recently, a new self-consistent population synthesis model for the cosmic x-ray background has been developed (Gilli et al. 2007), including the most recent determination of the AGN luminosity function evolution in the soft and hard x-ray bands. Gilli et al. consider a N_H distribution with a relatively small fraction of unabsorbed AGN (log $N_H < 22$), but with a growing contribution at higher column densities. Most of the emission originates from the Compton-thin AGN population. For the spectral templates a more complex model is assumed: The spectral index of the sources scatters in a Gaussian fashion around an average value of $< \Gamma >= 1.9$. Such spectral templates increase the contribution of Compton-thin AGN at 30 keV in the background spectrum by 30% (as compared to $\Gamma = const$ templates). However, the observed background spectrum still calls for a moderately and heavily obscured Compton-thick AGN population that contributes significantly around 30 keV. Moderately obscured Compton-thick AGN have $24 < \log N_H < 25$, whereas heavily obscured Compton-thick AGN have $\log N_H > 25$. To date only ≃40 local sources are candidates for Compton-thick AGN (Comastri 2004). However, the Compton-thick AGN are expected to become very numerous below current flux limits. This fact motivates for intensive studies with future x-ray missions such as *Simbol-X*, *Spectrum-X-Gamma*, *NeXT*, *Constellation-X* and *XEUS*.

Figure 9 shows the results from the most recent population synthesis model (Gilli et al. 2007). The model involving three AGN populations (unabsorbed Compton-thin, absorbed Compton-thin and Compton-thick sources) is in good agreement with the observed data points. Note that only the inclusion of the Compton-thick population (lowest bump peaking around 20–30 keV) reproduces the total cosmic XRB spectrum. The fraction of Compton-thick AGN predicted by this model is consistent with the actually observed numbers in the *Swift* BAT, *INTEGRAL* and CDF-S catalogs.

4. Conclusions

Black holes are the major scientific topic of this spring symposium. Many theoretical contributions have been presented at the meeting that help in understanding these mysterious objects. In contrast, x-ray astronomy offers the possibility to study black holes in action. Together with gravitational waves, x-rays are best suited because they allow insight into the direct surroundings of black holes.

This review demonstrates that x-ray astronomy is a powerful tool to understand black hole activity and their cosmological role in galaxy formation and galaxy evolution. Deep x-ray surveys probe black holes at cosmological distances. It is crucial to determine the redshift of these sources to trace back their cosmic history. This is done by optical follow-up observations with the best telescopes worldwide. Many x-ray surveys together provide a highly complete data set which can be exploited scientifically.

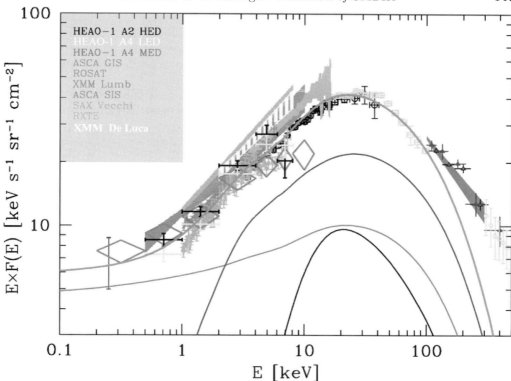

FIGURE 9. The observed cosmic XRB spectrum (data points) and the predicted contributions of different AGN families (solid lines), taken from Gilli et al. (2007). Total AGN plus galaxy clusters, type-2 Compton-thin AGN, type-1 Compton-thin AGN and type-2 Compton-thick AGN are drawn as solid lines (from top to bottom). Only the three populations together fit the observed XRB spectrum well. Compton-thin AGN contribute most.

The most relevant findings of recent AGN x-ray cosmology are the following:

• Observed x-ray spectra support the standard model of AGN, that is an accreting SMBH. However, AGN unification schemes have to consider luminosity effects and feedback mechanisms.

• The cosmic x-ray background is a total signal of individual active galaxies and galaxy clusters.

• An AGN classification scheme requires both optical (broad vs. narrow emission lines) and x-ray (absorbed versus unabsorbed spectrum) information.

• The fraction of type-2 to type-1 AGN decreases with luminosity. This indicates an interaction between the luminous galactic nucleus and its environment. Luminous AGN seem to be able to clean out their environment.

• The type-2 fraction shows a small evolution with redshift in the sense that higher redshift environments seem to have somewhat more absorption.

• Active galaxies do not evolve in a uniform manner. Depending on the luminosity (or black hole mass, respectively) SMBHs exhibit an anti-hierarchical growth, i.e., quasars enter the cosmic stage significantly earlier than Seyfert galaxies. In other words: the density evolution of AGN is luminosity dependent.

• X-ray luminosity functions are in perfect agreement with other wavebands.

• Compton-thin AGN contribute most to the cosmic x-ray background.

• However, Compton-thick AGN are an essential ingredient to reproduce the total cosmic x-ray background spectrum. They mainly contribute at photon energies around 30 keV.

• Accreted BHMFs from the x-ray background can be reconciled with dormant relic BHs in the local universe if one assumes that the BHs rotate.

• Theory and observation indirectly support the existence of progenitors such as mini-quasars at $z \simeq 11$ and GRBs at $z \simeq 20$. This should be tested by direct observations with *XEUS*.

X-ray astronomy is going to be successful in the future if the survey capacities are extended successively. Surveys with larger fields and deeper pencil beams are natural goals. Therefore, upcoming projects such as eRosita aboard *Spectrum-X-Gamma*, *Simbol-X*, *NeXT*, and *XEUS* are promising scientific drivers.

We thank Roberto Gilli for providing the model prediction in Figure 6. AM acknowledges the invitation to the STScI at Baltimore. It was a great experience to discuss current black hole research in an exciting and warm atmosphere.

REFERENCES

AKIYAMA, M., OHTA, K., & YAMADA, T. ET AL. 2000 *ApJ* **532**, 700.

BARCONS, X., CARRERA, F. J., CEBALLOS, M. J., ET AL. 2007 *A&A* **476**, 1191.

BARGER, A. J., COWIE, L. L., CAPAK, P., ET AL. 2003 *AJ* **126**, 632.

BARGER, A. J., COWIE, L. L., MUSHOTZKY, R. F., ET AL. 2005 *AJ* **129**, 578.

BRANDT, W. N. & HASINGER, G. 2005 *ARA&A* **43**, 829.

COMASTRI, A. 2004. In *Supermassive Black Holes in the Distant Universe* (ed. A. J. Barger). Ap&SS Library, vol. 308, p. 245. Kluwer.

ECKART, M. E., STERN, D., HELFAND, D., ET AL. 2006 *ApJS* **165**, 19.

FAN, X., STRAUSS, M. A., SCHNEIDER, D. P., ET AL. 2001 *AJ* **121**, 54.

FAN, X., STRAUSS, M. A., SCHNEIDER, D. P., ET AL. 2003 *AJ* **125**, 1649.

FIORE, F., BRUSA, M., COCCHIA, F., ET AL. 2003 *A&A* **409**, 79.

GIACCONI, R., GURSKY, H., PAOLINI, F. R., & ROSSI, B. B. 1962 *Phys. Rev. Lett.* **9**, 439.

GILLI, R., COMASTRI, A., & HASINGER, G. 2007 *A&A* **463**, 79.

GRAZIAN, A., FONTANA, A., DE SANTIS, C., ET AL. 2006 *A&A* **449**, 951.

HASINGER, G. 2004 *Nucl. Phys. B* **132**, 86.

HASINGER, G. 2008 *A&A* **490**, 905.

HASINGER, G., ALTIERI, B., ARNAUD, M., ET AL. 2001 *A&A* **365**, L45.

HASINGER, G., CAPPELLUTI, N., BRUNNER, H., ET AL. 2007 *ApJS* **172**, 29.

HASINGER, G., MIYAJI, T., & SCHMIDT, M. 2005 *A&A* **441**, 417.

HOPKINS, P. F., RICHARDS, G. T., & HERNQUIST, L. 2007 *ApJ* **654**, 731.

LA FRANCA, F., FIORE, F., COMASTRI, A., ET AL. 2005 *ApJ* **635**, 864.

LI, Y., HERNQUIST, L. ROBERTSON, B., ET AL. 2006 *ApJ* **665**, 187.

LOCKMAN, F. J., JAHODA, K., & MCCAMMON, D. 1986 *ApJ* **302**, 432.

MAINIERI, V., BERGERON, J., HASINGER, G., ET AL. 2002 *A&A* **393**, 425.

MAINIERI, V., ROSATI, P., TOZZI, P., ET AL. 2005 *A&A* **437**, 805.

MAIOLINO, R., SHEMMER, O., IMANISHI, M., ET AL. 2007 *A&A* **468**, 979.

MARCONI, A., RISALITI, G., GILLI, R., ET AL. 2004 *MNRAS* **351**, 169.

MARTINEZ-SANSIGRE, A., RAWLINGS, S., BONFIELD, D. G., ET AL. 2007 *MNRAS* **379**, L6.

MERLONI, A. 2004 *MNRAS* **353**, 1035.

MULCHAEY, J. S., KORATKAR, A., WARD, M. J., ET AL. 1994 *ApJ* **436**, 586.

RICHARDS, G., STRAUSS, M. A., FAN, X., ET AL. 2006 *AJ* **131**, 2766.

RISALITI, G., MAIOLINO, R., & SALVATI, M. 1999 *ApJ* **522**, 157.

SHINOZAKI, K., MIYAJI, T., ISHISAKI, Y., UEDA, Y., & OGASAKA, Y. 2006 *AJ* **131**, 2843.

SILVERMAN, J. D., GREEN, P. J., BARKHOUSE, W. A., ET AL. 2005 *ApJ* **624**, 630.

SIMPSON, C. 2005 *MNRAS* **360**, 565.

SOLTAN, A. 1982 *MNRAS* **200**, 115.

STEFFEN, A. T., BARGER, A. J., CAPAK, P., ET AL. 2004 *AJ* **128**, 1483.

STEFFEN, A. T., BARGER, A. J., COWIE, L. L., MUSHOTZKY, R. F., & YANG, Y. 2003 *ApJ* **569**, L23.

SZOKOLY, G. P., BERGERON, J., HASINGER, G., ET AL. 2004 *ApJS* **155**, 271.

SZOKOLY, G. P., ET AL. 2007 in preparation.

TOZZI, P., GILLI, R., MAINIERI, V., ET AL. 2006 *A&A* **451**, 457.

TREISTER, E. & URRY, C. M. 2006 *ApJ* **652**, L79.

UEDA, Y., AKIYAMA, M., OHTA, K., & MIYAJI, T. 2003 *ApJ* **598**, 886.

UEDA, Y., EGUCHI, S., Y., TERASHIMA, Y., ET AL. 2007 *ApJ* **664**, L79.

UEDA, Y., ISHISAKI, Y., TAKAHASHI, T., MAKISHIMA, K., & OHASHI, T. 2005 *ApJS* **161**, 185.

VOLONTERI, M., MADAU, P., QUATAERT, E., & REES, M. J. 2005 *ApJ* **620**, 69.

WALL, J. V., JACKSON, C. A., SHAVER, P. A., HOOK, I. M., & KELLERMANN, K. I. 2005 *A&A* **434**, 133.

WOLF, C., MEISENHEIMER, K., KLEINHEINRICH, M., ET AL. 2004 *A&A* **421**, 913.

ZHENG, W., MIKLES, V. J., MAINIERI, V., ET AL. 2004 *ApJS* **155**, 73.

Black-hole masses of distant quasars

By M. VESTERGAARD[1,2]

[1] Steward Observatory, Department of Astronomy, University of Arizona, 933 North Cherry
Avenue, Tucson, AZ, 85718, USA

[2] Current address: Department of Physics and Astronomy, Tufts University, Robinson Hall,
Medford, MA 02155

A brief overview of the methods commonly used to determine or estimate the black-hole mass
in quiescent or active galaxies is presented and it is argued that the use of mass-scaling relations
is both a reliable and the preferred method to apply to large samples of distant quasars. The
method uses spectroscopic measurements of a broad emission-line width and continuum luminosity and currently has a statistical 1σ uncertainty in the absolute mass values of about a factor
of 4. Potentially, this accuracy can be improved in the future. When applied to large samples of
distant quasars it is evident that the black-hole masses are very large, of order 1 to 10 billion $M_{\odot}$,
even at the highest redshifts of 4 to 6. The black holes must build up their mass very fast in the
early universe. Yet they do not grow much larger than that: a maximum mass of $\sim 10^{10}$ $M_{\odot}$ is
also observed. Preliminary mass functions of active black holes are presented for several quasar
samples, including the Sloan Digital Sky Survey. Finally, common concerns related to the application of the mass-scaling relations, especially for high redshift quasars, are briefly discussed.

1. Introduction: Mass-estimation methods for active galaxies and quasars

The *Hubble Space Telescope* has played a key role in our ability to detect supermassive
black holes in the centers of nearby galaxies and to determine their mass through its
high angular resolution. This has started a new, exciting era in which we are now able
to study these massive objects and how they affect their environment and its cosmic
evolution, and thereby get a more complete picture of structure formation and evolution
over the history of the universe.

There are different ways in which to determine or estimate the black-hole mass in active
nuclei, and they all differ somewhat from the methods applied to quiescent galaxies.
Therefore, to place these methods in perspective and to explain why the method of
"mass-scaling relations" is preferred for distant active nuclei, I will start by giving a brief
overview of the methods used to determine or estimate black-hole masses in quiescent
and active galaxies.

In galaxies containing a quiescent or dormant central black hole, its mass is measured
by determining the velocity dispersion of stars or gas close enough to the black hole
that their dynamics are dominated by its gravity (i.e., within the black hole's "radius of
influence") by means of the virial theorem. While this appears the most direct way to
measure the black-hole mass, this method cannot be applied to active black holes due to
the strong glare from the central source. The most robust mass-determination method
for active nuclei depends on its source type. Active galaxies are classified as Type 1
and Type 2, depending on whether the broad emission-line region can be viewed by us
(Type 1) or not (Type 2). Type 1 sources tend to be oriented more face-on, and Type 2
sources are typically more highly inclined, such that gas and dust in the host galaxy
or in the outskirts of the central engine block our direct view of the broad-line region.
For Type 2 sources, water mega-masers emitted by the molecular gas disk in the galaxy
(e.g., Miyoshi et al. 1995) is a very reliable and accurate way to determine the black-
hole mass. For less-inclined Type 1 sources, the mega-maser method cannot be used.

150

The most robust way to determine their black-hole masses is through the emission-line reverberation mapping method that relies entirely on the variability nature of the central engine (e.g., Peterson 1993; see also Peterson, this volume), and thus not on high angular-resolution observations as are needed for quiescent galaxies. At the present time, the monitoring data that has been collected allow the so-called one-dimensional reverberation mapping in which the broad emission-line time delay relative to continuum variations can be determined. This delay is the light travel time of photons from the central continuum region and is, hence, the distance of the broad line gas from the continuum region. The black-hole mass is determined from the virial theorem by also measuring the velocity dispersion of this variable gas. Ultimately, we would like to do a two-dimensional analysis which also yields a velocity-delay map. With the latter, we would be able to determine the black-hole mass zero-point independently of other methods. The one-dimensional method does not allow this, so at present we are limited to estimating the absolute-mass zero-point by other means. This is explained next.

For quiescent local galaxies, the masses of their black holes display a strong, well-established correlation with the velocity dispersion of stars in the galactic bulge, far beyond the gravitational reach of the black hole (e.g., Ferrarese & Merritt 2000; Gebhardt et al. 2000a), the so-called M_{BH}–σ_* relationship. Active black holes also display such a relationship (Gebhardt et al. 2000b; Ferrarese et al. 2001; Onken et al. 2004), but given the unknown geometry and structure of the broad emission-line region, it is not yet possible to independently determine the zero-point of the relationship. However, if one assumes that the active black holes in the local neighborhood have essentially built up their mass, we can assume that the two M_{BH}–σ_* relationships are, in fact, one and the same, and thereby empirically establish the absolute zero-point offset of the mass scale of active black holes (e.g., Onken et al. 2004). The assumption is fair, since most of the active nuclei in the reverberation-mapping sample have black holes accreting at low levels, between 0.1% to a few percent of the Eddington rate (Peterson et al. 2004).

It is not always possible to apply the primary methods of reverberation mapping or mega-masers to determine the black-hole mass. This is where "secondary or tertiary methods" become useful. They typically build on or approximate the primary methods. Table 1 gives an overview of these methods, which I name "Secondary" for convenience. For example, BL Lac-type objects are viewed at such small inclination angles that the continuum emission is very strongly Doppler amplified, and the source variability reflects details of the powerful radio jets we are looking right into, rather than that of the continuum and line-emitting regions. Also, BL Lacs typically have very weak or no broad emission lines, most likely due to the strong continuum boosting. For these types of sources, researchers have utilized the fact that they are hosted by elliptical galaxies. The physical parameters of elliptical galaxies (surface brightness Σ_e, effective radius, r_e, and stellar velocity dispersion σ_*) correlate such that they form a "plane," the Fundamental Plane (Djorgovski & Davis 1987; Dressler et al. 1987). To determine the central black-hole mass, one can measure Σ_e and r_e so to infer σ_* from the Fundamental Plane, and next use the M_{BH}–σ_* to infer the black-hole mass (e.g., Woo & Urry 2002; Falomo et al. 2004). Unfortunately, owing in part to the finite thickness of the Fundamental Plane, the inference from this plane relationship itself can add as much as a factor of 4 uncertainty (Woo & Urry 2002) to the mass estimates. This method can also be applied when only (host galaxy) imaging data are available. Alternatively, one can determine the luminosity of the galaxy (bulge) and from the M_{BH}–L(Bulge) relationship (e.g., Magorrian et al. 1998), a sister relationship to the M_{BH}–σ_* relation, infer the black-hole mass (McLure & Dunlop 2002). The lowest uncertainties that have been reported are a factor 3 to 4. Unfortunately, this method is prone to larger typical uncertainties when applied to active

	Low-*z*		High-*z*	
	Low-*L*	High-*L*	High-*L*	Best Accuracy
	LINERs, Sy 2s	QSOs, Sy 1s, BL Lacs	QSOs	(Dex)
Scaling Relations Via $M_{\rm BH}$–σ_*:	$\checkmark$	$\checkmark$	$\checkmark$	0.5–0.6
$-\ \sigma_*$	$\checkmark$	$\checkmark$	$\div$	0.3
$-$ [O III] FWHM	$\checkmark$	$\checkmark$	$\checkmark$	0.7
$-$ Fundamental Plane: $\sum_e, r_e$	$\checkmark$	$\checkmark$	$\div$	>0.7
Via $M_{\rm BH}$–$L_{\rm bulge}$ & scaling rel.: $-\ M_R$	$\checkmark$	$\checkmark$	$\div$	0.6–0.7

TABLE 1. Secondary mass-estimation methods

galaxies, owing to the difficulties in separating the luminosity of the galaxy bulge from the strong nuclear source (e.g., Wandel 2002). With spectroscopic measurements of the galaxy bulge σ_*, the black-hole mass can be inferred from the $M_{\rm BH}$–σ_* relation with a significantly smaller uncertainty (Table 1). However, as the Ca II triplet lines $\lambda\lambda$ 8498, 8542, 8662 shift into the water vapor lines for $z \gtrsim 0.068$, it becomes increasingly difficult and the measurement uncertainties increase (e.g., Ferrarese et al. 2001; Onken et al. 2004). Some authors have then resorted to determining σ_* from a collection of stellar features, although weaker, at shorter wavelengths (e.g., Barth et al. 2005).

Because reverberation mapping requires a lot of telescope time and resources, this method is difficult or impossible to apply to large samples of sources, especially for distant objects (e.g., Vestergaard 2004a; Kaspi 2001). It is also an issue that more luminous sources, owing to their larger size, vary on longer time scales and with smaller amplitudes, making reverberation-mapping analysis challenging. Even for the nearby PG quasars of order 10 years of monitoring data are required to constrain the black-hole mass well enough (e.g., Kaspi 2001). Instead, one method that has been applied to obtain mass estimates of large samples makes use of the fact that the width of the [O III] $\lambda5007$ line approximates the velocity dispersion of the stars in the galaxy (Nelson 2000; Nelson & Whittle 1996). The idea is to use the FWHM([O III]) as a substitute for σ_* and then the $M_{\rm BH}$–σ_* relationship to infer the central mass. However, the 1σ statistical uncertainty in this method is a factor of 5 (Boroson 2003).

One other secondary method seems to perform better overall: the method sometimes referred to as "mass-scaling relations." The method uses measurements of line width and continuum luminosity from a single spectrum of an active nucleus to estimate the black-hole mass. It is an approximation to the reverberation-mapping method, and therefore is also based on the virial theorem. The method relies strongly on another result from reverberation mapping, namely the radius–luminosity relationship, based on the size of the line-emitting region (as measured for a particular emission line) which can be estimated from the continuum luminosity. The advantages of this method are that it can both be applied to nearby and distant Type 1 active nuclei and to large samples with relative ease. The methods relying on accurate measurements of the host galaxy properties are limited to relatively low redshifts below z of about 0.5 to 1.0, since more distant host galaxies are difficult to characterize (e.g., Kukula et al. 2001). Mass-scaling relations presently provide an accuracy of the absolute-mass values of a factor of about 4. This is a method for which we can potentially even further improve the mass estimates in the future.

In this contribution, I will focus on mass-scaling relations as my preferred method to obtain mass estimates of black holes in large samples of distant quasars: (a) because of their easy application to a larger redshift range than the other methods, without going into the infrared regime; and (b) due to the lower uncertainties associated. Before I discuss this method, I will address the key component of this method—namely the broad-line region radius–luminosity relationship (Section 2).

Section 3 is dedicated to mass-scaling relationships and important related issues. The distribution of quasar black-hole masses over the history of the universe is described in Section 4, while the first mass functions based on various large quasar samples are presented in Section 5. Potential issues with mass-scaling relations as commonly highlighted in the literature are discussed in Section 6, before the summary and conclusions (Section 7).

A flat cosmology with $H_0 = 70$ km s^{-1}/Mpc and $\Omega_\lambda = 0.7$ is used throughout.

2. The radius–luminosity relationship

Peterson (this volume) has already introduced the radius–luminosity relationship: the relation between the characteristic size of the gaseous region emitting a particular broad emission line and the ionizing continuum luminosity that excites this gas. Therefore I will not discuss these relationships and their history in detail, but only emphasize a few issues that are important for their applications to estimating the black-hole mass for distant active nuclei.

2.1. The optical R–L relationship

Peterson discussed the efforts to establish the physically relevant relationship between the size of the Hβ-emitting region and the ionizing optical continuum luminosity without the contamination from stellar light in the host galaxy of the active nucleus (Bentz et al. 2006a). Keep in mind the reason that the first R–L relations published had a steeper slope is mostly due to the contamination from the host galaxy. The degree of contamination, even from nearby luminous quasars thought to entirely dominate their host-galaxy light, is in fact larger (Bentz et al. 2006a) than originally assumed (e.g., Kaspi et al. 2005). Therefore, the only radius–luminosity relation for R(Hβ) and L_λ(5100 Å) that should be applied is that of Bentz et al. (2006a), for which the slope is firmly established to be $1/2$, because this reflects the intrinsic, physical relationship.

2.2. The UV R–L relationships

Kaspi et al. (2005) established that a similar R–L relationship exists for continuum luminosities in the UV and x-ray regions using reverberation results and archival data. In particular, R(Hβ) scales with the monochromatic continuum luminosity at 1450 Å as $R(\mathrm{H}\beta) \propto L_\lambda(1450 \text{ Å})^{0.53}$. Restframe UV luminosities are most useful when studying distant quasars, since we can then use ground-based observations in the observed optical region. However, in this case it is not the R(Hβ)–L(UV) relationship that is relevant, but rather that pertaining to the UV emission line (Mg II or C IV) with which it is to be used (Section 3). Prior to 2007, no such relationship could be established for C IV, mostly due to the unavailability of a sufficient data base (see Peterson, this volume). The inherent assumption in the mass estimates based on C IV at the time was therefore that the R(C IV)–L(UV) relationship that had to exist, would be similar to that of the R(Hβ)–L(UV) relation (e.g., Vestergaard & Peterson 2006). The recent successes of Peterson et al. (2005) to measure a C IV time delay (or lag) in the low-luminosity Seyfert galaxy NGC 4395 and of Kaspi et al. (2007) to measure a C IV lag for a luminous quasar has finally extended the luminosity range (to over 7 orders of magnitude in luminosity) of previously measured (reliable) C IV lags, allowing us to confirm that the slope of the R(C IV)–L(UV) relation (of 0.53) is entirely consistent with that for Hβ. This confirms the previous assumptions made and shows that, as it happens, the recently updated mass-scaling relationships based on C IV (Vestergaard & Peterson 2006), which assumed R(C IV) $\propto L$(UV)$^{0.53}$, still hold without any modifications needed.

It is worth noting that an R–L relationship has not been established for the Mg II emission line, since only one reliable determination of the size of the Mg II-emitting regions exists at present (Metzroth et al. 2006). Mg II is the one emission line that has typically not been targeted in monitoring campaigns. Its isolation in the spectrum from the UV high-ionization lines and the optical Balmer lines has significantly contributed to this fact. The only reliable Mg II lag is based on improved reverberation data of NGC 4151, which show that Mg II is consistent with being emitted from a distance twice that of C IV for two epochs; yet within the uncertainties, the two lines are also consistent with being emitted from similar distances (Metzroth et al. 2006). Apart from measurement uncertainties, the limitation that no contemporaneous measurements of the Hβ and Mg II lags currently exists. However, new Hβ-monitoring data of NGC 4151 do suggest that Mg II and Hβ are emitted at similar distances from the continuum regions (Bentz et al. 2006b).

2.3. Do the R–L relations apply to distant luminous quasars?

There are a few reasons why we can comfortably apply these R–L relationships to quasars beyond our local neighborhood in which they are defined. First, quasar spectra look remarkably similar (to first order) at all redshifts. This has been well demonstrated by Dietrich et al. (2002), who show in their Figures 3 and 5 the strong similarity of the quasar broad-line spectra across the history of the universe when binning in either luminosity or redshift. While this complicates a thorough understanding of the broad-line region, it is a help in this case, because it shows that for all quasars the broad-line regions are essentially the same—whether you look at a quasar in your backyard or one at the furtherest reaches of the universe (e.g., Barth et al. 2003; Jiang et al. 2007). If the broad-line regions are similar, they will also obey the same radius–luminosity relationships. Second, a key point about the Kaspi et al. (2007) results is that the quasar for which the size of the C IV-emitting region could be determined is located at a redshift of 3! Hence, these results also indicate that high-redshift quasars do have similar broad-line regions that respond to luminosity changes as do nearby active nuclei. Third, when we apply the R–L relations to more distant quasars, we are, contrary what is sometimes stated, not extrapolating beyond the luminosity range for which these relations are defined. The R–L relation pertaining to the optical continuum luminosity L_λ(5100 Å) and R(Hβ) is defined over 5 orders of magnitude in luminosity: from 10^{41} erg s^{-1} to 10^{46} erg s^{-1} (Bentz et al. 2006a; see Figure 3 in Peterson's contribution in this volume). The UV R–L relation, relevant for R(C IV) and L_λ(1350 Å) or L_λ(1450 Å)—these luminosities can be used interchangeably (Vestergaard & Peterson 2006)—is defined over 7 orders of magnitude in luminosity: from $10^{39.3}$ erg s^{-1} to 10^{47} erg s^{-1} (Kaspi et al. 2007). Next, I will compare these characteristic luminosity ranges to the luminosities of known quasars of various surveys.

Figure 1 shows the luminosity distributions of a few quasar samples, including the Bright Quasar Survey (BQS) at $z \leqslant 0.5$, the Large Bright Quasar Survey (shown in two redshift ranges: $z \leqslant 0.5$ and $1.1 \leqslant z \leqslant 2.9$ that are discussed in later Sections), the color-selected sample from the Sloan Digital Sky Survey Fall Equatorial Stripe (Fan et al.

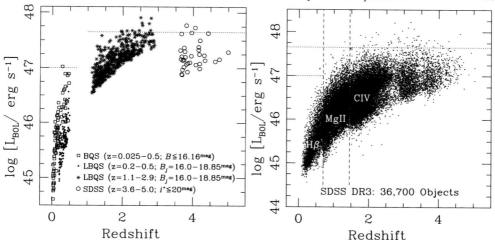

FIGURE 1. Distribution of bolometric luminosities for various quasar samples (labeled). The dotted lines mark the maximum luminosity for which the optical (applied at $z \lesssim 0.9$) and UV $R–L$ relationships are defined. Very few quasars, even at high z, have luminosities exceeding these, so the use of the $R–L$ relations is reasonable. Left: The Bright Quasar Survey (BQS), the Large Bright Quasar Survey (LBQS), and the color-selected sample presented by Fan et al. (2001), as labeled. Right: The Sloan Digital Sky Survey Data Release 3 quasar catalog. Vertical dashed lines indicate the subsets where different emission lines (labeled) are used for the mass estimates (left to right: Hβ, Mg II, C IV).

2001), and the SDSS Data Release 3 quasar catalog (Schneider et al. 2005). These samples are discussed further in Sections 4 and 5. I converted the monochromatic power-law continuum luminosities to bolometric luminosities, L_{Bol}, in order to show the luminosities of the individual samples in the same diagrams. The continuum luminosities are determined through a decomposition of the quasar spectra into a power-law continuum component, an iron-emission component using the best and most complete iron templates in the UV (Vestergaard & Wilkes 2001) and the optical (Veron et al. 2004), and a line-emission component. The L_{Bol} values are estimated by a simple scaling factor applied to the monochromatic luminosities of 4.62, 4.65, 5.8, 9.74, and 10.5 (Vestergaard 2004b, and based on Richards et al. 2006) for the $L_\lambda(1350\ \text{Å})$, $L_\lambda(1450\ \text{Å})$, $L_\lambda(3000\ \text{Å})$, $L_\lambda(4400\ \text{Å})$, and $L_\lambda(5100\ \text{Å})$ luminosities, respectively.

The luminosity distributions in Figure 1 show that, as expected, more distant quasars are also more luminous. The lower flux limits of the different surveys uncovering these quasar samples are evident by the sometimes sharp cutoff of the distribution toward lower luminosities. In each panel I have indicated the maximum luminosity, in L_{Bol} units, for which the $R–L$ relationships are defined for the optical continuum luminosity, $L_\lambda(5100\ \text{Å})$, and for the UV continuum luminosities, $L_\lambda(1350\ \text{Å})$ and $L_\lambda(1450\ \text{Å})$. They are 10^{47} and $10^{47.65}$ erg s^1, respectively, given the bolometric scaling factors introduced above. It is clear from Figure 1 that very few quasars, even the distant, most luminous ones, have luminosities beyond the luminosity ranges where the $R–L$ relations are defined. This holds both at low and high redshift. Statements that the mass-scaling relationships may not be valid for distant luminous quasars because of a need to extrapolate far beyond the validity of the $R–L$ relations for these sources are clearly not supported.

The R–L relationships have tremendous power—providing they are reliable, of course—as discussed here and by Peterson (this volume), we have good reasons to believe that they are. They allow us to determine black-hole mass estimates of thousands of quasars on a relatively short timescale from a single spectrum of each source. This is in sharp contrast to the large amount of time and resources required to determine the mass using reverberation mapping. However, R–L relationships do not marginalize monitoring campaigns and the reverberation method. In fact, reverberation mapping is still very important, because it provides the time lags and sizes of the various line-emitting regions and the more directly and more robustly determined masses to which the scaling relations are anchored.

3. Single-epoch mass estimates based on scaling relations

Scaling relations are approximations to the mass estimates based on the reverberation mapping method, as discussed by Peterson in this proceedings. Thus, similar to reverberation-based masses, the single-epoch mass estimates rely on the virial theorem, where $M_{\rm BH} = V^2 R/G$, G is the gravitational constant and R is the distance from the black hole to the gas with velocity dispersion, V. It is thus assumed that the broad-line emitting gas that is used to probe the black-hole mass is gravitationally dominated by the black-hole potential. It has by now been firmly established to hold for most, if not all, active galaxy broad-line emitting gas (Peterson, this volume; Peterson & Wandel 1999, 2000; Onken & Peterson 2002; Metzroth et al. 2006).

Specifically, these (mass-) scaling relations use the broad-line width as a measure of proxy for the velocity dispersion of the broad emission line gas. The line width is parameterized either as the line dispersion $\sigma_{\rm line}$ (the second moment of the line profile) or the full-width-at-half-maximum, FWHM. As discussed later and by Peterson et al. (2004) and Collin et al. (2006), the line dispersion is the preferred line-width measure for its robustness. The continuum luminosity is used to estimate the distance to the line-emitting gas via the important R–L relationship introduced in Section 2, since, clearly, no time delays can be determined without extensive multi-epoch observations. In summary, the mass-scaling relations take the form of either:

$$M_{\rm BH} = {\rm C1} \times {\rm FWHM(line)}^2 \times L_\lambda^\beta \ ,$$

or

$$M_{\rm BH} = {\rm C2} \times \sigma_{\rm line}^2 \times L_\lambda^\beta \ ,$$

where C1 and C2 are the mass zero-points for each of the equations and the index β is essentially 0.5 for continuum luminosities (Section 2). The specifics of the most recent updates to these relationships for C IV and Hβ are presented by Vestergaard & Peterson (2006), to which the interested reader is referred.

The possibility of using measurements from a single observation of an active nucleus to estimate the mass of its central black hole was first introduced by Wandel, Peterson, & Malkan (1999) in the optical wavelength region using the Hβ emission line and the ionizing luminosity estimated from photoionization theory. Vestergaard (2002) expanded this into the UV-wavelength region using the C IV emission line by calibrating single-epoch measurements of FWHM(C IV) and L(1350 Å) to black-hole masses obtained with reverberation mapping data. This was followed by work using FWHM(C IV) and L(1450 Å; Warner et al. 2003) and FWHM(Mg II) and L(3000 Å; McLure & Jarvis 2002). In the latter study, Mg II is assumed to be emitted from the same distance as Hβ, such that the Mg II FWHM is substituting the Hβ line width. With the availability of UV-scaling relations, it became more easily possible to estimate black-hole masses of distant quasars residing at the highest observable redshifts, and for large samples thereof.

3.1. *An empirical determination of the f-factor*

In 2004, the reverberation-mapping database underwent a major revision (Peterson et al. 2004), which resulted in improved determinations of black-hole mass and the radius of the variable Hβ-emitting gas for the 36 active nuclei monitored at that time. As summarized in that paper (and Peterson, this volume), an f-factor enters the virial mass estimates owing to the unknown geometry and structure of the broad emission-line region. The appropriate value of f is largely unknown, but is expected to be of order unity. However, Onken et al. (2004) empirically determined the average value of the f-factor to be 5.5 for the reverberation sample of active nuclei. For those sources with reliable (and completely independent) determinations of both black-hole mass and host-galaxy bulge stellar-velocity dispersion, Onken et al. (2004) made the assumption that intrinsically the M–σ relationships for active and quiescent galaxies are the same. The distribution of black-hole masses and stellar-velocity dispersion for the active galaxies show the same slope, but an offset in the zero-point, suggesting that the assumed f-factor at the time was not representative. Prior to 2004, we had to make assumptions about the value of f so to get an estimate of the black-hole mass. The most simple kinematical structure, an isothermal gas distribution, with equal gas velocities in all directions, was often adopted, yielding an f value of 0.75. It is, in fact, a very important point that the first values of f adopted were entirely *arbitrary*, but were accepted due to the lack of a more qualified guess at this time. The work by Onken et al. (2004) has made all other previously assumed values of f obsolete. It is worth emphasizing, since this is often misunderstood, that this empirical determination of $f = 5.5$ is valid for random inclinations of the broad emission-line region with respect to our line of sight. The average value of f was determined such that some sources will have the mass either overestimated or underestimated, but on average for the reverberation sample, the mass zero-point is consistent with the M–σ relationship for quiescent galaxies. I will discuss in Section 6.1 that the effects of source inclination can be minimized simply by using the line dispersion to measure the line width, or by making an f-factor correction to the mass estimates based on the FWHM.

It is important to note that if one seeks to determine the black-hole mass of a *single* object based on a single-epoch spectrum—as opposed to the masses for a sample of objects for which the mass-scaling relationships discussed in Section 3.2 can be used— then one should use $f = 3.85$ along with the line dispersion to determine the mass (Collin et al. 2006). Since a single-epoch spectrum resembles the mean spectrum much more than the RMS spectrum, and given that the mean and rms profiles are often different, the appropriate f-factor is also different. In other words, to recover the reverberation black-hole mass of one of the reverberation-mapped active galaxies using the line dispersion measured from a single spectrum, $f = 3.85$ is needed. But, if only the FWHM of the line is available, one should use Eq. 7 of Collin et al. (2006) to determine the appropriate f-factor, which depends on the FWHM value.

3.2. *Improved mass-scaling relations for Hβ and C IV*

Since the Peterson et al. (2004) and Onken et al. (2004) papers significantly revised the reverberation data on which the Vestergaard (2002) scaling relations are anchored, a revision of the latter was necessary. Using a larger database than the 2002 calibrations, Vestergaard & Peterson (2006) present updated and improved scaling relationships for both the C IV and Hβ emission lines, calibrated to the revised reverberation-mapping masses. Equations using either the FWHM or line dispersion are presented for C IV. Note that since the scaling relationships are calibrated directly to the reverberation masses—deemed the most accurate mass determinations of the objects in the reverberation-mapping sample—the f-factor is absorbed in the zero-points of the mass estimates. The Vestergaard & Peterson (2006) study shows that, although the scatter of the single-epoch mass estimates relative to the reverberation masses has decreased somewhat owing to the improved data base used, there is still some scatter remaining, amounting to a 1σ statistical uncertainty in the absolute values of the mass estimates of a factor 3.5 to 5, depending on the emission-line and line-width measure. This includes the estimated uncertainty in the absolute zero-point of the reverberation masses of a factor of 2.9 or less, as estimated by Onken et al. (2004). This means that there is a 68% chance that the mass estimate of a single object is accurate to within a factor 3.5–5, but there is a 95% chance it is accurate to within a factor of about 6–7.

The remaining scatter is possibly due to several factors. For one, the intrinsic scatter in the R–L relationship dominates the uncertainties in the mass estimates. That scatter is likely related to the detailed geometric and kinematical structure of the broad-line region that we are unable to account for. In addition, active nuclei do vary in luminosity. The R–L and virial relationships indicate that as the nuclear luminosity changes, so will the size of the line-emitting region, and its location will dictate the gas velocity for a given black-hole mass. As a result, the R and V values should slide along the virial relationship, $R \propto V^{-1/2}$, as the luminosity changes. This has been clearly demonstrated for the Seyfert 1 galaxy, NGC 5548 (Peterson & Wandel 1999, 2000) and for several broad emission lines. And the mass estimates should always be consistent. However, deviations from a tight relationship may be expected in a couple of situations. For example, if the broad-line region has not yet adjusted to the luminosity change such that the actual R and the measured V may not accurately reflect the measured L. With BLR sizes of hundreds of days for luminous quasars (e.g., Kaspi et al. 2000, 2007) this can perhaps never be avoided. Another possibility for deviation is when the density distribution of the gas in the broad-line region is not smooth or "typical" (i.e., based on the R–L relation, which yields an ensemble average). The different emission lines could be emitted from regions lying closer or farther from one another than what is "typical." According to the Locally Optimized Cloud model of the broad-line region (Baldwin et al. 1995), the predominant part of the flux in a particular emission line is emitted from the part of the broad-line gas that is most efficient, at a given time, of emitting this line. The different emission lines will thus be emitted from different subregions in the broad-line region determined by the detailed physical conditions, such as the gas density and the incident photon flux. And since the broad-line region is a dynamic place, the locations of most efficient C IV-line emission relative to that of Hβ may change in time and as the changing luminosity allows us to "probe" different part of the broad-line region. In other words, there may always be some scatter appearing in the R–L relationships, and thus always some intrinsic uncertainties in the mass estimates that will limit how well we can establish the black-hole mass, especially for individual objects. However, the hope is that as we learn more about the broad-line region, we are able to make appropriate adjustments to these relationships, so the accuracy of the black-hole mass estimates and of distant quasars will improve.

One additional factor that can affect mass estimates is the relative source inclination of the broad-line region to our line of sight. I discuss this separately in Section 6.1.

Unfortunately, the Vestergaard & Peterson (2006) calibrations were published before Collin et al. (2006) demonstrated that the Hβ FWHM line widths may, in fact, provide a biased estimate of the broad-line gas velocity dispersion, depending on the inclination of the active nucleus with respect to our line of sight. So, while the Vestergaard & Peterson (2006) scaling relations are anchored in the reverberation masses for which inclination effects are reduced, the relations based on FWHM do not directly take into account the possible inclination effects that can affect the measured FWHM values. This issue, for both Hβ- and C IV-scaling relations, will be addressed in a forthcoming paper.

3.3. *Mass-scaling relations for Mg II*

The existing relations involving the Mg II emission line presented by McLure & Jarvis (2002) can no longer be used because: (a) these mass estimates turn out to be inconsistent with those based on Hβ or C IV by up to a factor of 5 (Dietrich & Hamann 2004), and (b) the relationship has not been updated to the new reverberation-mass scale. I have instead established new relationships using a high signal-to-noise subset of SDSS quasars for which Mg II and one other broad emission line (Hβ or C IV) is also present. These new relations make no prior assumptions of where Mg II is emitted. The relationships and the details of this analysis will be presented in a forthcoming paper (Vestergaard & Osmer 2009).

3.4. *Relationship involving the Hβ luminosity*

For sources where the luminosity at 5100 Å is not representative of the nuclear luminosity due to contamination from the host-galaxy light or the high-energy tail of the powerful radio emission in radio-loud sources that cannot easily be corrected, the Hβ-line luminosity can be used to approximate the size of the Hβ-emitting region (e.g., Wu et al. 2004). This is valid because the Hβ-line flux is directly proportional to the ionizing luminosity, which is the determining factor for the size of the broad-line region. Notably, since the C IV-emission line is partially collisionally excited, this assumption is not valid for C IV. This is also emphasized by the Baldwin Effect (an inverse correlation between the C IV equivalent width and the continuum luminosity): the continuum and C IV-line luminosities do not scale with one another. However, there is also no real need for an alternate luminosity measure—since neither the host-galaxy light, the high-energy tail of the radio emission, or even the UV iron emission between Ly α and C IV make a significant contribution to the 1350 Å or 1450 Å continuum luminosity (e.g., Elvis et al. 1994). A recalibration of the relationship involving the Hβ-line width and Hβ-line luminosity is presented by Vestergaard & Peterson (2006).

This relationship is not applied to the quasar samples discussed in this contribution.

4. Black-hole masses in distant quasars

4.1. *Quasar samples*

In this Section I discuss the results of applying the mass-scaling relations to large samples of quasars: namely, the Bright Quasar Survey (BQS; Schmidt & Green 1983), the Large Bright Quasar Survey (LBQS; Hewett et al. 1995), the Sloan Digital Sky Survey (SDSS) Data Release 3 quasar catalog (Schneider et al. 2005), the color-selected sample of Fan et al. (2001) based on the SDSS Fall Equatorial Stripe, the $z \approx 2$ sample of radio-loud and radio-quiet quasars studied by Vestergaard et al. (2000) and Vestergaard (2000, 2003, 2004b), and several $z \approx 4$ samples from the literature. These samples and the data, with exception of the LBQS and the SDSS DR3 Quasar samples, are discussed in detail by Vestergaard (2004b). Details of the analysis of the LBQS and SDSS DR3 samples will be presented elsewhere (Vestergaard & Osmer 2009). However, it is appropriate to point out that the spectral measurements are made after modeling and subtraction of the Fe II and Fe III emission and, in the case of the DR3 sources below $z = 0.5$, the stellar light from the host galaxy; the LBQS quasars were selected to be strictly point sources, which limits the contribution from the host galaxy. The Mg II and Hβ lines, in particular, are embedded in broad bands of iron emission, which can be very strong in some sources. If the iron emission is not (well) fitted and subtracted, it will bias the line-width measurements. In particular, for Mg II one will systematically obtain smaller line widths because half of the broad Mg II-emission-line flux is blended with the iron emission (Vestergaard & Wilkes 2001). To minimize measurement errors due to noise, bad pixels, and narrow absorption, the line-width measurements were made on model fits to the broad emission line that are representative of the broad profiles.

The black-hole mass estimates are based on the updated calibrations of Vestergaard & Peterson (2006) for Hβ and C IV; this work also lists the most current mass determinations and estimates of the BQS sample. For Mg II, the new relations discussed in Section 3 were applied. I am using the scaling relations based on the FWHM line widths, because most of the data are not of sufficient quality to allow the line dispersion to be significantly measured. Since the line dispersion is an integral measure, it is less affected by (residual) narrow-line emission, but is much more affected by flux in the extreme wings of the line profiles, as well as the placement of the underlying continuum level. For these data that are not of very high signal-to-noise, contrary to reverberation-mapping data, the use of the FWHM allows us to probe a much larger subset of the quasar samples. As discussed in Section 6.1, corrections can be applied to the mass estimates based on FWHM measurements to reduce potential effects of source inclination. While this is the best approach, it has not been done in the current work (for the sake of consistency), because a correction only exists for the Hβ line and not yet for the C IV and Mg II lines.

4.2. *Does radio-loudness depend on black-hole mass?*

The sample selection of the $z \approx 2$ quasar sample makes it an ideal dataset for studying the properties of quasars with respect to radio-loudness, since each individual radio-loud quasar is matched in redshift and luminosity to a radio-quiet quasar in the sample. The distribution of masses in Figure 2 shows that when luminosity and redshift differences are eliminated, there is no difference in black-hole mass for quasars of different radio-loudness, as the BQS otherwise suggests (Laor 2000); that result is probably due to selection effects.

4.3. *Distributions with redshift*

The distributions of the black-hole mass estimates for the different quasar samples discussed in Section 4.1 are shown in Figure 2. The distributions at the low-mass end are strongly affected by the flux limits of the surveys by which the quasars were found. At low redshift, say $z < 0.5$, we can probe less luminous sources than at higher redshifts. Since the BQS is a "bright" quasar sample in the nearby universe, it has black-hole masses spanning 3 orders of magnitudes. It contains some quite luminous sources like 3C 273, which has a black-hole mass of a few billion solar masses. At redshifts higher than 1, we can typically only probe down to a few times 10^8 $M_\odot$. The lack of data points below this mass is therefore an artifact of our sample selection. We do expect the quasar population to extend much below the current flux limit cutoff.

It is characteristic that for the quasars we do detect at high redshift, say at $z \gtrsim 1$, their black-hole masses are typically very large—of order a billion solar masses or more. This holds even beyond a redshift of 3 to 3.5, at which the quasar space density is declining, and for the most distant quasars known at $z \gtrsim 6$. This shows that black holes must build up their mass very quickly, since, for example, at a redshift of 6.3 the universe was less than 900 Myrs old. The first star formation is thought to occur at redshifts between 6

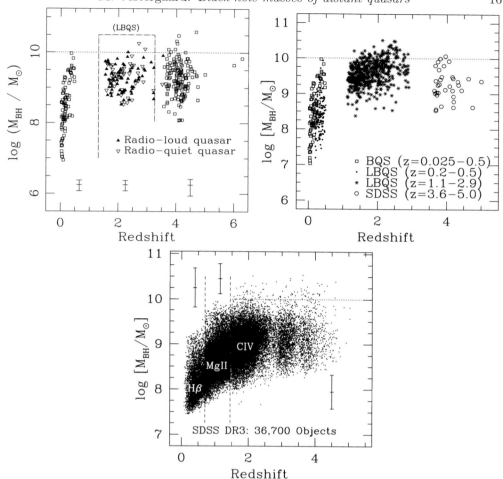

FIGURE 2. Distribution of black-hole mass estimates for different quasar samples. These mass estimates are based on the Vestergaard & Peterson (2006) single-epoch scaling relations and are computed in a flat cosmology with $H_0 = 70$ km s^{-1}/Mpc and $\Omega_\lambda = 0.7$. A mass of $10^{10} M_\odot$ is indicated by the dotted line. Upper left panel: The Bright Quasar Survey (BQS), the intermediate redshift sample studied by Vestergaard et al. (2000), Vestergaard (2003, 2004b), and various samples of $z \gtrsim 4$ samples from the literature studied by Vestergaard (2004b). The measurement uncertainty (not including the statistical accuracy of the mass estimates) is indicated at the bottom of the panel. Upper right panel: The BQS, The Large Bright Quasar Survey, and the Color-Selected Sample in the Fall Equatorial Stripe (Fan et al. 2001). Lower panel: The Sloan Digital Sky Survey Data Release 3 Quasar Catalog. Vertical dashed lines indicate the subsets where different emission lines (labeled) are used for the mass estimates (left to right: Hβ, Mg II, C IV). The error bars in each region indicate the typical uncertainty in the mass estimates and includes the statistical uncertainty in the mass zero-point of about a factor of 4. The decreased densities of objects at redshifts of about 2.8 and 3.5 is due to the increased difficulty in selecting quasars at these redshifts, as in color space the quasar tracks overlap with the stellar locus.

and 9, based on the presence of heavy elements in the spectra of quasars at $z \approx 5$ (Dietrich et al. 2003a,b) and $z \approx 6$ (e.g., Barth et al. 2003; Maiolino et al. 2005; Jiang et al. 2007; Dwek et al. 2007) and chemical evolution models (e.g., Friaça & Terlevich 1998; Matteucci & Recchi 2001). While the black holes may form earlier, this does show that given the short time available, supermassive black holes must evolve rapidly.

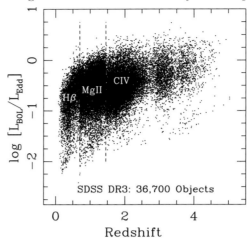

FIGURE 3. Distribution of Eddington luminosity ratios, $L_{\rm Bol}/L_{\rm Edd}$ for the SDSS DR3 quasar catalog.

It is noteworthy that there appears to be a maximum black-hole mass for quasars at 10 billion solar masses. This mass value is marked in the diagrams of Figure 2 for reference. There are very few black holes with masses above 10^{10} $M_\odot$, even for very large samples like the SDSS. While SDSS does have an upper flux limit to the survey, these limits do not affect the quasar selection at high redshift. The number of objects with mass estimates above 10^{10} $M_\odot$ are statistically consistent with the uncertainties in the black-hole mass estimates. Therefore it is not unreasonable to conclude that there appears to be a limit to how massive actively accreting black holes can become, and that this limit lies at or near 10^{10} $M_\odot$. At this point, it is unclear whether this limit is due to the dark matter halo in which the black hole resides, whether it has to do with details of the accretion physics, or whether the black hole is simply exhausting its accessible fuel supply. Of course, it could be due to a combination of all three possibilities.

The distributions indicate that there is a decline in black-hole mass from a redshift of about 1.5 to the present. The SDSS survey does have a bright flux limit that deselects the brightest (and hence the most luminous) objects in the nearby universe. That is the reason that at $z < 0.5$ no SDSS quasars have a black-hole mass above about 10^9 $M_\odot$ at $z \approx 0$, while the BQS has a number of objects above this mass. Nonetheless, even for the bright quasar samples of BQS and LBQS, one can see a (rapid) decline in black-hole mass from a redshift of 0.5 to the present. This does not mean, of course, that black holes with mass above 10^9 $M_\odot$ do not exist in the local universe; the giant elliptical galaxy M87 is known to have a black-hole mass of a few billion solar masses. It simply shows that the activity of the most massive black holes rapidly declines from $z = 0.5$ to the present. They simply "turn off" and become increasingly quiescent.

4.4. *The Eddington luminosity ratio distribution with redshift*

With both the black-hole mass and the bolometric luminosities readily available, we can infer the Eddington luminosity ratio, $L_{\rm Bol}/L_{\rm Edd}$, a crude estimate of the black-hole accretion rate. The distribution of Eddington ratios for the SDSS DR3 quasars is shown in Figure 3; as seen for the luminosity and mass distributions, the lower flux limits define the lower boundary to the distribution. As expected, most of the quasars appear to radiate between 10% and 100% of their Eddington luminosities. Even in the

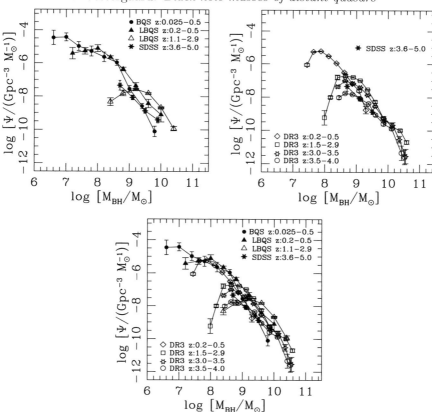

FIGURE 4. Mass functions of actively accreting black holes at different redshifts and for different quasar samples, as labeled. The "SDSS" sample at a redshift of 3.6 to 5.0 is the color-selected sample of Fan et al. (2001). The SDSS DR3 mass functions are preliminary. The downturn at the low-mass end is not real, but caused by uncorrected incompleteness in these bins. The bottom panel shows how the mass functions shown in the top panels compare to each other.

low redshift ($z < 0.5$) universe there are many highly accreting quasars. Although some objects appear to have super-Eddington accretion rates up to a value of 3 ($0.5 \, \mathrm{dex}$), this is not likely to be real. For one, the statistical uncertainties are entirely consistent with no significant population of sources with super-Eddington luminosities. In addition, the $L_{\mathrm{Bol}}/L_{\mathrm{Edd}}$ estimate is based on the assumption of Bondi accretion (i.e., spherical accretion), which is not very likely to hold for a black hole fed by an accretion disk. Moreover, the $L_{\mathrm{Bol}}/L_{\mathrm{Edd}}$ estimates are crude and have uncertainties at least of a factor of about 3.5 due to the uncertainties in the mass estimates.

The distribution of the $L_{\mathrm{Bol}}/L_{\mathrm{Edd}}$ values with redshift shows that at the highest red-shifts the quasars must be highly accreting to be detected. As a result, all known $z \gtrsim 4$ quasars are accreting at or near their Eddington limits.

5. Mass functions of actively accreting black holes

In the spirit of showing preliminary results from work in progress, I briefly present and discuss our first cut at the mass functions of actively accreting black holes of quasars at a range of redshifts. The mass functions for the different quasar samples are shown in

Figure 4. Please note that the downturn of the mass functions at the low-mass end is an artifact of incompleteness in these bins that is not yet corrected for.

The upper left panel shows the BQS, LBQS, and the color-selected sample of Fan et al. (2001). The BQS (filled circles) and low redshift subset of the LBQS (filled triangles) show consistent shape and normalization. The high-z LBQS subset (open triangles) lies slightly higher than its low-z subset (at its high-mass end) and lies even higher than the color-selected sample (filled stars) located at higher redshifts yet. However, this is entirely consistent with the observed space density of quasars, which peaks at a redshift of about 2 to 3 and declines above and below these redshifts (e.g., Peterson 1997). This diagram also seems to indicate that the comoving volume density of 1 to 10 billion solar-mass black holes at redshifts of 4 to 5 (the color-selected sample) approximately resembles that of the nearby bright quasars in the BQS (at $z < 0.5$).

The upper right panel shows the mass functions for the SDSS DR3 quasar catalog divided into various redshift bins as labeled. For reference, the color-selected sample from the left panel is also shown. Again, we see a decrease in the normalization of the mass function beyond a redshift of 3. In the bottom panel, the individual mass functions from the top panels are overplotted to show their relative normalizations. In particular, the low-z DR3 subset (open diamonds) is entirely consistent with the other two low-z subsets of the BQS and the LBQS. A more detailed discussion of the finalized mass functions are presented by Vestergaard et al. (2009, in preparation) and Vestergaard & Osmer (2009).

5.1. *Constraints on black-hole growth*

The ultimate goal is to use the black-hole mass estimates and the estimates of the mass functions of actively accreting black holes to constrain how black holes grow and evolve over cosmic time. This will be done by comparing these empirical results with theoretical models of how black holes can grow. While a more detailed analysis is yet to be done, it is interesting to do a simple comparison with one such theoretical model. Steed & Weinberg (2003) modeled how the distribution of black-hole masses change with redshift depending on the dominant growth mechanism. In Figure 5, I compare the distribution of quasar black-hole masses in SDSS DR3 with two such growth scenarios from Steed & Weinberg. The empirical data suggest that the black holes are more likely grow in a fashion in which the accretion rate depends on the black-hole mass rather than one in which the black holes obtain their mass in a short accretion phase, equal to the quasar lifetime. A more detailed discussion will be presented elsewhere.

5.2. *The space density of very massive black holes*

In Section 4 and Figure 2, it was shown that the typical black-hole mass at high redshift is a few billion solar masses and extends to 10 billion solar masses. Since the most-massive black hole (in the elliptical galaxy M87) measured in the local neighborhood is only about $3 \times 10^9 \ M_\odot$, and the measurement method does not exclude the determination of more-massive black holes, one may question whether the single-epoch mass estimates, discussed here, may indeed be overestimates. I address this issue further in Section 6.4 but it is instructive to also look at the high-redshift space density of black holes more massive than M87. Since the mass functions contain information on the comoving space density of the black holes as a function of their mass, I have integrated the current version of the quasar black-hole mass function above a mass of $3 \times 10^9 \ M_\odot$ for the subsets with quasars at and above a redshift of 4 (i.e., the mass functions shown by the open circles and filled stars in Figure 4). This measurement shows that such massive black holes are so rare that we need to probe the local neighborhood out to a distance of

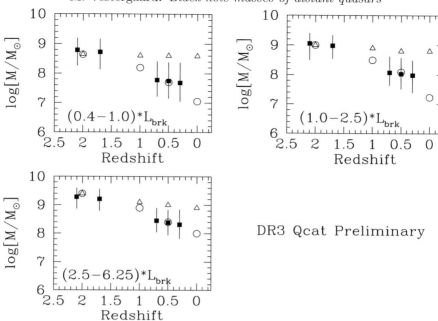

FIGURE 5. Distributions of masses of actively accreting black holes at different redshifts for the SDSS DR3 quasar catalog. The filled squares in each panel show the observed distribution with redshift for different luminosity ranges, as labeled. The filled squares show the mean value, while the vertical bars show the 10% to 90% ranges. The break luminosity, L_{brk}, is the characteristic luminosity at the knee of the quasar luminosity function. Theoretical model predictions are indicated by open symbols: Open triangles: Short accretion phase build-up; Open circles: mass-dependent accretion build-up.

$\sim$290 Mpc from earth in order to detect just *one* such massive black hole. This is a volume about 25 times larger than what has been studied at present ($R \approx 100$ Mpc; Ferrarese & Merritt 2000; Tremaine et al. 2002). Owing to the uncertainty in the mass estimates, the mass functions may intrinsically be steeper than our estimates in Figure 4 show. In that case, the space density of the most massive black holes will be even lower, and an even larger local volume is needed. Strictly speaking, based on the uncertainties in the mass estimates, the mass functions could be shallower and thus the frequency of very massive black holes could be higher. We may be able to test this in the future. Figure 1 of Ferrarese (2003) indicates that with a diffraction-limited 8 m telescope, in principle, it is technically feasible to detect $10^{10}\ M_\odot$ black holes out to distances of about 1500 Mpc. With a diffraction-limited 30 m we can possibly probe a volume of several thousand Mpc.

In conclusion, it seems the explanation for the lack of very massive black holes in the local neighborhood is simply that of their extreme rarity and the relatively small volume that we have probed so far. However, as briefly discussed in Section 6.4, there are now suggestions that the brightest cluster galaxies may have black holes with masses of order $10^{10}\ M_\odot$. It is, however, still to be clarified whether or not these black holes are the quiescent equivalent of the very massive black holes found in distant quasars.

6. Potential issues with scaling relationships

There are a few issues that may potentially present problems for the single-epoch mass estimates, including those based on scaling relationships, in the sense that the current estimates may in some sense be skewed or systematically biased. The issues most commonly mentioned in the literature include the effects of the relative inclination of the broad-line region to our line of sight, the potential danger of high-ionization outflows in quasars which may affect the C IV line emission, the unexpected narrow C IV profile for some active nuclei, and the question whether the single-epoch mass methods may overestimate the actual black-hole mass.

6.1. *Source inclination issues*

There are clear indications that the broad emission-line region may have the plane-like geometry because the measured gas velocities are larger for more inclined sources. This has been established for the Hβ line by Wills & Browne (1986) and for the base of the C IV line by Vestergaard et al. (2000) for radio-loud sources where the ratio of the radio-core flux to the radio-lobe flux yields a crude inclination measure: as the relative inclination of the radio jet increases, the core flux becomes less dominant relative to the radio lobe flux; notably, the direction of the jet is expected to be normal to the central engine's accretion disk. In both studies, the authors saw a lack of objects with small inclinations (i.e., near face-on) and large line widths. The largest line widths are only seen in the most inclined sources. This "Zone of Avoidance" distribution indicates that the gas velocities in a plane parallel or nearly parallel to that of the accretion disk are larger than perpendicular to this plane.

Since active nuclei are expected to be randomly oriented (for Type 1 sources: within the inclination ranges that defines this subset: $\lesssim 45°$; Barthel 1989), and the current mass determinations of active nuclei (both the reverberation-mapping method and mass-scaling relations) do not specifically account for source inclination, inclination effects can contribute to the uncertainties in the black-hole mass of individual active nuclei. For ensemble determinations, this effect is expected to average out owing to the way the mass zero-point was obtained (see Section 3.1).

Source inclination has been suggested to explain the scatter of the reverberation masses around the M_{BH}–σ_* relationship (e.g., Wu & Han 2001; Zhang & Wu 2002). Collin et al. (2006) investigated this issue and found that this only appears to work in a statistical sense and is not supported by individual sources for which the source inclination is known or well constrained (see Peterson, this volume). However, Collin et al. did establish that the FWHM measurement tends to be more sensitive to source inclination than the line dispersion. This is consistent with the finding of Peterson et al. (2004) that the line dispersion is the most robust line-width measure for mass determinations. Based on the reverberation-mapping data base, Collin et al. provide f-factor corrections to mass estimates based on the FWHM(Hβ) values if the line-dispersion measurement cannot be obtained. This allows mass estimates for which the effects of the relative source inclination is decreased.

However, this correction is currently only available for the Hβ emission line. Corrections applicable to the C IV and Mg II emission lines will be addressed in a future publication.

6.2. *High-ionization outflows*

There are general concerns that active nuclei typically contain high-ionization outflows that potentially affect the C IV line emission, and therefore the mass estimates based thereon. These concerns are based on several observations to suggest this. Leighly (2000) finds for the subset classified as narrow-line Seyfert 1 galaxies (NLS1s), that for increasing nuclear luminosity the C IV emission line displays an increasing blue asymmetry. Observations of NLS1s in the far-UV (Yuan et al. 2004) support the interpretation that these sources have high-ionization outflows. I Zw 1 is a typical example of a NLS1 with a broad blue asymmetric C IV profile. It has a FWHM(Hβ) $\approx$ 1200 km s^{-1}, while its C IV line is measured to have FWHM of $\sim$4400 km s^{-1} (Baskin & Laor 2005). Since NLS1s typically have Hβ line widths below 2000 km s^{-1}, C IV line widths several times that of the Hβ line are not very representative of the velocities of broad line gas dominated by black-hole gravity. This is the background for the discouragement of using C IV for mass estimates for NLS1s, and in general if the C IV profile has a triangular and blue asymmetric shape similar to that of I Zw 1 (Vestergaard 2004b).

Because NLS1s are luminous sources with highly accreting (albeit small) black holes, there is a concern that quasars that are also luminous and highly accreting may also display such high-ionization outflows that would make mass estimates based on C IV biased or even invalid (e.g., Bachev et al. 2004; Shemmer et al. 2004; Baskin & Laor 2005). Based on the $z > 1$ samples shown in the upper left panel of Figure 2 I previously established (Vestergaard 2004b) that none of the high-z quasar spectra, at least in those samples, have C IV profiles resembling that of I Zw 1. For the sources with the strongest asymmetries, I compared their luminosities and mass estimates with those of the remaining quasars and found no indication at all that the mass estimates were affected. The mass differences were at the 0.1 dex level and some subsets with asymmetries even showed a lower mass. Assuming that the quasar samples of that study are typical of the quasar population at those redshifts, it is fair to conclude that high-z quasars with C IV line profiles resembling that of I Zw 1 must be rare.

However, it is still possible that the C IV emission line is associated with outflowing gas. It is well known that the C IV emission line is blueshifted relative to its restframe wavelength of 1549 Å (e.g., Wilkes 1984; Espey et al. 1989; Tytler & Fan 1992). Based on about 3800 SDSS quasars, Richards et al. (2002) established that as the blueshift increases (up to a maximum of about 2000 km s^{-1}), the line-equivalent width decreases, the profile shape changes, and the continuum luminosity increases. When comparing the line profiles of subsets of quasars binned by C IV blueshift, the authors concluded that the observed blueshift is generated by an increasing deficit of the red side of the profile. Their Figure 4 does suggest a significant broadening of the C IV profile with increasing blueshift. Nonetheless, to determine the effect on the single-epoch mass estimates, a comparison of the FWHM of the profiles with the most and the least blueshift is needed. Based on the entries in Table 2 of Richards et al. one can infer that, intriguingly, there is only a 15% difference in the FWHM measurements, which translates into a mere 30% effect in the mass estimates. Regardless of the cause of the blueshift, since most quasars display a smaller blueshift than 2000 km s^{-1} and the current mass estimates based on C IV scaling relations have uncertainties of order of a factor of 4, the effects of the C IV blueshift is insignificant at the present time. However, as the mass estimates improve sufficiently this effect needs to be accounted for.

6.3. *The unexpected relative line widths of C IV and Hβ*

Motivated by the blueshifts and potential blue profile asymmetries of C IV, Baskin & Laor (2005) investigated the C IV- and Hβ-profile differences using non-contemporaneous optical- and UV-literature data of the BQS quasar sample. They find, in particular, that for objects with FWHM(Hβ) > 4000 km s^{-1}, the C IV profile is not broader than Hβ. Existing monitoring data and the fact that C IV is a high-ionization line indicate that C IV is emitted from a region closer to the black hole than Hβ. The Baskin & Laor study shows that FWHM(C IV) is not always larger than FWHM(Hβ), and in fact the former rarely comply with being a factor $\sqrt{2}$ broader, as expected based on the virial relationship, if Hβ is emitted from a distance twice that of C IV. As it turns out, this issue is much less severe when problem data such as low quality *IUE* data, strongly absorbed profiles, and

NLS1s are eliminated from the database (see Vestergaard & Peterson 2006 for details). However, it does remain that FWHM(C IV) does not increase as fast as FWHM(Hβ) and an inverse correlation between FWHM(C IV)/FWHM(Hβ) and FWHM(Hβ) is seen extending to ± 0.2 dex in the FWHM ratio. Part of this difference can be attributed to measurement uncertainties.

How well established is the factor 2 difference in the C IV and Hβ lags? Of the few sources with reliable lags for both lines, three sources (NGC 3783, NGC 5548, and NGC 7469) display C IV lags that are about a factor 2 shorter than those of Hβ for the same epochs (Peterson et al. 2004). For NGC 4151 there are no contemporaneous epochs for the two lines, but the data are also consistent with this result (Bentz et al. 2006b; Metzroth et al. 2006). But for two sources the measured C IV and Hβ lags argue otherwise. Existing data for Fairall 9 display a C IV lag that is 1.5 times larger than that of Hβ and has large errors (Peterson et al. 2004). Photoionization theory expects the C IV and Lyα line gas to be emitted from similar distances, so judging from the Lyα lag, a more reliable C IV lag may be at 2/3 the lag of Hβ. For 3C 390.3 the C IV line is narrower than Hβ and the C IV lag is almost twice that of Hβ but within the large errors, the two lines could be emitted from similar distances (Peterson et al. 2004). So judging by the most recent analysis of the reverberation data base of Peterson et al. (2004) there are good reasons to expect Hβ to be emitted from a distance twice that of C IV. But the data on Fairall 9 and 3C 390.3 suggest that perhaps not all sources have broad-line regions that are as neatly organized. As argued in Section 3.2, the broad-line region is a dynamic place and a neat onion-skin–like emission-line gas distribution is, in fact, not expected. This is supported by photoionization theory. In their Figure 3 Korista et al. (1997) present the strength of different emission lines as a function of gas density and incident ionizing luminosity. A comparison of the diagrams for C IV and Hβ shows a significant overlap of parameter space for which sufficiently strong Hβ and C IV can be emitted. Hence, their line regions are not necessarily mutually exclusive, and they can indeed be located closer than the factor 2 difference in distance. In fact, the non-contemporaneous nature of most UV and optical data being compared and the dynamic nature of the broad-line region, which also likely has a clumpy gas distribution, is likely part of the explanation for the C IV- and Hβ-line widths not always scaling perfectly with one another.

After all, the most compelling argument in favor of C IV-based mass estimates is that reverberation results show the virial products (RV^2) are consistent for *all* the measured emission lines, which includes C IV (Peterson & Wandel 1999, 2000; Onken & Peterson 2002).

6.4. *Are the masses overestimated?*

When the scaling relationships are applied to samples of high-redshift quasars, the estimated typical black-hole mass is of order a billion solar masses and extends to a few tens of billions of solar masses (Section 4). The validity of these large black-hole masses have been questioned (e.g., Netzer 2003) for a couple of reasons. Firstly, black-hole masses of 10 billion solar masses or more are not found in the local universe among quiescent black holes assumed to have been active in the past. The most massive black hole detected in the local universe is that in the giant elliptical galaxy, M87, with a mass of 3×10^9 $M_\odot$ (Harms et al. 1994; Macchetto et al. 1997). Second, from the well-established M_{BH}–σ_* relationship one can infer that a 10 billion solar-mass black hole will reside in a galaxy with a bulge that has stellar velocity dispersions above 400 km s^{-1} (e.g., Tremaine et al. 2002). Such massive galaxies are not observed in the local universe. It is therefore fair to question whether the single-epoch mass estimates from the scaling relations systematically over-predicts the mass of the black hole by about an order of magnitude. This would bring the highest quasar black-hole mass estimates in full consistency with the black-hole masses in quiescent galaxies measured so far in the local universe. The methods to determine their black-hole masses are by many considered more reliable since they rely on stellar velocities that are less prone to non-gravitational forces, such as radiation pressure, that is expected to be present and strong in the presumably violent environment near the accreting black hole.

There are different considerations supporting that mass-scaling relationships do not systematically overestimate the black-hole mass. A full discussion can be found elsewhere (Vestergaard 2004b). Here I will summarize the main points. First, it is argued in Section 2.3 that the radius–luminosity relationship is fully applicable to high-z quasars. Hence, there is no indication that the use of this relationship should bias us in any way. Second, there is no indication that the kinematics of the broad-line region of high-redshift quasars are dominated by forces other than black hole gravity. The argument in favor is a combination of the fact that this is certainly found to hold for nearby active nuclei and quasars (Peterson & Wandel 1999, 2000; Onken & Peterson 2002; Kollatschny 2003) and quasar spectra look very similar at all redshifts (e.g., Dietrich et al. 2002). In addition, as noted in Section 2.3, the monitoring data of the $z \approx 3$ quasar for which Kaspi et al. (2007) was able to measure a C IV emission-line lag also support the notion that high-z quasar broad-line regions are very similar to those at lower redshift. Third, it is argued in Section 6.2 that high-ionization outflows are not important for a very large fraction of quasars. Fourth, one can demonstrate that it is unlikely that the mass estimates are systematically too large using results from photoionization modeling (e.g., Korista et al. 1997). For the sake of this argument, assume an estimated black-hole mass of $10^9 \, M_\odot$ that is a factor 10 overestimated. The virial theorem shows that a typical quasar with FWHM(C IV) of 4500 km s^{-1} will emit the Ly α- and C IV-lines from a distance of only $\sim$33 light-days from the central source. At this location for a typical quasar luminosity of $L_{\rm Bol}$ of order 10^{47} erg s^{-1}, the ratio of photons-to-gas particles is so high that not only are these lines emitted inefficiently, but especially the C III]-line emission cannot be generated. The prominence of the C III] in quasar spectra argue for the presence of low-density gas subjected to lower ionizing flux. The key is that both the Ly α and the C IV lines are much more efficiently emitted from these regions, that necessarily are located much further from the central source in accordance with the predicted distance from the R–L relationship.

In conclusion, there are no obvious indications based on our existing knowledge of the broad-line region that the mass estimates based on scaling relations are systematically overestimated. Even the crudeness of using the FWHM of a single-epoch emission line that contains a contribution from non-varying emission line gas does not significantly or systematically affect the mass estimates (Vestergaard 2004b).

In closing, it is interesting to note that recent findings show that a natural "saturation" of the bulge stellar-velocity dispersion occurs above $\sigma_* \approx 400$ km s^{-1} (Lauer et al. 2007). For the brightest cluster galaxies and other high σ_* galaxies, σ_* is thus not an accurate indicator of the central black-hole mass. On the other hand, the galaxy luminosity indicates black-hole masses of order $10^{10} \, M_\odot$. The apparent contradiction of the large quasar black-hole masses is therefore no longer that evident. Moreover, as I showed in Section 5.2, $10^{10} \, M_\odot$ active black holes are so rare that we should not expect to see any in the local volume that has been probed so far.

7. Summary and conclusions

It is argued that mass-scaling relations are the preferred method to estimate black-hole masses of distant active galaxies and quasars owing to the ease with which the method can be applied to large samples, its higher accuracy relative to alternative comparable methods, and the reliability of the mass estimates. Furthermore, the accuracy can potentially be improved. This is important because we seek to use these mass estimates to understand how black holes grow and affect galaxies and their evolution—those that host the active black holes and those in their neighborhood. At present, the statistical 1σ uncertainty in the absolute-mass values amounts to a factor of 3.5 to 5, depending on the emission line and line-width measure applied; the C IV relations have lower uncertainties.

Distributions of black-hole masses of distant quasars at a range of epochs is presented. The estimated masses are very large, a billion $M_\odot$ or more, even at redshifts of 4 to 6; the black holes can clearly grow and mature very quickly in the early universe. However, a maximum mass of 10^{10} $M_\odot$ is also observed, which shows that black holes reach their ultimate growth limit then and thus must shut off their activity. Preliminary black-hole mass functions of various large quasar samples is also presented and discussed; a full analysis will be presented elsewhere (Vestergaard et al. 2008; Vestergaard & Osmer 2009; Vestergaard et al. 2009, in preparation). Furthermore, I outlined and briefly discussed some potential issues associated with applying mass-scaling relations to luminous high-redshift quasars and concluded that none of these jeopardize this method or significantly affects the mass estimates. Also, there are no indications that the mass estimates are systematically too high.

I am grateful for financial support of this work by NASA through grants HST-AR-10691 and HST-GO-10417 from the Space Telescope Science Institute, and by NSF grant AST-0307384 to the University of Arizona. I thank Xiaohui Fan and Bradley Peterson for discussions related to this work.

REFERENCES

BACHEV, R., ET AL. 2004 *ApJ* **617**, 171.
BALDWIN, J., ET AL. 1995 *ApJ* **455**, L119.
BARTH, A., ET AL. 2003 *ApJ* **594**, L95.
BARTH, A., ET AL. 2005 *ApJ* **619**, L151.
BARTHEL, P. D. 1989 *ApJ* **336**, 606.
BASKIN, A. & LAOR, A. 2005 *MNRAS* **356**, 1029.
BENTZ, M., ET AL. 2006a *ApJ* **644**, 133.
BENTZ, M., ET AL. 2006b *ApJ* **651**, 775.
BOROSON, T. A. 2003 *ApJ* **585**, 647.
COLLIN, S., ET AL. 2006 *A&A* **456**, 75.
DIETRICH, M., ET AL. 2002 *ApJ* **581**, 912.
DIETRICH, M., ET AL. 2003a *ApJ* **589**, 722.
DIETRICH, M., ET AL. 2003b *ApJ* **596**, 817.
DIETRICH, M. & HAMANN, F. 2004 *ApJ* **611**, 761.
DJORGOVSKI, S. & DAVIS, M. 1987 *ApJ* **313**, 59.
DRESSLER, A., ET AL. 1987 *ApJ* **313**, 42.
DWEK, E., ET AL. 2007 *ApJ* **662**, 927.
ELVIS, M., ET AL. 1994 *ApJS* **95**, 1.
ESPEY, B., ET AL. 1989 *ApJ* **342**, 666.
FALOMO, R., ET AL. 2004. In *Coevolution of Black Holes and Galaxies* (ed. L. C. Ho). Carnegie Observatories Astrophysics Series, Vol. 1. Carnegie Observatories; http://www.ociw.edu/ociw/symposia/series/symposium1/proceedings.html/.

FAN, X., ET AL. 2001 *AJ* **121**, 31.

FERRARESE, L. 2003. In *Hubble's Science Legacy: Future Optical-Ultraviolet Astronomy from Space* (eds. K. R. Sembach, J. C. Blades, G. D. Illingworth, & R. C. Kennicutt, Jr.). ASP Conference Series, Vol. 291, p. 196. ASP.

FERRARESE, L. & MERRITT, D. 2000 *ApJ* **539**, L9.

FERRARESE, L., ET AL. 2001 *ApJ* **555**, L79.

FRIAÇA, A. & TERLEVICH, R. 1998 *MNRAS* **298**, 399.

GEBHARDT, K., ET AL. 2000a *ApJ* **539**, L13.

GEBHARDT, K., ET AL. 2000b *ApJ* **543**, L5.

HARMS, R. J., ET AL. 1994 *ApJ* **435**, L35.

HEWETT, P., ET AL. 1995 *AJ* **109**, 1498.

JIANG, L., ET AL. 2007 *ApJ* **134**, 1150.

KASPI, S. 2001. In *Probing the Physics of Active Galactic Nuclei* (eds. B. M. Peterson, R. W. Pogge, & R. S. Polidan). ASP Conference Proceedings, Vol. 224, p. 347. ASP.

KASPI, S., ET AL. 2000 *ApJ* **533**, 631.

KASPI, S., ET AL. 2005 *ApJ* **629**, 61.

KASPI, S., ET AL. 2007 *ApJ* **659**, 997.

KOLLATSCHNY, W. 2003 *A&A* **407**, 461.

KORISTA, K., ET AL. 1997 *ApJS* **108**, 401.

KUKULA, M., ET AL. 2001 *MNRAS* **326**, 1533.

LAOR, A. 2000 *ApJ* **543** , L111.

LAUER, T., ET AL. 2007 *ApJ* **662**, 808.

LEIGHLY, K. 2000 *NewAR* **44**, 395.

MACCHETTO, D., ET AL. 1997 *ApJ* **489**, 579.

MAGORRIAN, J., ET AL. 1998 *AJ* **115**, 2285.

MAIOLINO, R., ET AL. 2005 *A&A* **440**, 51.

MATTEUCCI, F. & RECCHI, S. 2001 *ApJ* **558**, 351.

MCLURE, R. & DUNLOP, J. 2002 *MNRAS* **331**, 795.

MCLURE, R. & JARVIS, M. 2002 *MNRAS* **337**, 109.

METZROTH, K. G., ONKEN, C. A., & PETERSON, B. M. 2006 *ApJ* **647**, 901.

NELSON C. 2000 *ApJ* **544**, L91.

NELSON, C. & WHITTLE, M. 1996 *ApJ* **465**, 96.

NETZER, H. 2003 *ApJ* **583**, L5.

ONKEN, C. A. & PETERSON, B. M. 2002 *ApJ* **572**, 746.

ONKEN, C. A., ET AL. 2004 *ApJ* **615**, 645.

PETERSON, B. M. 1993 *PASP* **105**, 247.

PETERSON, B. M. 1997 *An Introduction to Active Galactic Nuclei*. Cambridge University Press.

PETERSON, B. M. & WANDEL, A. 1999 *ApJ* **521**, L95.

PETERSON, B. M. & WANDEL, A. 2000 *ApJ* **540**, L13.

PETERSON, B. M., ET AL. 2004 *ApJ* **613**, 682.

PETERSON, B. M., ET AL. 2005 *ApJ* **633**, 799.

RICHARDS, G. T., ET AL. 2002 *AJ* **124**, 1.

RICHARDS, G. T., ET AL. 2006 *ApJS* **166**, 470.

SCHMIDT, M. & GREEN, R. F. 1983 *ApJ* **269**, 352.

SCHNEIDER, D., ET AL. 2005 *AJ* **130**, 367.

SHEMMER, O., ET AL. 2004 *ApJ* **614**, 547.

STEED, A. & WEINBERG, D. H. 2003; astro-ph/0311312.

TREMAINE, S., ET AL. 2002 *ApJ* **574**, 740.

TYTLER, D. & FAN, X. 1992 *ApJS* **79**, 1

VERON, M.-P., ET AL. 2004 *A&A* **417**, 515.

VESTERGAARD, M. 2000, *PASP* **112**, 1504.

VESTERGAARD, M. 2002 *ApJ* **571**, 733.

VESTERGAARD, M. 2003 *ApJ* **599**, 116.

VESTERGAARD, M. 2004a. In *AGN Physics with the Sloan Digital Sky Survey* (eds. G. T. Richards & P. B. Hall). ASP Conference Series, Vol. 311, p. 69. ASP.

VESTERGAARD, M. 2004b *ApJ* **601**, 676.

VESTERGAARD, M., FAN, X., TREMONTI, C. A., OSMER, P. S., & RICHARDS, G. T. 2008 *ApJ* **674**, L1.

VESTERGAARD, M. & OSMER, P. S. 2009 *ApJ* **699**, 800.

VESTERGAARD, M. & PETERSON, B. M. 2006 *ApJ* **641**, 689.

VESTERGAARD, M. & WILKES, B. J. 2001 *ApJS* **34**, 1.

VESTERGAARD, M., WILKES, B. J., & BARTHEL, P. D. 2000 *ApJ* **538**, L103.

WANDEL, A. 2002 *ApJ* **565**, 762.

WANDEL, A., PETERSON, B. M., MALKAN, M. 1999 *ApJ* **526**, 579.

WARNER, C., ET AL. 2003 *ApJ* **596**, 72.

WILKES, B. J. 1984 *MNRAS* **207**, 73.

WILLS, B. & BROWNE, I. 1986 *ApJ* **302**, 56.

WOO, J.-H. & URRY, C. M. 2002 *ApJ* **579**, 530.

WU, X. & HAN, J. L. 2001 *ApJ* **561**, L59.

WU, X.-B., ET AL. 2004 *A&A* **424**, 793.

YUAN, Q., BROTHERTON, M., GREEN, R. F., & KRISS, G. A. 2004. In *Recycling Intergalactic and Interstellar Matter* (eds. P.-A. Duc, J. Braine, & E. Brinks). IAU Symposium Series, Vol. 217, p. 364. ASP.

ZHANG, T.-Z. & WU, X.-B. 2002 *Chin. J. Astron. Astrophys.* **2**, 487.

The accretion history of super-massive black holes

By K A T E   B R A N D[1]
AND **THE NDWFS BOÖTES SURVEY TEAMS**

[1]Space Telescope Science Institute, 3700 San Martin Drive, Baltimore, MD 21218

How did the mass of 10^8–10^{10} $M_\odot$ super-massive black holes at the center of massive galaxies in the local Universe build up? Did the bulk of the growth happen in an optically luminous AGN phase? Or did a substantial fraction of SMBH growth occur in a dusty, obscured phase, visible as a luminous infrared galaxy? Has there been substantial SMBH growth in a low luminosity or radiatively inefficient regime after the more luminous AGN phase? These are particularly important questions, given the tight relationship between the mass of galaxy bulges and their SMBHs, suggesting that the formation and evolution of galaxies and SMBHs are intimately linked. We use the multi-wavelength data in the NDWFS Boötes field to address this issue. We have performed an x-ray stacking analysis of ~20,000 red galaxies at $z = 0.2$–1 to show that the average nuclear accretion rates in these sources are low and decreasing with time. Given the long timescale, significant SMBH mass growth could occur in this regime. We also investigate the nature of an extreme, obscured population of AGN-dominated luminous infrared galaxies which are likely to host SMBHs undergoing a period of rapid and substantial growth.

1. Introduction

In the present day Universe, most (if not all) massive galaxies contain super-massive black holes (SMBHs). How did the mass of these SMBHs grow as a function of time? The correlation between the mass of SMBHs and the galaxy bulge in which they reside (Magorrian et al. 1998; Gebhardt et al. 2000; Ferrarese & Merritt 2000) suggests that the processes which govern the build up of galaxies and SMBHs are related. Thus, determining how SMBHs grow via accretion may also help us understand galaxy evolution.

Assuming an accretion efficiency in transforming mass to light, we can trace the build-up of the mass of SMBHs via their AGN activity. We can then compare the total inferred mass to the mass density of SMBHs in the local Universe (known as Soltan's argument; Soltan 1982). This should show whether the AGN identified in existing surveys are responsible for the major build-up of SMBH mass via accretion. Current estimates of the accreted SBMH mass via hard x-ray emission ($\rho_{\rm BH,acc} \approx 3 \pm 1 \times 10^5$ $M_\odot$ Mpc^{-3}; Barger et al. 2005) broadly agree with estimates of the mass density of SMBHs in the local Universe ($\rho_{\rm BH} \approx 4.6 \pm 1.9 \times 10^5$ $M_\odot$ Mpc^{-3}; Marconi et al. 2004). However, the many assumptions needed to determine these numbers mean that there is still room for substantial growth in other regimes which are challenging to identify in optical and x-ray surveys.

Are there important populations of AGN which have been largely missed by previous optical and x-ray surveys, but in which SMBHs could have grown a substantial fraction of their mass? There has been recent evidence from *Chandra* and *Spitzer* that both hidden AGN and star-formation activity may occur in red galaxies out to $z \sim 1$ (e.g., Rodighiero et al. 2007; Davoodi et al. 2006). With the advent of *Spitzer*, it is becoming clear that there may be a large population of powerful but obscured AGN at high redshift (e.g., Martínez-Sansigre et al. 2005). In these proceedings, we present work which considers the possibilities that SMBHs could have grown substantially in a low luminosity and/or radiatively inefficient regime at $z < 1$, and in an intrinsically luminous but dusty phase at $z \approx 2\text{--}3$.

2. The data

We use the multi-wavelength data from the NOAO Deep Wide-Field Survey (NDWFS; Jannuzi & Dey 1999) Boötes field. This field covers a contiguous 9.3 square degree area and has a plethora of existing multi-wavelength imaging and spectroscopy. As part of the NDWFS Boötes survey, the field has been mapped in B_W, R, and I bands to median 3σ point source depths of ≈ 27.7, 26.7, and 26.0 (Vega) respectively, and to $K = 19.6$ (Vega) in the near-infrared. The *Chandra X-ray Observatory* has imaged the entire field in the 0.5–7 keV energy range with the Advanced CCD Imaging Spectrometer (ACIS; Murray et al. 2005). The x-ray catalog is presented in Kenter et al. (2005) and Brand et al. (2006). *Spitzer* has imaged the field at 24, 70, and 160 μm using the Multiband Imaging Photometer for *Spitzer* (MIPS; Rieke et al. 2004) to 5σ rms depths of 0.4 mJy, 25 mJy, and 150 mJy respectively, yielding a catalog of $\approx 22{,}000$ sources. *Spitzer* has also imaged the field with the Infrared Array Camera (IRAC) to 5σ depths of 6.4, 8.8, 51, and 50 microJy at 3.6, 4.5, 5.8, and 8μm respectively (Eisenhardt et al. 2004).

3. The growth of SMBHs in a low luminosity regime at $z < 1$

In this section, we use an x-ray stacking technique to measure the nuclear accretion history in red galaxies at $z < 1$. We use this to estimate the possible growth of SMBHs in a low luminosity regime.

3.1. *Selection of a population of red galaxies*

Our procedure for selecting a red galaxy sample is very similar to that of Brown et al. (2007). We consider B_W, R, and I imaging over almost the entire 9.3 deg^2 of the NDWFS Boötes field. We determined photometric redshifts for all galaxies using the ANNz artificial neural networks code of Collister & Lahav (2004), calibrated with a training set of 20,000 $I < 20$ spectroscopic redshifts from the AGN and Galaxy Evolution Survey (AGES; Kochanek et al., in prep.) and several hundred $R < 24.5$ Keck spectroscopic redshifts.

We select all galaxies with $0.2 < z < 1.0$ and with an evolving optical absolute magnitude cut;

$$M_{B_W} > -15.5 - 5 \log h - 0.87z \quad . \tag{3.1}$$

We then use a rest-frame selection criterion that selects red galaxies along the color-magnitude relation. We use a slightly more conservative cut than that of Brown et al. (2007), to minimize contamination from dusty star-forming galaxies:

$$U - V > 1.55 - 0.25 - 0.04(M_V - 5 \log h + 20.0) - 0.42(z - 0.05) + 0.07(z - 0.05)^2 \quad . \tag{3.2}$$

Figure 1 shows the galaxy sample divided into red, intermediate ("green valley"), and blue galaxies. There are $\approx 26{,}000$ red galaxies. In Figure 2, we plot the photometric and spectroscopic redshifts of a sub-sample of the red galaxies. The 1σ uncertainties are ≈ 0.03 and 0.1 in redshift at $I{=}19.5$ and 22 respectively (Brown et al. 2007).

3.2. *Optical line diagnostics*

We investigate the optical line diagnostics of red galaxies in our sample to determine whether their optical line emission is dominated by obscured AGN or star-formation activity. Of the $\approx 26{,}000$ red galaxies in our sample, 3500 have optical spectroscopy from

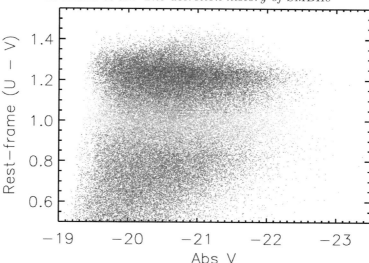

FIGURE 1. Color-magnitude diagram for NDWFS galaxies with $0.2 < z < 1.0$. Our red galaxies are shown as dark gray dots with large $U - V$ colors. The blue star-forming and intermediate "green valley" galaxies are shown as medium and light gray dots, respectively.

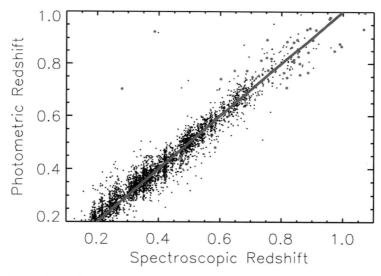

FIGURE 2. Comparison of spectroscopic and photometric redshifts for the red galaxy sample. Spectroscopic redshifts are from AGES (black) and Keck optical spectroscopy (gray).

the AGN and Galaxy Evolution Survey (AGES; Kochanek et al., in prep.). In Figure 3, we show optical line diagnostic diagrams (e.g., Baldwin et al. 1981) for all of the red galaxies with the relevant optical emission lines. Figure 3a shows that red galaxies tend to have [O III]/Hβ versus [N II]/Hα line ratios characteristic of AGN-dominated sources or 'transition' sources with both AGN and star-formation activity (e.g., Kauffmann et al. 2003; Kewley et al. 2006). The [O III]/Hβ versus [S II]/Hα diagnostic diagram (Figure 3b) shows a broader range of properties, but again, has a larger fraction of AGN-dominated (Seyfert or LINER) sources than that of blue galaxies. Yan et al. (2006) show that the ratio of the [O II]–Hα equivalent widths provides a complimentary classification tool to

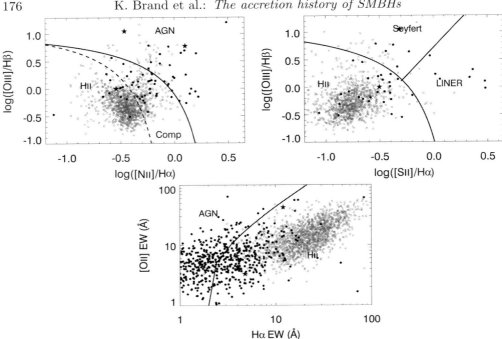

FIGURE 3. Optical line diagnostic diagrams for NDWFS galaxies. Empty gray circles denote blue galaxies and filled black circles denote red sequence galaxies. Black stars denote x-ray detected red galaxies in the XBoötes survey. Top left: the log([O III]/Hβ) vs. log([N II/Hα) line diagnostic diagram. We overplot the Kewley et al. (2006) extreme starburst classification line (solid line) and the empirically derived line of Kauffmann et al. (2003) to distinguish between star-forming galaxies and AGN. Top right: the log([O III]/Hβ) vs. log([S II/Hα) line diagnostic diagram. The solid lines denote the Kewley et al. (2006) classification lines. Bottom: the [O II] vs. Hα equivalent widths. The solid line denotes the classification line of Yan et al. (2006). Most diagnostics show that red galaxies with narrow optical emission lines are primarily AGN-dominated or a mixture of AGN and star-forming galaxies.

standard line ratio diagnostics. This diagnostic is useful because it only requires measurements of the typically strong [O II] and Hα lines. Figure 3c shows that the red and blue galaxies separate into two regions of the plot: red galaxies with high [O II]/Hα and blue galaxies with low [O II]/Hα ratios. All three of these diagnostics show that red galaxies tend to have line ratios more characteristic of AGN-dominated sources or composite sources with both AGN and star-formation activity. This is consistent with the results of Yan et al. (2006), who find that 94% of red galaxies have line ratios resembling AGN.

3.3. *X-ray stacking*

Brand et al. (2005) performed an x-ray stacking technique on the red galaxies in a 1.4 deg^2 region of the NDWFS Boötes field. Here, we use the same techniques, but for a larger sample of red galaxies over almost the entire 9.3 deg^2 of the Boötes field, to determine their mean x-ray luminosities down to very faint levels. By stacking the x-ray observations of galaxies, one can increase the effective observation depth on the mean object by a factor of the sample size: a sample of 10^4 galaxies effectively increases the depth of our 5 ks *Chandra* exposure to that of a 50 Msec observation on the mean object.

The total x-ray emission from galaxies may contain contributions from a variety of sources: stellar objects, such as low-mass and high-mass x-ray binaries (hereafter LMXBs

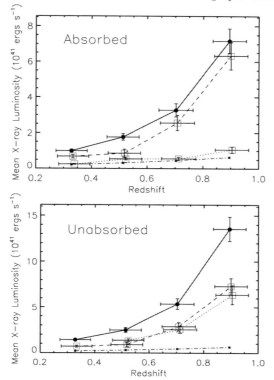

FIGURE 4. Redshift evolution of the mean x-ray luminosity of red galaxies. The x-ray luminosity was calculated assuming a power-law spectrum with a photon index, $\Gamma = 1.7$. The hydrogen column density was then calculated from the mean x-ray hardness ratio. We show both the mean absorbed (top) and unabsorbed intrinsic (bottom) x-ray luminosities. The total, hard, and soft luminosities are represented by the solid, dashed, and dotted lines respectively. The expected contribution from low and high-mass x-ray binaries is show by the dash-dotted line.

and HMXBs); diffuse hot gas; and AGN. The LMXB population is long-lived and traces the old stellar population (and hence the stellar mass) in galaxies; in contrast, HMXBs have short lifetimes and the luminosity of this population reflects the star-formation rate in galaxies. We use the relations of Kim et al. (2004) and Grimm, Gilfanov, & Sunyaev (2005) to determine the expected x-ray contribution of LMXBs and HBXBs to the total x-ray luminosity for each red galaxy. The average combined x-ray luminosity from both LMXBs and HMXBs is expected to be $\approx 2 \times 10^{40}$ ergs s^{-1}. The stacked x-ray luminosity is at least a factor of ≈ 5–10 higher than that expected from stellar sources alone, suggesting that the AGN dominates the x-ray emission in these sources. This is consistent with their optical line diagnostics. We note that the mean x-ray luminosity is a factor of ~ 10–100 times smaller than the x-ray luminosities of typical Seyfert galaxies. This implies that although the x-ray luminosity is dominated by AGN emission, the mean accretion rate and/or accretion efficiency must be very low.

The large mean hardness ratio of our sample of red galaxies suggests that the x-ray spectrum is hard in comparison to typical unobscured AGN and is best fit by either an absorbed ($N_H \sim 1$–5×10^{22} cm^{-2}) $\Gamma = 1.7$ power-law or an unabsorbed $\Gamma = 0.7$ power-law consistent with models in which the fueling of the SMBH occurs at low accretion efficiencies (e.g., ADAF models; Rees et al. 1982; Di Matteo et al. 2000). If we assume the former model, the hardness ratios imply that the hydrogen column density increases from

$N_H \sim 1 \times 10^{22}$ cm^{-2} at $z = 0.2$ to $N_H \sim 5 \times 10^{22}$ cm^{-2} at $z = 1.0$. In Figure 4, we show both the absorbed and unabsorbed (i.e., corrected for absorption) x-ray luminosities as a function of redshift. The x-ray luminosity increases significantly as a function of redshift. The absorbed x-ray luminosity increases with redshift as, $L_{x,abs} \propto (1+z)^{5.7\pm0.9}$. The unabsorbed x-ray luminosity shows a sharper increase with redshift, $L_{x,unabs} \propto (1+z)^{6.6\pm1.0}$, because the sources are more heavily obscured at higher redshifts.

These results suggest a global decline in the mean AGN activity of normal early-type galaxies from $z \sim 1$ to the present, which indicates that since the quasar epoch we are witnessing the sharp tailing off of the accretion activity onto super-massive black holes in early type galaxies.

3.4. *SMBH growth*

If we assume that we know the efficiency at which energy is converted to radiation and that the x-ray luminosity dominates the bolometric luminosity of the AGN, we can use the scaling of Barger et al. (2001) to determine the black hole accretion rate for a given x-ray luminosity. We can then integrate the mean accretion rate as a function of look-back time to determine the typical build-up of SMBH masses within red galaxies. We assume a typical accretion efficiency of $\epsilon = 0.1$ and bolometric correction between the hard x-ray and bolometric luminosities of 35 (e.g., Barger et al. 2005). We infer that the total growth expected for a SMBH within a red galaxy between $z \sim 1$ and the present is $\approx 8 \times 10^7\ M_\odot$ (and $\approx 5 \times 10^7\ M_\odot$ when we only consider the sources that are not individually detected in the x-ray). Vasudevan et al. (2009) suggest that AGN with Eddington ratios below ~ 0.1 have mean bolometric corrections of ~ 20. If we use these values, the total inferred SMBH growth is $\approx 4 \times 10^7\ M_\odot$.

The Magorrian relation (Magorrian et al. 1998) implies that our sample of red galaxies host SMBHs with a mean mass of $M \approx 1 \times 10^8\ M_\odot$ at $z \sim 1$. We therefore conclude that SMBHs within red galaxies could have accreted 30–40% of their final mass between $z = 1$ and the present if they are accreting at typical efficiencies.

This calculation is highly uncertain due to a number of input parameters that are poorly constrained. The largest uncertainty is likely to be in the assumed accretion efficiency which could be substantially smaller ($\epsilon \approx 0.001$) if the accretion rates are small enough for the flow to switch into a radially inefficient mode such as an advection-dominated accretion flow. If this was the case, then a large and arguably implausible fraction of the SMBH mass could have been built up between $z \approx 1$ and the present. It is also possible that SMBH growth occurs primarily within very active (optically luminous) AGN phases, and that sources going through these phases temporarily fall out of our red galaxy classification (because their optical SEDs become dominated by AGN emission). In this case, we will also have under-estimated their SMBH mass build-up due to our assumption of a steady-state accretion.

4. The growth of SMBHs in a luminous, obscured phase at $z \sim 2$–3

In this section, we investigate a population of dusty, obscured AGN at high redshifts to determine how much an SMBH could have grown in an obscured, but bolometrically luminous, phase that may have been missed by previous optical and x-ray surveys.

4.1. *The nature of optically faint, luminous infrared galaxies*

We have uncovered a population of sources that are bright at 24 μm but optically very faint ($R - [24] > 14$; Dey et al. 2008). This has proved to be an effective method in identifying powerful but heavily obscured AGN at high redshifts. The sources are generally faint in the x-ray and their very faint (typically $R > 25$) optical counterparts means that spectroscopic follow-up is extremely challenging. Mid-IR spectroscopic follow-up with *Spitzer* IRS of ~ 50 of these sources with $f_{24} > 0.8$ mJy has revealed that they generally lie at $z \sim 2$–3 and are best fit by local AGN-dominated rather than starburst-dominated templates (Houck et al. 2005; Weedman et al. 2006; Higdon et al. 2008). Near-IR spectroscopy of a small sub-sample of 10 of these sources confirms that they are AGN-dominated sources (Brand et al. 2007).

4.2. Rapid but obscured SMBH growth at $z \sim 2$–3

Optically faint, luminous infrared galaxies constitute an important population of AGN, with space densities comparable to or greater than that of optically luminous type I AGN (Brown et al. 2006; Brand et al. 2007). More work is required to determine the redshift distribution and luminosity function of the population, and hence to reliably estimate the total build up of SMBHs in this regime. We can, however, make a rough calculation of the nuclear accretion rate of a typical source to see if substantial SMBH growth could occur in sources going through a dusty obscured growth phase. If we consider a source with a 24 μm flux of $f_{24} = 1$ mJy at $z = 2$ (which appears to be fairly typical from our mid-IR and near-IR spectroscopic follow-up of these sources), and use local AGN templates to extrapolate from the 24 μm luminosity to the total infrared luminosity (e.g., Houck et al. 2005), the source must have an infrared luminosity of $L_{8-1000 \ \mu m} \approx 0.6 \times 10^{13} \ L_\odot$. We assume an accretion efficiency of $\epsilon = 0.1$ and, because these sources are highly obscured, assume that all the bolometric luminosity comes out in the infrared. This luminosity then corresponds to a mass accretion rate of $\approx 3.5 \ M_\odot \ yr^{-1}$. The entire SMBH mass could be built up in only 3×10^7 years (less than 1% of the galaxies' lifetime) if these accretion rates were sustained over this time period. This is clearly an important SMBH growth phase and the extent to which it has been missed by previous surveys will become clearer with the compilation of larger spectroscopic samples of infrared-selected sources.

5. Conclusions

Although the x-ray–derived accreted-SMBH mass density agrees broadly with current estimates of the local SMBH mass density, there is sufficient uncertainty in these numbers that there is room for substantial SMBH growth in previously unidentified phases (particularly if the mean accretion efficiency, $\epsilon > 0.1$).

We have used the multi-wavelength data from the NDWFS Boötes field to identify two populations in which substantial SMBH growth may have occurred, but which may have been largely missed by previous optical and x-ray surveys:

- Low luminosity AGN in red galaxies at $z < 1$

Using an x-ray–stacking technique, we have measured the mean x-ray luminosity from red galaxies with $0.2 < z < 1$. Optical line diagnostics show that they generally have optical line ratios characteristic of AGN-dominated sources. The x-ray emission is also likely dominated by AGN given its luminosity and hardness ratio. We measure a sharp decline ($z \propto (1+z)^{6.6 \pm 1.0}$) in their absorption-corrected luminosity, suggesting a global decline in the mean AGN activity of red galaxies from $z = 1$ to the present. Although the nuclear accretion rates are small, given that ~ 8 Gyrs have passed between $z = 1$ and the present, SMBHs could have accreted 30–40% of their final mass in this regime.

- Luminous but dusty AGN at $z \sim 2$–3

We have investigated the nature of a population of optically faint, luminous infrared galaxies ($R - [24] > 14$). Follow-up mid- and near-IR spectroscopic observations of a subset of the sources with $f_{24} > 0.8$ mJy shows that they are dusty AGN-dominated sources at $z \sim 2$–3. Their space densities may be as large as that of optically luminous type I AGN, yet they have been largely missed in previous x-ray and optical surveys. Although optically faint, these sources must have huge bolometric luminosities, corresponding to a phase of rapid SMBH growth.

We thank our colleagues on the NDWFS, MIPS, IRAC, AGES, and XBoötes teams. KB acknowledges the support provided by the Giacconi fellowship. This work is based in part on observations made with the *Spitzer Space Telescope*, which is operated by the Jet Propulsion Laboratory, California Institute of Technology under a contract with NASA. Support for this work was provided by NASA through awards issued by JPL/Caltech. The *Spitzer MIPS* survey of the Boötes region was obtained using GTO time provided by the Spitzer Infrared Spectrograph Team (James Houck, P.I.) and by M. Rieke.

REFERENCES

BALDWIN, J. A., PHILLIPS, M. M., & TERLEVICH, R. 1981 *PASP* **93**, 5.

BARGER, A. J., COWIE, L. L., BAUTZ, M. W., BRANDT, W. N., GARMIRE, G. P., HORN-SCHEMEIER, A. E., IVISON, R. J., & OWEN, F. N. 2001 *AJ* **122**, 2177.

BARGER, A. J., COWIE, L. L., MUSHOTZKY, R. F., YANG, Y., WANG, W.-H., STEFFEN, A. T., & CAPAK, P. 2005 *AJ* **129**, 578.

BRAND, K., ET AL. 2005 *ApJ* **626**, 723.

BRAND, K., ET AL. 2006 *ApJ* **641**, 140.

BRAND, K., ET AL. 2007 *ApJ* **663**, 204.

BROWN, M. J. I., ET AL. 2006 *ApJ* **638**, 88.

BROWN, M. J. I., ET AL. 2007 *ApJ* **654**, 858.

COLLISTER, A. A. & LAHAV, O. 2004 *ApJS* **154**, 48.

DAVOODI, P., ET AL. 2006 *MNRAS* **371**, 1113.

DEY, A., ET AL. 2008 *ApJ* **677**, 943.

DI MATTEO, T., ET AL. 2000 *MNRAS* **311**, 507.

EISENHARDT, P. R., ET AL. 2004 *ApJS* **154**, 48.

FERRARESE, L., & MERRITT, D. 2000 *ApJ* **539**, L9.

GEBHARDT, K., ET AL. 2000 *ApJ* **539**, L13.

GRIMM, H.-J., GILFANOV, M., & SUNYAEV, R. 2005 *MNRAS* **339**, 793.

HIGDON, J. L., HIGDON, S. J. U., WILLNER, S. P., BROWN, M. J. I., STERN, D., LE FLOC'H, E., & EISENHARDT, P. 2008 *ApJ* **688**, 885.

HOUCK, J. R., ET AL. 2005 *ApJ* **622**, L105.

JANNUZI, B. T. & DEY, A. 1999. In *Photometric Redshifts and the Detection of High Redshift Galaxies* eds. R. Weymann, L. Storrie-Lombardi, M. Sawicki, & R. Brunner). ASP Conf. Ser. 191, p. 111. ASP.

KAUFFMANN, G., ET AL. 2003 *MNRAS* **346**, 1055.

KENTER, A., ET AL. 2005 *ApJS* **161**, 9.

KEWLEY, L. J., GROVES, B., KAUFFMANN, G., & HECKMAN, T. 2006 *MNRAS* **372**, 961.

KIM, D. W., ET AL. 2004 *ApJS* **150**, 19.

KOCHANEK, C. S., ET AL. 2010; in prep.

MAGORRIAN, J., ET AL. 1998 *AJ* **115**, 2285.

MARCONI, A., RISALITI, G., GILLI, R., HUNT, L. K., MAIOLINO, R., & SALVATI, M. 2004 *MNRAS* **351**, 169.

MARTÍNEZ-SANSIGRE, ET AL. 2005 *Nature* **436**, 666.

MURRAY, S. S., ET AL. 2005 *ApJS* **161**, 1.

REES, M. J., PHINNEY, E. S., BEGELMAN, M. C., & BLANDFORD, R. D. 1982 *Nature* **295**, 17.

RIEKE, G. H., ET AL. 2004 *ApJS* **154**, 25.

RODIGHIERO, G., ET AL. 2007 *MNRAS* **376**, 416.

SOLTAN, A. 1982 *MNRAS* **200**, 115.

VASUDEVAN, R. V., MUSHOTZKY, R. F., WINTER, L. M., & FABIAN, A. C. 2009 *MNRAS* **399**, 1553.

WEEDMAN, D. W., ET AL. 2006 *ApJ* **651**, 101.

YAN, R., NEWMAN, J. A., FABER, S. M., KONIDARIS, N., KOO, D., & DAVIS, M. 2006 *ApJ* **648**, 281.

Strong field gravity and spin of black holes from broad iron lines

By A. C. FABIAN

Institute of Astronomy, University of Cambridge, Madingley Road, Cambridge CB3 0HA, UK

Accreting black holes often show iron line emission in their x-ray spectra. When this line emission is very broad or variable, it is likely to originate from close to the black hole. The theory and observations of such broad and variable iron lines are briefly reviewed here. In order for a clear broad line to be found, one or more of the following have to occur: high iron abundance, dense disk surface and minimal complex absorption. Several excellent examples are found from observations of Seyfert galaxies and Galactic Black Holes. In some cases there is strong evidence that the black hole is rapidly spinning. Further examples are expected as more long observations are made with *XMM-Newton*, *Chandra* and *Suzaku*. The x-ray spectra show evidence for the strong gravitational redshifts and light bending expected around black holes.

1. Introduction

Most of the radiation from luminous accreting black holes is released within the innermost 20 gravitational radii (i.e., $20r_g = 20\,GM/c^2$). In such an energetic environment, iron is a major source of x-ray line emission, with strong emission lines in the 6.4–6.9 keV band. Observations of such line emission provides us with a diagnostic of the accretion flow and the behavior of matter and radiation in the strong gravity regime very close to the black hole (Fabian et al. 2000; Reynolds & Nowak 2003; Fabian & Miniutti 2009; Miller 2007).

The rapid x-ray variability found in many Seyfert galaxies is strong evidence for the emission orginating at small radii. The high frequency break in their power spectra, for example, corresponds to orbital periods at $\sim 20r_g$ and variability is seen at still higher frequencies (Uttley & McHardy 2004; Vaughan et al. 2004). Key evidence that the very innermost radii are involved comes from Soltan's (1982) argument relating the energy density in radiation from active galactic nuclei (AGN) to the local mean mass density in massive black holes, which are presumed to have grown by accretion which liberated that radiation. The agreement found between these quantities requires that the radiative efficiency of accretion be 10% or more (Yu & Tremaine 2002; Marconi et al. 2005). This exceeds the 6% for accretion onto a non-spinning Schwarzschild black hole and inevitably implies that most massive black holes are rapidly spinning, with accretion flows extending down to a just few r_g. Moreover, this is where most of the radiation in such accretion flows originates.

The x-ray spectra of AGN are characterized by several components: a hard power law which may turn over at a few hundred keV, a soft excess and a reflection component (Figure 1). This last component is produced from surrounding material by irradiation by the power law. It consists of backscattered x-rays, fluorescence and other line photons, bremsstrahlung and other continua from the irradiated surfaces. Examples of reflection spectra from photoionized slabs are shown in Figure 2. At moderate ionization parameters ($\xi = F/n \sim 100$ erg cm s^{-1}, where F is the ionizing flux and n the density of the surface) the main components of the reflection spectrum are the Compton hump peaking at ~ 30 keV, the iron line at 6.4–6.9 keV (depending on ionization state) and a collection of lines and reradiated continuum below 1 keV. When such a spectrum is produced from the

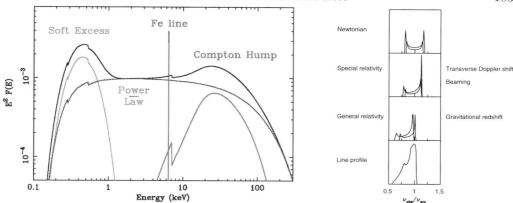

FIGURE 1. Left: The main components of the x-ray spectrum from an unobscured accreting BH: soft x-ray emission from the accretion disk (light gray); power law from Comptonization of the soft x-rays in a corona above the disk (dotted line); reflection continuum and narrow Fe line due to reflection of the hard x-ray emission from the disk and dense gas (dark gray). Right: The profile of an intrinsically narrow emission line is modified by the interplay of Doppler/gravitational energy shifts, relativistic beaming, and gravitational light bending occurring in the accretion disk (from Fabian et al. 2000). The upper panel shows the symmetric double-peaked profile from two annuli on a non-relativistic Newtonian disk. In the second panel, the effects of transverse Doppler shifts (making the profiles extend to lower energies) and of relativistic beaming (enhancing the blue peak with respect to the red) are included. In the third panel, gravitational redshift is turned on, shifting the overall profile to the red side and reducing the blue peak strength. The disk inclination fixes the maximum energy at which the line can still be seen, mainly because of the angular dependence of relativistic beaming and of gravitational light-bending effects. All these effects combined give rise to a broad, skewed line profile which is shown in the last panel, after integrating over the contributions from all the different annuli on the accretion disk. Detailed computations are given by Fabian et al. (1989), Laor (1991), Dovčiak et al. (2004) and Beckwith & Done (2004).

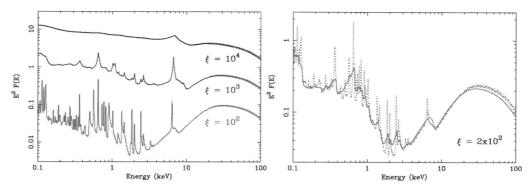

FIGURE 2. Left: Computed x-ray reflection spectra as a function of the ionization parameter ξ (from the code by Ross & Fabian 2005). The illuminating continuum has a photon index of $\Gamma = 2$ and the reflector is assumed to have cosmic (solar) abundances. Right: Relativistic effects on the observed x-ray reflection spectrum (solid line). We assume that the intrinsic rest-frame spectrum (dotted) is emitted in an accretion disk and suffers all the relativistic effects shown in Figure 1.

innermost parts of an accretion disk around a spinning black hole, the outside observer sees it smeared and redshifted (Figure 2) due to doppler and gravitational redshifts.

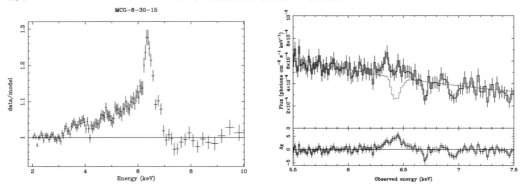

FIGURE 3. Left: The broad iron line in MCG−6-30-15 from the *XMM* observation in 2001 (Fabian et al. 2002b) is shown as a ratio to the continuum model. Right: *Chandra* HEG spectrum with ionized absorber model for the red wing (Young et al. 2005). The model predicts absorption between 6.4 and 6.5 keV which is not seen.

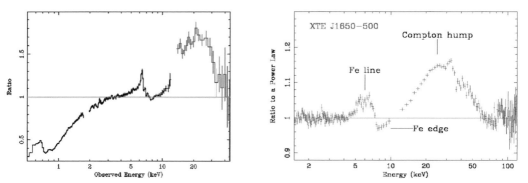

FIGURE 4. Left: Broad iron line and Compton hump in MCG−6-30-15 from *Suzaku* (Miniutti et al. 2007). Right: Broadband *BeppoSAX* spectrum of XTE J1650–500 (as a ratio to the continuum). The signatures of relativistically-blurred reflection are clearly seen (Miniutti et al. 2004).

2. Observations

All three main parts of the reflection spectrum have now been seen from AGN and Galactic Black Holes (GBH). The broad iron line and reflection hump are clearly seen in the Seyfert galaxy MCG−6-30-15 and in the GBH J1650−400 (Figure 4). More recently it has been realized that the soft excess in many AGN can be well explained by smeared reflection (Crummy et al. 2006). It had been noted by Gierlinski & Done (2004) that the soft excess posed a major puzzle if thermal quasi-blackbody disk emission since the required temperature was always about 150 eV, irrespective of black hole mass, luminosity etc. Explaining it as a feature due to smeared atomic lines resolves this puzzle. (An alternative interpretation involves smeared absorption features, Gierlinski & Done 2004).

Broad iron lines and reflection components are seen in both AGN (e.g., Tanaka et al. 1995; Nandra et al. 1997) and GBH (Martocchia & Matt 1996; Miller et al. 2002abc, 2003, 2004ab, 2005) with examples shown in Figures 5, 6 and 7. A recent exciting development are the reports that broad iron lines are present in 50% of all *XMM* AGN observations where the data are of high quality (more than 150,000 counts, Figure 5; Guanazzi et al. 2006; Nandra et al. 2006, 2007). They are not found in all objects or in all accretion states. There are many possible reasons for this, including overionization of the surface,

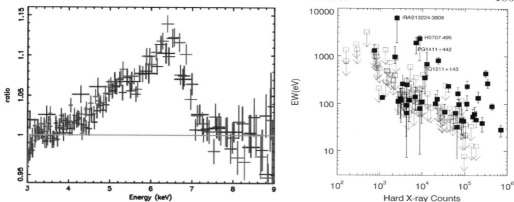

FIGURE 5. Left: *XMM* spectrum of the broad iron line in GX 339-4 (Miller et al. 2006). Right: Detectability of broad iron lines using archival *XMM-Newton* data (Guainazzi et al. 2006). Heavy symbols show the EW of broad iron lines when detected; more than 50% of the spectra having more than 150,000 counts show a significant iron line.

low iron abundance, and beaming of the primary power-law away from the disk (e.g., if the power-law originates from the mildly-relativistic base of the jet).

In many cases there is a narrow iron-line component due to reflection from distant matter. Absorption due to intervening gas, warm absorbers and outflows from the AGN, as well as the interstellar medium in both our Milky Way galaxy and the host galaxy must be accounted for. Moreover, if most of the emission emerges from within a few gravitational radii and the abundance is not high, then the extreme blurring can render the blurred reflection undetectable (Fabian & Miniutti 2009).

In order to distinguish between the various spectral components, both emission and absorption, we can use higher spectral resolution, broader bandwidth and variability. An example of the use of higher spectral resolution is the work of Young et al. (2005) with the *Chandra* high energy gratings. Observations of MCG−6-30-15 fail to show absorption lines or features associated with iron of intermediate ionization (Figure 3). Such gas could cause some curvature of the apparent continuum mimicking a very broad line. A broader bandwidth is very useful in determining the slope of the underlying continuum. This has been shown using *BeppoSAX* (e.g., Guainazzi et al. 1999; Fabian et al. 2002ab) and now in several sources with *Suzaku* (Figure 5, Miniutti et al. 2007; Reeves et al. 2006).

The iron line in MCG−6-30-15 predicted on the basis of the large observed reflection hump (Figure 4) requires that it be broad in order to match the observed intensity (the narrow core is too small: Miniutti et al. 2007).

3. Spin

The extent of the blurring of the reflection spectrum is determined by the innermost radius of the disk (Figure 5). Assuming that this is the radius of marginal stability enables the spin parameter a of the hole to be measured. Objects with a very broad iron line like MCG−6-30-15 are inferred to have high spin $a/M > 0.95$ (Dabrowski et al. 1997; Brenneman & Reynolds 2006). Some (Krolik & Hawley 2002) have argued that magnetic fields in the disk can blur the separation between innermost edge of the disk and the inner plunge region so that the above assumption is invalid. This probably makes little difference for the iron line, however, since the low ionization parameter of most observed reflection requires that the disk matter is very dense. The density of matter in the plunge

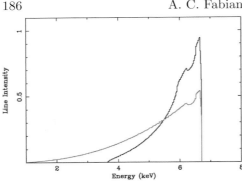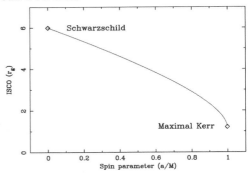

FIGURE 6. Left: The line profiles dependence from the inner disk radius is shown for the two extremal cases of a Schwarzschild BH (black, with inner disk radius at 6 r_g) and of a Maximal Kerr BH (gray, with inner disk radius at $\simeq 1.24\ r_g$). Right: The innermost stable circular orbit around a black hole plotted as a function of its spin (Bardeen et al. 1972).

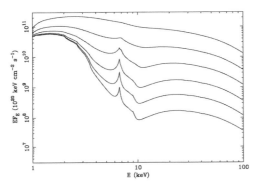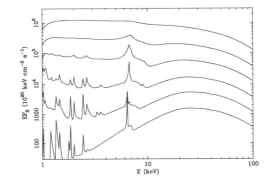

FIGURE 7. Left: Model reflection spectra for a slab of gas heated from below by blackbody emission and irradiated from above by a powerlaw of varying intensity, appropriate for a disk around a stellar mass black hole (Ross & Fabian 2007). Right: Model reflection spectra for a slab with irradiation only, appropriate for an AGN (Ross & Fabian 2005).

region drops very rapidly to low values (Reynolds & Begelman 1997) and only very strong magnetic fields, much larger than are inferred in disks, can stop this steep decline in density. Any reflection from the plunge region will be very highly ionized and so produce little iron emission.

The ionization parameter of the gas, ξ, is related through continuity and luminosity (assuming luminosity is $0.1\dot{M}c^2$) to the volume filling fraction, f, and radial inflow velocity, v_r, as follows

$$\xi = 3 \times 10^5 \left(\frac{f}{0.1}\right)\left(\frac{v_r/c}{0.1}\right) \quad , \tag{3.1}$$

so if $\xi < 3000$ in order to produce strong iron emission (Figure 2) then

$$\left(\frac{f}{0.1}\right)\left(\frac{v_r/c}{0.1}\right) < 100 \quad . \tag{3.2}$$

In other words, the inflowing gas has to be moving slowly in a thin disk for iron emission to be detectable. This does not occur in the plunge region.

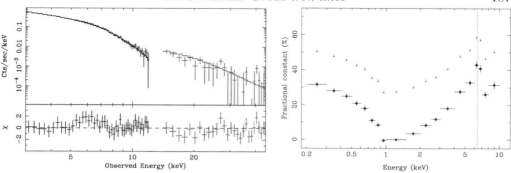

FIGURE 8. Left: Ratio of the *Suzaku* difference spectrum of MCG−6-30-15, obtained by subtracting the low flux state from the high flux state (Miniutti et al. 2007). Right: The spectrum of the constant component in MCG−6-30-15, expressed as a fraction of the total flux, (Vaughan & Fabian 2004). The shape strongly resembles a reflection component.

4. Variability

In the best objects where a very broad line is seen (e.g., MCG−6-30-15 Fabian et al. 2002ab; NGC 4051 Ponti et al. 2006), the reflection appears to change little despite large variations in the continuum. The spectral variability can be decomposed into a highly variable power-law and a quasi-constant reflection component. This behavior is also borne out by a the difference (high-low) spectrum (Figure 8), which is power-law in shape, and by the reflection-like shape of the spectrum resulting from the intercept in flux-flux plots (Figure 8).

This behavior was initially puzzling, until the effects of gravitational light bending were included (Fabian & Vaughan 2003; Miniutti et al. 2004; Miniutti & Fabian 2004). Recall that the extreme blurring in these objects means that much of the reflection occurs within a few r_g of the horizon of the black hole. The enormous spacetime curvature there means that changes in the position of the primary power-law continuum have a large effect on the flux seen by an outside observer (Martocchia & Matt 1996; Martocchia et al. 2002ab). An intrinsically constant continuum source can then appear to vary by large amounts just by moving about in this region of extreme gravity. The reflection component, which comes from the spatially fixed accretion disk, appears relatively constant in flux in this region. Consequently, the observed behavior of these objects may just be a consequence of strong gravity. Indeed, the strength of the observed reflection in MCG−6-30-15 requires strong anisotropy, such as expected from light bending effects.

Some of the Narrow-Line Seyfert 1 galaxies such as 1H0707, IRAS13224 and 1H0439 appear to share this behavior (Fabian et al. 2002a, 2004, 2005). Sharp drops around 7 keV, which are sometimes seen in these sources, may be interpreted as due to absorption from something only partially covering the source. (If the covering was total, then no strong soft emission would be seen, contrary to observation.) Alternatively, they may be the blue edge of a strong, broad, iron emission line (e.g., Figure 9).

5. Discussion

Clear examples of relativistically broadened iron lines are seen in some AGN and GBH in some states. Such objects must have dense inner accretion disks in order that the gas is not overionized. Detection of a line is helped greatly if the iron abundance is super-Solar and if there is little extra absorption due to very strong warm absorbers or winds. Where

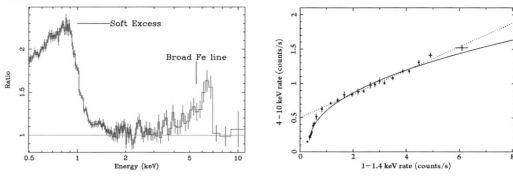

FIGURE 9. Left: Ratio of the spectrum of the NLS1 galaxy 1H0707 to a power law. Spectral fits with either a very broad iron line or a partial covering with a steep edge are equally good for this object (Fabian et al. 2004; Boller et al. 2002). Right: Flux-flux plot for NGC 4051 (Ponti et al. 2006) showing the interplay between reflection which dominates the 4–10 keV band and the power-law component which dominates the 1–1.4 keV band.

broad lines are seen and can be modelled well, then the spin of the central black hole can be reliably determined.

The study of absorption and emission variability of iron-K lines is in its infancy, with some interesting and tantalizing results produced so far. The inner regions of accretion flows are bound to be structured which give rise to variations. Some may be due to motion or transience in the corona or primary power-law source while others may reflect structure, e.g., spiral waves, on the disk itself, intercepting primary radiation from much smaller radii.

The number of broad lines detected is increasing and will continue to expand with the improved broad band coverage from *Suzaku*. Hints that broad lines are common in fainter objects such as in the Lockman Hole (Streblyanska et al. 2005) and Chandra Deep Fields (Brusa et al. 2005) could indicate that the conditions necessary for strong line production—perhaps high metallicity—are common in typical AGN at redshifts 0.5–1.

We have an excellent tool with which to observe the innermost regions of accretion disks immediately around spinning black holes. The effects of redshifts and light bending expected from strong gravity in this regime are clearly evident. This can and should be exploited by future x-ray missions. To make significant progress, we need large collecting areas. The count rate in the broad iron line of MCG–6-30-15 is about 2 ph m^{-2} s^{-1}, which means that square meters of collecting area are required around 6 keV in order to look for reverberation effects. GBH are much brighter, but the orbital periods of matter close to the black hole are much, much smaller, so reverberation is difficult here. Instead, variations with mass accretion rate and source state are accessible.

I am grateful to many colleagues for work on broad iron lines, including Giovanni Miniutti, Jon Miller, Chris Reynolds, Randy Ross, and Josefin Larsson. The Royal Society is thanked for continued support. This brief overview is an update of one I presented at the Prague IAU.

REFERENCES

BARDEEN, J., PRESS, W. H., & TEUKOLSKY, S. A. 1972 *ApJ* **178**, 347.
BECKWITH, K. & DONE, C. 2004 *MNRAS* **352**, 353.
BOLLER, TH., ET AL. 2002 *MNRAS* **329**, L1.
BRENNEMAN, L. & REYNOLDS, C. S. 2006 *ApJ* **652**, 1028.

BRUSA, M., GILLI, R., & COMASTRI, A. 2005 *ApJ*, **621**, L5.

CRUMMY, J., FABIAN, A. C., GALLO, L., & ROSS, R. R. 2006 *MNRAS*, **365**, 1067.

DABROWSKI, Y., FABIAN, A. C., IWASAWA, K., LASENBY, A. N., & RYNOLDS, C. S. 1997 *MNRAS* **288**, L11.

DOVČIAK, M., KARAS, V., & YAQOOB, T. 2004 *ApJSS*, **153**, 205.

FABIAN, A. C., BALLANTYNE, D. R., MERLONI, A., VAUGHAN, S., IWASAWA, K., & BOLLER, TH. 2002a *MNRAS* **331**, L35.

FABIAN, A. C., IWASAWA, K., REYNOLDS, C. S., YOUNG, A. J. 2000 *PASP* **112**, 1145.

FABIAN, A. C. & MINIUTTI, G. 2009. In *Kerr Spacetime: Rotating Black Holes in General Relativity* (eds. D. L. Wiltshire, M. Visser & S. M. Scott. p. 236. Cambridge Univ. Press.

FABIAN, A. C., MINIUTTI, G., GALLO, L., BOLLER, TH., TANAKA, Y., VAUGHAN, S., & ROSS, R. R. 2004 *MNRAS* **353**, 1071.

FABIAN, A. C., MINIUTTI, G., IWASAWA, K., & ROSS, R. R. 2005 *MNRAS* **361**, 795.

FABIAN, A. C., REES, M. J., STELLA, L., & WHITE, N. E. 1989 *MNRAS* **238**, 729.

FABIAN, A. C. & VAUGHAN, S. 2003 *MNRAS* **340**, L28.

FABIAN, A. C., VAUGHAN, S., NANDRA, K., IWASAWA, K., BALLANTYNE, D. R., LEE, J. C., DE ROSA, A., TURNER, A., & YOUNG, A. J. 2002b *MNRAS* **335**, L1.

GIERLIŃSKI, M. & DONE, C. 2004 *MNRAS* **349**, L7.

GUAINAZZI, M., ET AL. 1999 *A&A* **341**, L27.

GUAINAZZI, M., BIANCHI, S., & DOVCIAK, M. 2006 *AN* **327**, 1032.

KROLIK, J. H. & HAWLEY, J. F. 2002 *ApJ* **573**, 754.

LAOR, A. 1991 *ApJ* **376**, 90.

MARCONI, A., ET AL. 2005 *MNRAS* **351**, 169.

MARTOCCHIA, A. & MATT, G. 1996 *MNRAS* **282**, L53.

MARTOCCHIA, A., MATT, G., & KARAS, V. 2002a *A&A* **383**, L23.

MARTOCCHIA, A., MATT, G., KARAS, V., BELLONI, T., &FEROCI, M. 2002b *A&A* **387**, 215.

MILLER, J. M. 2007 *ARAA* **45**, 441.

MILLER, J. M., FABIAN, A. C., IN'T ZAND, J. J. M., REYNOLDS, C. S., WIJNANDS, R., NOWAK, M. A., & LEWIN, W. H. G. 2002b *ApJ* **577**, L15.

MILLER, J. M., FABIAN, A. C., NOWAK, M. A., & LEWIN, W. H. G. 2005. In *Proc. of the 10th Marcel Grossman Meeting*, (eds. M. Novello, S. Perez Bergliaffa & R. Ruffini). p. 1296. World Scientific Publishing.

MILLER, J. M., FABIAN, A. C., REYNOLDS, C. S., NOWAK, M. A., HOMAN, J., FREYBERG, M. J., EHLE, M., BELLONI, T., WIJNANDS, R., VAN DER KLIS, M., CHARLES, & P. A., LEWIN, W. H. 2004b *ApJ* **606**, L131.

MILLER, J. M., FABIAN, A. C., WIJNANDS, R., REMILLARD, R. A., WOJDOWSKI, P., SCHULZ, N. S., DI MATTEO, T., MARSHALL, H. L., CANIZARES, C. R., POOLEY, D., & LEWIN, W. H. G. 2002c *ApJ* **578**, 348.

MILLER, J. M., FABIAN, A. C., WIJNANDS, R., REYNOLDS, C. S., EHLE, M., FREYBERG, M. J., VAN DER KLIS, M., LEWIN, W. H. G., SANCHEZ-FERNANDEZ, C., & CASTRO-TIRADO, A. J. 2002a *ApJ* **570**, L69.

MILLER, J. M., HOMAN, J., STEEGHS, D., RUPEN, M., HUNSTEAD, R. W., WIJNANDS, R., CHARLES, P. A., & FABIAN, A. C. 2006 *ApJ* **653**, 525.

MILLER, J. M., MARSHALL, H. L., WIJNANDS, R., DI MATTEO, T., FOX, D. W., KOMMERS, J., POOLEY, D., BELLONI, T., CASARES, J., CHARLES, P. A., FABIAN, A. C., VAN DER KLIS, M., & LEWIN, W. H. G. 2003 *MNRAS* **338**, 7.

MILLER, J. M., RAYMOND, J., FABIAN, A. C., HOMAN, J., NOWAK, M. A., WIJNANDS, R., VAN DER KLIS, M., BELLONI, T., TOMSICK, J. A., SMITH, D. M., CHARLES, P. A., & LEWIN, W. H. G. 2004a *ApJ* **601** 450.

MINIUTTI, G. & FABIAN, A. C. 2004 *MNRAS* **349**, 1435.

MINIUTTI, G., FABIAN, A. C., ANABUKI, N., CRUMMY, J., FUKAZAWA, Y., GALLO, L., HABA, Y., HAYASHIDA, K., HOLT, S., KUNIEDA, H., LARSSON, J., MARKOWITZ, A., MATSUMOTO, C., OHNO, M., REEVES, J. N., TAKAHASHI, T., TANAKA, Y., TERASHIMA, Y., TORII, K., UEDA, Y., USHIO, M., WATANABE, S., YAMAUCHI, M., & YAQOOB, T. 2007 *PASJ* **59**, S315.

MINIUTTI, G., FABIAN, A. C., & MILLER, J. M. 2004 *MNRAS* **351**, 466.

NANDRA, K., GEORGE, I. M., MUSHOTZKY, R. F., TURNER, T. J., & YAQOOB, T. 1997 *ApJ* **476**, 70.

NANDRA, K., O'NEILL, P. M., GEORGE, I. M., & REEVES, J. N. 2007 *MNRAS* **382**, 194.

NANDRA, K., O'NEILL, P. M., GEORGE, I. M., REEVES, J. N., & TURNER, T. J. 2006 *AN* **327**, 1039.

PONTI, G., ET AL. 2006 *MNRAS* **368**, 903.

REEVES, J., ET AL. 2006 *PASJ* **59**, S301.

REYNOLDS, C. S. & BEGELMAN, M. C. 1997 *ApJ* **488**, 109.

REYNOLDS, C. S. & NOWAK, M. A. 2003 *PhR* **377**, 389.

ROSS, R. R. & FABIAN, A. C. 2005 *MNRAS* **358**, 211.

ROSS, R. R. & FABIAN, A. C. 2007 *MNRAS* **381**, 1697.

SOLTAN, A. 1982 *MNRAS* **200**, 115.

STREBLYANSKA, A., HASINGER, G., FINOGUENOV, A., BARCONS, X., MATEOS, S., & FABIAN, A. C. 2005 *A&A* **432**, 395.

TANAKA, Y., ET AL. 1995 *Nature* **375**, 659.

UTTLEY, P. & MCHARDY, I. M. 2004 *Progr. Th. Phys.* **S155**, 170.

VAUGHAN, S. & FABIAN, A. C. 2004 *MNRAS* **348**, 1415.

VAUGHAN, S., IWASAWA, K., FABIAN, A. C., & HAYASHIDA, K. 2004 *MNRAS* **356**, 524.

YOUNG, A. J., LEE, J. C., FABIAN, A. C., REYNOLDS, C. S., GIBSON, R. R., & CANIZARES, C. R. 2005 *ApJ* **631** 733.

YU, Q. & TREMAINE, S. 2002 *MNRAS* **335**, 965.

Birth of massive black hole binaries

By MONICA COLPI,[1] MASSIMO DOTTI,[2]
LUCIO MAYER,[3] AND STELIOS KAZANTZIDIS[4]

[1]Department of Physics G. Occhialini, University of Milano Bicocca, Milano, Italy

[2]Department of Physics, University of Insubria, Como, Italy

[3]Institute of Theoretical Physics, Zurich, Switzerland

[4]Kavli Institute for Particle Astrophysics and Cosmology, Department of Physics,
Stanford University, Stanford, CA 94305, USA

If massive black holes (BHs) are ubiquitous in galaxies and galaxies experience multiple mergers during their cosmic assembly, then BH binaries should be common, albeit temporary, features of most galactic bulges. Observationally, the paucity of active BH pairs points toward binary lifetimes far shorter than the Hubble time, indicating rapid inspiral of the BHs down to the domain where gravitational waves lead to their coalescence. Here, we review a series of studies on the dynamics of massive BHs in gas-rich galaxy mergers that underscore the vital role played by a cool, gaseous component in promoting the *rapid formation of the BH binary*. The BH binary is found to reside at the center of a massive self-gravitating nuclear disk resulting from the collision of the two gaseous disks present in the mother galaxies. Hardening by gravitational torques against gas in this grand disk is found to continue down to sub-parsec scales. The eccentricity decreases with time to zero and when the binary is circular, accretion sets in around the two BHs. When this occurs, each BH is endowed with its own small-size ($\lesssim 0.01$ pc) accretion disk comprising a few percent of the BH mass. Double AGN activity is expected to occur on an estimated timescale of $\lesssim 1$ Myr. The double nuclear point-like sources that may appear have typical separation of $\lesssim 10$ pc, and are likely to be embedded in the still ongoing starburst. We note that a potential threat of binary stalling, in a gaseous environment, may come from radiation and/or mechanical energy injections by the BHs. Only short-lived or sub-Eddington accretion episodes can guarantee the persistence of a dense cool gas structure around the binary necessary for continuing BH inspiral.

1. Introduction

Dormant black holes (BHs) with masses in excess of $\gtrsim 10^6\,M_\odot$ are ubiquitous in bright galaxies today (Kormendy & Richstone 1995; Richstone 1998). Like quasars, these massive BHs are relics of an earlier active phase, and appear to be a clear manifestation of the cosmic assembly of galaxies. The striking correlations observed between the BH masses and properties of the underlying hosts (Magorrian et al. 1998; Ferrarese & Merritt 2000; Gebhardt et al. 2000; Graham & Driver 2007) unambiguously indicate that BHs evolve in symbiosis with galaxies, affecting the environment on a large scale and self-regulating their growth (Silk & Rees 1998; King 2003; Granato et al. 2004; Di Matteo, Springel, & Hernquist 2005).

According to the current paradigm of structure formation, galaxies often interact and collide as their dark matter halos assemble in a hierarchical fashion (Springel, Frenk, & White 2006), and BHs incorporated through mergers into larger and larger systems are expected to evolve concordantly (Volonteri, Haardt, & Madau 2003). In this astrophysical context, close BH *pairs* form as a natural outcome of binary galaxy mergers (Kazantzidis et al. 2005).

In our local universe, one outstanding example is the case of the ultra-luminous infrared galaxy NGC 6240, an ongoing merger between two gas-rich galaxies (Komossa et al. 2003; for a review on binary black holes see also Komossa 2006). *Chandra* images have revealed the occurrence of two nuclear x-ray sources, 1.4 kpc apart, whose spectral distribution is consistent with being two accreting massive BHs embedded in the diluted emission of a starburst. Similarly, Arp 299 (Della Ceca et al. 2002; Ballo et al. 2004) is an interacting system hosting an obscured active nucleus, and possibly a second less luminous one, several kpc away. A third example is the elliptical galaxy 0402+369, where the cleanest case of a massive BH *binary* has been recently discovered. Two compact variable, flat-spectrum active nuclei are seen at a projected separation of only 7.3 pc (Rodriguez et al. 2006). Arp 299, NGC 6240, and 0402+369 may just highlight different stages of the BH dynamical evolution along the course of a merger, with 0402+369 being the latest, most evolved phase (possibly related to a dry merger). Energy and angular momentum losses due to gravitational waves are not yet significant in 0402+392, so that stellar interactions and/or material and gravitational torques are still necessary to bring the BHs down to the domain controlled by General Relativity.

From the above considerations and observational findings, it is clear that binary BH inspiraling down to coalescence is a major astrophysical process that can occur in galaxies. It is accompanied by a gravitational wave burst so powerful it can be detected out to very large redshifts with current planned experiments like the *Laser Interferometer Space Antenna* (*LISA*; Bender et al. 1994; Vitale et al. 2002). These extraordinary events will provide not only a firm test of General Relativity, but also a view, albeit indirect, of galaxy clustering (Haehnelt 1994; Jaffe & Backer 2003; Sesana et al. 2005). With *LISA*, BH masses and spins will be measured with such accuracy (Vecchio 2004) that it will be possible to trace the BH mass growth across all epochs. Interestingly, *LISA* will explore a mass range between $10^3 M_\odot$ and $10^7 M_\odot$ that is complementary to that probed by the distant massive quasars ($> 10^7 M_\odot$), providing a complete census of the BHs in the universe.

Both minor as well as major mergers with BHs accompany galaxy evolution in environments that involve either gas-rich (wet), as well as gas-poor (dry) galaxies. Thus, the dynamical response of galaxies to BH pairing should differ in many ways, according to their properties. Exploring the expected diversities in a self-consistent cosmological scenario is a major challenge and only recently, with the help of high-resolution N-body/SPH simulations, it has become possible to "start" addressing a number of compelling issues. Galaxy mergers cover cosmological volumes (a few to hundred kpc aside), whereas BH mergers probe volumes of only few astronomical units or less. Thus, tracing the BH dynamics with scrutiny requires N-Body/SPH force resolution simulations spanning more than nine orders of magnitude in length. For this reason, two complementary approaches have been followed in the literature. A statistical approach (based either on Monte Carlo realizations of merger trees or on N-Body/SPH large-scale simulations) follows the collective growth of BHs inside dark matter halos. Supplemented by semi-analytical modeling of BH dynamics (Volonteri et al. 2003) and/or by sub-grid resolution criteria for accretion and feedback (Springel & Hernquist 2003; Springel, Di Matteo, & Hernquist 2005a), these studies have proved to be powerful in providing estimates of the expected coalescence rates, and in tracing the overall cosmic evolution of BHs, including their feedback on the galactic environment (Di Matteo, Springel, & Hernquist 2005; Di Matteo et al. 2008). The second approach, that we have been following, looks at individual binary collisions as it aims at exploiting in detail the BH dynamics and some bulk physics from the galactic scale down to and within the BH sphere of influence. Both approaches, the collective and the individual, are necessary and complementary, the main challenge being the implementation of realistic input physics in the dynamically active environment of a merger.

Following a merger, how can BHs reach the gravitational wave inspiral regime? The overall scenario was first outlined by Begelman, Blandford, & Rees (1980) in their seminal study on the dynamical evolution of BH pairs in pure stellar systems. They indicated three main roots for the loss of orbital energy and angular momentum: (I) dynamical friction against the stellar background acting on each individual BH; (II) hardening via 3-body scatterings off single stars when the BH binary forms; (III) gravitational wave back reaction.

Early studies explored phase (I), simulating the collisionless merger of spherical halos (Makino & Ebisuzaki 1996; Milosavljević & Merritt 2001; Makino & Funato 2004). Governato, Colpi, & Maraschi (1994) in particular first noticed that when two equal-mass halos merge, the twin BHs nested inside the nuclei are dragged effectively toward the center of the remnant galaxy by dynamical friction and form a close pair, but that the situation reverses in unequal-mass mergers, where the less massive halo, tidally disrupted, leaves its "naked" BH wandering in the outskirts of the main halo. Thus, depending on the halo mass ratio and internal structure, the transition from phase (I) to phase (II) can be prematurely aborted or drastically scaled back. Similarly, the transit from phase (II) to phase (III) is not always secured, as the stellar content inside the "loss cone" may not be rapidly refilled with fresh low-angular momentum stars to harden the binary down to separations where gravitational wave driven inspiral sets in (see, e.g., Milosavljević & Merritt 2001; Yu 2002; Berczik, Merritt, & Spurzem 2005; Sesana, Haardt, & Madau 2007). For an updated review on the last parsec problem and its possible solution (see Merritt 2006a; Gualandris & Merritt, this volume).

Since BH coalescences are likely to be events associated with mergers of (pre-)galactic structures at high redshifts, it is likely that their dynamics occurred in gas-dominated backgrounds, NGC 6240 being the most outstanding case visible in our local universe. Other processes of BH binary hardening are expected to operate in presence of a dissipative gaseous component that we will highlight and study here.

Kazantzidis et al. (2005) first explored the effect of gaseous dissipation in mergers between gas-rich disk galaxies with central BHs, using high resolution N-Body/SPH simulations. They found that the merger triggers large-scale gas dynamical instabilities that lead to the gathering of cool gas deep in the potential well of the interacting galaxies. In minor mergers, this fact is essential in order to bring the BHs to closer and closer distances before the less massive galaxy, tidally disrupted, is incorporated in the main galaxy. Moreover, the interplay between strong gas inflows and star formation leads naturally to the formation, around the two BHs, of a grand, massive ($\sim 10^9\ M_\odot$) gaseous disk on a scale smaller than ~ 100 pc. It is in this equilibrium circum-nuclear disk that the dynamical evolution of the BHs continues, after the merger has subsided. Escala et al. (2005, hereinafter ELCM05; see also Escala et al. 2004) were the first to study the role played by gas in affecting the dynamics of massive ($\sim 10^8\ M_\odot$) twin BHs in equilibrium Mestel disks of varying clumpiness. In both these approaches (i.e., in the large scale simulations of Kazantzidis et al. and in the equilibrium disk models of ELCM05) it was clear that gas temperature is a key physical parameter and that hot gas brakes the BHs inefficiently. Instead, when the gas is allowed to cool, the drag becomes efficient: the large enhancement of the local gas density relative to the stellar one leads to the formation of prominent density wakes that are decelerating the BHs down to the scale where they form a "close" binary. Later, binary hardening occurs under mechanisms that are only partially explored, and that are now subject of intense investigation. The presence of a cool circumbinary disk and of small-scale disks around each individual BH appears to be critical for their evolution down to the domain of gravitational wave-driven inspiral. In this context, there is no clear "stalling problem" that emerges from current hydrodynamical simulations, but this critical phase needs a more through, coherent analysis.

The works by Kazantzidis et al. (2005) and ELCM05 have provided our main motivation to study the process of BH pairing along two lines: In gas-rich binary mergers, line (1) aims at studying the transit from state (A) of pairing when each BH moves individually inside the time-varying potential of the colliding galaxies, to state (B) when the two BHs dynamically couple their motion to form a binary. The transit from (A) → (B) requires exploring a dynamic range five orders of magnitude in length from the cosmic scale of a galaxy merger of 100 kpc down to the parsec scale for BHs of a million solar masses (i.e., BHs in the *LISA* sensitivity domain). After all transient inflows have subsided and a new galaxy has formed, the BH binary is expected to enter phase (C), where it hardens under the action of gas-dynamical and gravitational torques. Research line (2) aims at studying the braking of the BH binary from (B) → (C) and further in, exploring the possibility that during phase (C) two disks form and grow around each individual BH. As first discussed by Gould & Rix (2000), the binary may later enter a new phase (D), controlled by the balance of viscous and gravitational torques in a circum-binary disk surrounding the BHs, in a manner analogous to the migration of planets in circum-stellar disks (a scenario particularly appealing when the BH mass ratio is less than unity). Phase (D) likely evolves into (E) when gravitational wave inspiral terminates the BH binary evolution.

There are a number of key questions to address:

(i) How does the transition from state (A) → (B) depend on the gas thermodynamics? How do BHs bind?

(ii) In the grand nuclear disk inside the remnant galaxy, how do eccentric orbits evolve? Do they become circular or highly eccentric?

(iii) During the hardening through phase (B) and (C), do the BHs collect substantial amounts of gas to form cool individual disks?

(iv) Can viscous torques drive the binary into the gravitational wave decaying phase?

(v) Is there a threat of a *stalling* problem when transiting from (C) → (D) or from (D) → (E)? And, for which mass ratios and ambient conditions?

2. Dynamics of BHs in disk-galaxy mergers

In this section, we track the large-scale dynamics of two massive BHs during the merger between two gas-rich (equal mass) disk galaxies, and later focus on the process leading to the formation of a Keplerian BH binary.

2.1. Modeling galaxy mergers

We start simulating, with the N-Body/SPH code *Gasoline* (Wadsley, Stadel, & Quinn 2004), the collision between two galaxies, similar to the Milky Way, comprising a stellar bulge, a disk of stars and gas, and a massive, extended spherical dark-matter halo with NFW density profile (Navarro, Frenk, & White 1996; Klypin, Zhao, & Somerville 2002). The halo has a virial mass $M_{\rm vir} = 10^{12} M_\odot$, concentration parameter $c = 12$ and dimensionless spin parameter $\lambda = 0.031$ consistent with current structure formation models. The disk of mass $M_{\rm disk} = 0.04 M_{\rm vir}$ has a surface density distribution that follows an exponential law with scale length of 3.5 kpc and scale height 350 pc. The spherical bulge (Hernquist 1993) has mass $M_{\rm bulge} = 0.008 M_{\rm vir}$ and scale radius of 700 pc. Initially, the dark matter halo has been adiabatically contracted to respond to the growth of the disk and bulge, resulting in a model with a central density slope close to isothermal. Each galaxy consists of 10^5 stellar disk particles, 10^5 bulge particles, and 10^6 halo particles. The gas fraction, $f_{\rm g}$, is 10% of the total disk mass and is represented by 10^5 particles (10^6 in a refined simulation). To each of the galaxy models we added a softened particle, initially at rest, to represent the massive BH at the center of the bulge. The BH mass is

$M_{BH} = 2.6 \times 10^6 \ M_\odot$, according to M_{BH}–σ relation. In the major merger, the BHs are twin BHs (i.e., the mass ratio $q_{BH} = 1$).

Different encounter geometries were explored in the simulations by Kazantzidis et al. (2005): they comprise prograde or retrograde coplanar mergers, as well as mergers with galactic disks inclined relative to the orbital plane. The simulation presented in this proceeding refers to a coplanar prograde encounter. This particular choice is by no means special for our purpose, except that the galaxies merge slightly faster than in the other cases, thus limiting our computational burden. The galaxies approach each other on parabolic orbits with pericentric distances equal to 20% of the galaxy's virial radius, typical of cosmological mergers (Khochfar & Burkert 2006). The initial separation of the halo centers is twice their virial radii and their initial relative velocity is determined from the corresponding Keplerian point-mass orbit.

We include radiative cooling from a primordial mixture of hydrogen and helium, and adopt the star-formation algorithm by Katz (1992) where gas particles, in dense, cold Jeans-unstable regions and in convergent flows, spawn N-body particles at a rate proportional to the local dynamical time (with star-formation efficiency of 0.1). Radiative cooling is switched off at a relatively high floor temperature of 20,000 K to account for turbulent heating, non-thermal pressure forces, and the presence of a warm interstellar medium. In this large-scale simulation, the force resolution is $\sim$100 pc.

The computational volume is later refined during the final stage of the merger with the technique of static particle splitting (Kaufmann et al. 2006) in order to achieve a resolution of 2 pc. The fine-grained region is large enough to guarantee that the dynamical timescale of the entire coarse-grained region is much longer than that of the refined region, so that gas particles from the coarse region will reach the fine region on a timescale longer than the actual time span probed in this work. In the refined simulations, stars and dark matter particles essentially provide a smooth background potential, while the computation focuses on the gas component which dominates by mass in the nuclear region. A volume of 30 kpc in radius is selected, while the two galaxy cores are separated by only 6 kpc. Inside this region, the simulation is carried on with as many as 2×10^6 gas particles. The mass resolution in the gas component, originally of $2 \times 10^4 \ M_\odot$, now becomes $\sim$3000 $M_\odot$ after splitting.

A starburst with a peak star-formation rate of $\sim$30 $M_\odot$ yr^{-1} takes place when the cores finally merge, and it is in this environment that the BHs couple to form a binary. The short dynamical timescale involved in this process, compared to the starburst duration ($\sim$10^8 yr), suggests to model the thermodynamics and radiation physics simply via an effective equation of state. Calculations that include radiative transfer show that the thermodynamic state of a metal-rich gas heated by a starburst can be approximated by an equation of state of the form $P = (\gamma - 1)\rho u$ with $\gamma = 7/5$ (Spaans & Silk 2000). The specific internal energy u evolves with time as a result of PdV work and shock heating modeled via the standard Monaghan artificial viscosity term. Shocks are generated even when a self-gravitating disk with strong spiral arms forms. The highly dynamical regime modeled here is different from that considered in the next section, which could be evolved using an adiabatic equation of state. With this prescription we treat the gas as a one-phase medium, whose mean density and internal energy (the sum of thermal and turbulent energy) correspond to the mean density and line width seen in observed nuclear disks (Downes & Solomon 1998).

2.2. *Large-scale dynamics and BH pairing*

The galaxies first experience few close fly-bys before merging. In these early phases of the collision, the cuspy potentials of both galaxies are deep enough to allow for the survival

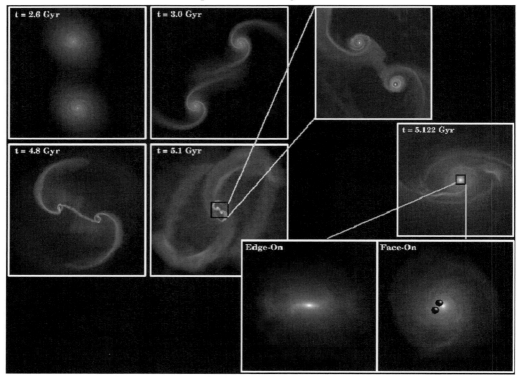

FIGURE 1. The different stages of the merger between two identical disk galaxies seen face-on. The color-coded density maps of the gas component are shown using a logarithmic scale, with brighter colors for higher densities. The four panels to the left show the large-scale evolution at different times. The boxes are 120 kpc on a side (top) and 60 kpc on a side (bottom) and the density ranges between 10^{-2} atoms cm^{-3} and 10^2 atoms cm^{-3}. During the interaction, tidal forces tear the galactic disks apart, generating spectacular tidal tails and plumes. The panels to the right show a zoom-in of the very last stage of the merger, about 100 million years before the two cores have fully coalesced (upper panel), and 2 million years after the merger (middle panel), when a massive, rotating nuclear gaseous disk embedded in a series of large-scale ring-like structures has formed. The boxes are now 8 kpc on a side and the density ranges between 10^{-2} atoms cm^{-3} and 10^5 atoms cm^{-3}. The two bottom panels, with a gray color scale, show the detail of the inner 160 pc of the middle panel; the nuclear disk is shown edge-on (left) and face-on (right), and the two BHs are also shown in the face-on image.

of the baryonic cores where the BHs reside. As the two dark matter halos sink into one another under the action of dynamical friction, strong spiral patterns appear in both the stellar and the gaseous disks. Non-axisymmetric torques redistribute angular momentum, and as much as 60% of the gas originally present in the disks of the parent galaxies is funneled inside the inner few hundred parsecs of each core. This is illustrated in the upper-right panel of Figure 1, where the enlarged color coded density map of the gas is shown, after 5.1 Gyr from the onset of the collision. *Each of the two BHs are found to be surrounded by a rotating stellar and gaseous disk of mass $\sim 4 \times 10^8$ $M_\odot$ and of a size of a few hundred parsecs.* The two disks and BHs are just 6 kpc apart. In the meantime, a starburst of ~ 30 $M_\odot$ yr^{-1} has invested the central region of the merger.

It is tempting to imagine that an episode of accretion onto each BH starts at this time, similar to that observed in NGC 6240 (see Colpi et al. 2007, for a discussion on accretion excited along the course of a large-scale merger). The *double* AGN activity in NGC 6240

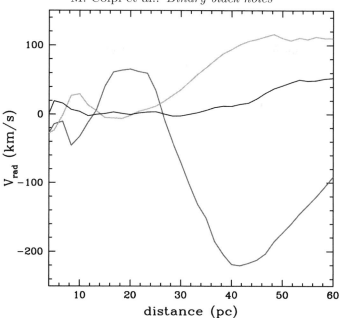

FIGURE 2. Radial velocities within the nuclear disk ($\gamma = 7/5$) starting at $t = 5.1218$ Gyr (dark gray line), and then after another 10^5 yr (light gray line) and 2×10^5 yr (black line). Remarkable inflow and outflow regions are the result of streaming motions within the bar and spiral arms arising in the disk during the phase of non-axisymmetric instability sustained by its self-gravity. At later times the instability saturates due to self-regulation, and the radial motions also level down (green line).

occurs just on a similar scale, as the x-ray nuclei are 1 kpc apart in projection on the sky. It is remarkable that high-resolution near-infrared images at the Keck II telescope, combined with radio and x-ray positions, have revealed the habitat of the two active BHs in this ultra-luminous system. Each active BH appears to be at the center of a rotating stellar disk surrounded by a cloud of young star clusters lying in the plane of each disk (Max, Canalizo, & de Vries 2007). This hints to a consumption of a fraction of the gas disk into stars along the course of the major merger and to BH fueling by the winds of these young stars.

As the interaction proceeds in the simulation, the two baryonic disks around each BH get closer and closer, and eventually merge in a single structure: a massive circumnuclear disk. This is illustrated in Figure 1, in the middle-right panel. The two BHs are now at a relative separation comparable to the softening length of $\sim$100 pc. At this stage, we stop the simulation and start the one with increased resolution.

2.3. *Formation of a circumnuclear disk*

The gaseous cores merge in a single nuclear disk with mass of 3×10^9 $M_\odot$ and size of $\lesssim 100$ pc. This *grand disk* is more massive than the sum of the two nuclear cores formed earlier because radial gas inflows occur in the last stage of the galaxy collision. A strong spiral pattern in the disk produces remarkable radial velocities whose amplitude declines as the spiral arms weaken over time. Just after the merger, when non-axisymmetry is strongest, radial motions reach amplitudes of $\sim$100 km s^{-1} (see Figure 2). This phase lasts only a couple of orbital times, while later the disk becomes smoother as spiral shocks increase the internal energy which in turn weakens the spiral pattern. Inward radial velocities of order 30–50 km s^{-1} are seen for the remaining few orbital times.

The disk is surrounded by several rings and by a more diffuse, rotationally-supported envelope extending out to more than a $\sim$ kpc from the center (Figure 1). A background of dark matter and stars distributed in a spheroid is also present but the gas component is dominant in mass within $\sim$300 pc from the center.

The grand disk is rotationally supported ($v_{rot} \sim 300$ km s^{-1}) and also highly turbulent, having a typical velocity $v_{turb} \sim 100$ km s^{-1}. Multiple shocks generated as the cores merge are the main source of this turbulence. The disk is composed by a very dense, compact region of size about 25 pc which contains half of its mass (the mean density inside this region is $> 10^5$ atoms cm^{-3}). The outer region instead, from 25 to 75–80 pc, has a density 10–100 times lower, and is surrounded by even lower-density rotating rings that extend out to a few hundred parsecs. The disk scale height also increases from inside out, ranging from 20 pc to nearly 40 pc. The volume-weighted density within 100 pc is in the range 10^3–10^4 atoms cm^{-3}, comparable to that of observed nuclear disk (Downes & Solomon 1998). This suggests that the degree of dissipation implied by our equation of state is a reasonable assumption despite the simplicity of the thermodynamical scheme adopted.

2.4. *Birth of a BH binary*

The BHs have been dragged together with their cores toward the dynamical center of the merging galaxies under the action of dynamical friction, and now move inside the grand disk.

They keep sinking down from about 40 pc to a few pc, our resolution limit. We find that *in less than a million years after the merger, the two holes are gravitationally bound to each other, as the mass of the gas enclosed within their separation is less than the mass of the binary. It is the gas that controls the orbital decay, not the stars.* Dynamical friction against the stellar background would bring the two BHs this close only on a much longer timescale, $\sim 5 \times 10^7$ yr (Mayer et al. 2007, supporting online material). This short sinking timescale comes from the combination of (1) the fact that gas densities are much higher than stellar densities in the center, and (2) that in the mildly supersonic regime the drag against a gaseous background is stronger than that in a stellar background with the same density (Ostriker 1999). Adding star formation is unlikely to change this conclusion as in our low-resolution galaxy merger simulations, the starburst timescale of $\sim 10^8$ yr is much longer than the binary formation timescale.

2.5. *Effect of thermodynamics of the BH sinking*

We tested how a smaller degree of dissipation in the gas affects the structure and dynamics of the nuclear region by increasing γ to 5/3. This would correspond to a purely adiabatic evolution. The radiative injection of energy from an active nucleus is a good candidate for a strong heating source that our model does not take into account (Spaans & Silk 2000; Klessen, Spaans, & Jappsen 2007). An AGN would not only act as a source of radiative heating, but would also increase the turbulence in the gas by injecting kinetic energy (Springel, Di Matteo, & Hernquist 2005b) in the surrounding medium, possibly suppressing gas cooling. Before the two galaxy cores merge, double (or single) AGN activity can in principle alter the thermal state of the gas.

We have run another refined simulation with $\gamma = 5/3$ to explore this extreme situation. In this case we find that a turbulent, pressure supported cloud of a few hundred parsecs arises from the merger rather than a disk. The mass of gas is lower within 100 pc relative to the $\gamma = 7/5$ case because of the adiabatic expansion following the final shock at the

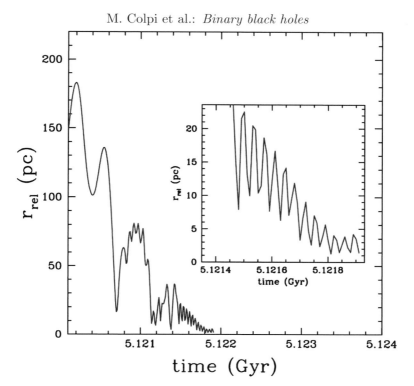

FIGURE 3. Orbital separation of the two BHs as a function of time during the last stage of the galaxy merger shown in Figure 1. The orbit of the pair is eccentric until the end of the simulation. The two peaks at scales of tens of parsecs at around $t = 5.1213$ Gyr mark the end of the phase during which the two holes are still embedded in two distinct gaseous cores. Until this point the orbit is the result of the relative motion of the cores combined with the relative motion of each BH relative to the surrounding core, explaining the presence of more than one orbital frequency. The inset shows the details of the last part of the orbital evolution, which takes place in the nuclear disk arising from the merger of the two cores. The binary stops shrinking when the separation approaches the softening length (2 pc).

merging of the cores. The nuclear region is still gas dominated, but the stars/gas ratio is >0.5 in the inner 100 pc. This suggests that the $\gamma = 5/3$ simulation does not describe the typical nuclear structure resulting from a dissipative merger.

The BH duet does not form a binary owing to inefficient orbital decay, and maintains a separation of $\sim$100–150 pc, as shown in Figure 4. The gas is hotter and more turbulent; the sound speed $c_{\rm s} \sim 100$ km s^{-1} and the turbulent velocity $v_{\rm turb} \sim 300$ km s^{-1} are of the same order of $v_{\rm BH}$, the velocity of the BHs, and the density around them is $\sim$5 times lower than in the $\gamma = 7/5$ case. Stars and gas will drive the BHs closer to form a binary, but on an estimated dynamical friction time of several 10^7 yr (Mayer et al. 2007, supporting online material).

3. BH inspiral in equilibrium rotating nuclear disks

3.1. *Initial conditions*

In this section, we present an independent series of simulations (carried out with *GAD-GET*; Springel, Yoshida, & White 2001) that trace the dynamics of a BH pair (with $q_{\rm BH} = 1, 1/4, 1/10$) orbiting inside a self-gravitating, rotationally supported disk composed

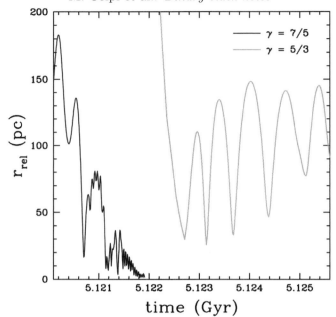

FIGURE 4. Orbital separation of the binary BHs. The black line shows the relative distance as a function of time for $\gamma = 7/5$ as shown in Figure 2, while the gray line shows it for $\gamma = 5/3$.

either of gas, gas and stars, or just stars (Dotti, Colpi, & Haardt 2006; Dotti et al. 2007). The main parameters of the simulations are summarized in Table 1.

In all the simulations there are two BHs, a stellar bulge, and a massive rotationally supported nuclear disk composed by stars, gas, or both (except A3). The disk has a mass $M_{\text{Disk}} = 10^8 \, M_\odot$, an extension of 100 pc, a vertical thickness of 10 pc, and follows a Mestel surface density profile:

$$\Sigma_{\text{Disk}}(R) = \frac{K_{\text{Disk}}}{R} \quad , \tag{3.1}$$

where R is the radial distance projected into the disk plane, and K_{DisK} is determined by the total mass of the disk. The vertical profile of the disk is initially set homogeneous. The rotational velocity of the gas v_{rot} is constant through the disk and this implies that fluid elements are rotating differentially with an angular velocity $\Omega_{\text{rot}} = v_{\text{rot}}/R$. The spheroidal stellar bulge is modeled with 10^5 collisionless particles, initially distributed as a Plummer sphere with mass density profile:

$$\rho_{\text{Bulge}}(r) = \frac{3}{4\pi} \frac{M_{\text{Bulge}}}{b^3} \left(1 + \frac{r^2}{b^2}\right)^{-5/2} \quad , \tag{3.2}$$

where $b \, (= 50 \text{ pc})$ is the core radius, r the radial coordinate, and $M_{\text{Bulge}}(= 6.98 M_{\text{Disk}})$ the total mass of the spheroid. With such choice, the mass of the bulge within 100 pc is five times the mass of the disk, as suggested by Downes & Solomon (1998). We relax our initial composite model (bulge, disk and, if present, the central BH) for ≈ 3 Myrs, until the bulge and the disk reach equilibrium. Given the initial homogeneous vertical structure of the disk, the gas initially collapses on the disk plane exciting small waves that propagate through the system. The parameters varying in the simulations are:

• The disk mass fraction in stars. We run four different sets of simulations assuming a purely gaseous disk (runs A, B, C, and G), and a disk in which 1/3 (runs D), 2/3 (runs E),

run	central BH	f_*†	$M_{\mathrm{BH},1}$‡	$M_{\mathrm{BH},2}$‡	M_{Disk}‡	M_{Bulge}‡	e	Q	res¶
A1					100		0		
A2	no	0	1	1	100	698	0.9	1.8	1
A3					0		0.9		
B1							0		
B2	no	0	5	1	100	698	0.9	1.8	1
B3‖							0.9		
C1††				4					
C2	yes	0	4	1	100	698	0.7	3	1
C3				0.4					
D1				4					
D2	yes	1/3	4	1	100	698	0.7	3	1
D3				0.4					
E1				4					
E2	yes	2/3	4	1	100	698	0.7	3	1
E3				0.4					
F1				4					
F2	yes	1	4	1	100	698	0.7	3	1
F3				0.4					
G1‡‡	yes	0	4	4	100	698	0.4	3	0.1

† f_*: disk mass fraction in stars.

‡ BH masses are in units of $10^6\,M_\odot$.

¶ Force resolution in pc.

‖ The secondary lighter BH in run B3 has a retrograde orbit.

†† simulation C1 was run two times, setting the secondary BH on a prograde and a retrograde orbit.

‡‡ Simulation C1 re-run using the particle splitting technique to improve force resolution.

TABLE 1. Run parameters

and, finally, all gas particles (runs F) are turned into collisionless particles, respectively. For each disk model with fixed star fraction, we evolved the initial condition in isolation until equilibrium is reached. We do not convert any gaseous particle in stars when we follow the dynamics of the BHs, so that the disk stellar fraction remains constant;

• The disk mass. In all the simulations the disk mass is $M_{\mathrm{Disk}} = 10^8\,M_\odot$, apart from run A3, in which we study the BH pair evolution in a pure stellar bulge;

• The Toomre parameter Q of the disk. In our simulations with a pure gaseous disk (runs A, B, C, and G), we set a initial internal energy profile $u(R) = K_{\mathrm{Th}}\,R^{-2/3}$, where K_{Th} is a constant defined so that the Toomre parameter of the disk,

$$Q = \frac{kc_{\mathrm{s}}}{\pi G \Sigma} \qquad (3.3)$$

is $\gtrsim 1.8$ for runs A and B (cold disk runs), and $\gtrsim 3$ for runs C and G (hot disk runs) everywhere. In equation (3.3), k is the local epicyclic frequency and c_{s} is the local sound speed of the gas. The internal energy of the gas is evolved adiabatically neglecting radiative cooling/heating processes. Our choice of $Q > 1.8$ everywhere prevents fragmentation of the disk, and, when $Q > 3$ formation of large-scale over-densities, such as spiral arms or bars. We do not model any turbulent motion in the disk, but we consider the internal energy as a form of unresolved turbulence, and, as a consequence, c_{s} as local turbulent velocity. In the simulations with a non-null stellar fraction of the disk mass (runs D, E, and F) we set the local velocity dispersion of the disk stars equal to the local sound speed. With this procedure Q is the same than in a pure gaseous disk;

• The masses of the two BHs. As shown in Table 1, we explore various mass ratios q_{BH} from 1 to 1/10;

• The BH orbits. The BHs are placed on coplanar orbits that can either be prograde or retrograde, circular or eccentric (with $e = 0.4, 0.7, 0.9$). In run A and B the two BHs are placed at 55 pc from the dynamical center of the disk. In runs C, D, E, and F the primary BH is at the center of the disk;

• The initial eccentricity of the orbiting BH. In all the runs of A and B classes a BH is initially moving on a circular orbit, while the second BH can have an orbital eccentricity (e) of 0 (circular motion, runs A1 and B1) or 0.9. In runs C, D, E, F, and G a BH is initially placed at rest in the center of the structure (see point 1) while the orbiting BH is initially moving on eccentric orbits with $e = 0.7$ in runs C, D, E, and F, $e = 0.4$ in run G1;

• The spatial resolution of the simulations. In our low-resolution simulations (runs A, B, C, D, E, and F), we model our relaxed Mestel disk with 235,331 particles, and use a number of neighbors of 50. This number defines a subsample of particles used by the code to evaluate the local hydrodynamical parameters. This number defines, in the center of the disk, a hydrodynamical force resolution (usually defined as smoothing length) of ≈ 1 pc. We set also equal to 1 pc our gravitational softening, the parameter setting the spatial resolution of the gravitational acceleration evaluation. Evaluating the two forces, hydrodynamical and gravitational, with the same resolution prevents spurious local condensation of gas (occurring when the smoothing length exceeds the gravitational softening) or outflows (in the opposite case). The stellar gravitational softening is set equal to the gaseous one. In our high-resolution simulation (run G1) we re-sample the output of simulation C1 thanks to the technique of particle splitting (Kitsionas & Whitworth 2002). With this technique we locally increase the number of gaseous particles of one order of magnitude reducing accordingly the smoothing length and the gravitational softening. The new spatial resolution in the central regions is 0.1 pc.

3.2. *Orbital decay on circular orbits*

In simulations A1 and B1, the BHs move initially on circular prograde orbits inside the disk. The initial separation, relative to the center of mass, is ≈ 55 pc.

We plot in Figure 5 the density map of the gas surrounding the unequal mass BHs in run B1 at a selected time. Both BHs are exciting prominent density wakes whose extent depends on the amount of disk mass perturbed by the orbiting BHs, which is a function of the BH masses, as can be noted in the figure. The presence of two wakes near each BH is due to the different angular velocity pattern of the gas in the disk. Given our choice for the bulge and disk profile, in the BH comoving frame, the disk inside the BH orbit is moving counterclockwise while the outer regions are moving clockwise, as shown in the right panel of Figure 5. In a non-rotating frame, two wakes develop around each BH, one outside the orbit and one inside. The motion of the BH is highly supersonic, and this explains the coherent structure and shape of their wakes (Ostriker 1999). The interaction between the BH and it inner wake increases the BH angular momentum, while the outer brakes the motion. Given our choice of the density/internal energy profiles, and the velocity field in the disk, the outer wake is more effective in removing orbital energy and angular momentum from each BH compared to the positive (accelerating) torque exerted by the wake excited in the inner part of the disk. The net effect is the braking of the BH pair, so that a binary can form on a timescale of $\sim 10^7$ yr.

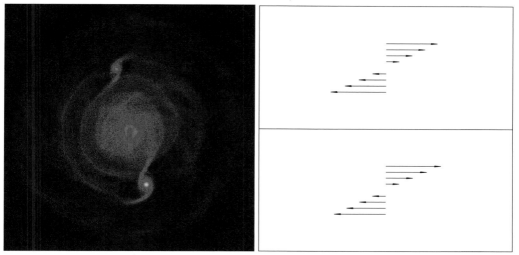

FIGURE 5. Left: face-on projection of the disk for run B1 at time 1 Myr. The color coding shows the z-averaged gas density on a logarithmic scale between 100 (gray) and $3 \times 10^5 \, M_\odot \, pc^{-3}$ (white). The two orbiting bright dots highlight the position of the two BHs (the top BH is the lighter one) that are moving counterclockwise. They excite prominent wakes along their trails. Right panels: relative angular-velocity pattern between the BHs and the gaseous disk for different radii. The upper (lower)-right panel is centered in the upper (lower) BH radius. The separations between two arrows are 1 pc.

3.3. *Orbital decay along eccentric orbits*

Figure 6 shows the BH relative separation s and eccentricity as a function of time (upper and lower panel respectively) for equal mass, initially eccentric binaries of runs C1, D1, E1, and F1, where a primary BH is initially at rest in the dynamical center of the disk. Regardless the fraction of star-to-gas disk particles, the secondary BH spirals in, at the same pace, reaching ~ 1 pc after ~ 5 Myr. The velocity dispersion of the disk-stars is similar to the gas sound speed, and the two components share the same differential rotation. This is why dynamical friction on the secondary BH, caused by stars and gas, is similar. As the orbit decays, the eccentricity decreases to $e \lesssim 0.2$. This value is not a physical lower limit, but rather a numerical artifact due to the finite resolution, as will be shown in run G1.

Here, we show that circularization occurs regardless the nature of the disk particles (gas and/or stars). Note that circularization takes place well before the secondary feels the gravitational potential of the primary, so the BH mass ratio does not play any role in the process.

To show how the circularization process works, let us consider two different snapshots of run C1. In Figure 7 we plot the gas densities in the disk at the time corresponding to the first passage at the apocenter. In this simulation, the initially orbiting BH is co-rotating with the disk, with a speed equal, in modulus, to the velocity corresponding to a circular orbit at that initial position, but with an orbital eccentricity of 0.7. Near the apocenter the BH is moving slower than the gas in the disk, as highlighted by the two vectors in Figure 7, and in a frame comoving with the BH, gas blows ahead of the BH. The gas average rotational velocity decreases due to the gravitational interaction with the BH, and the back-reaction on the BH is a temporary "increase" of its orbital energy and angular momentum. This excites a wake in the forward direction, i.e., at the apocenter the density wake is in front of the BH, as shown in Figure 7.

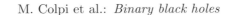

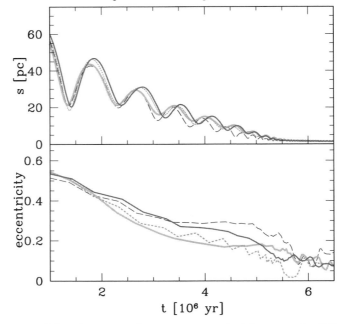

FIGURE 6. Equal mass BHs. Upper panel: separations s (pc) between the BHs as a function of time. Lower panel: eccentricity of the BH binary as a function of time. Dashed, gray, black and dotted lines refer to stellar to total disk mass ratio of 0, 1/3, 2/3 and 1 (run C1, D1, E1, and F1) respectively.

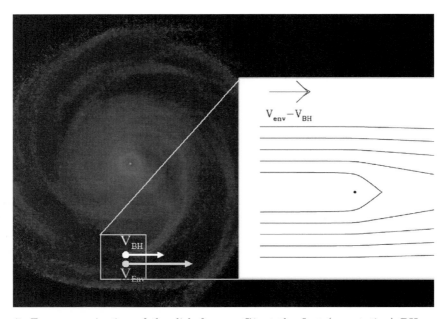

FIGURE 7. Face-on projection of the disk for run C1 at the first (co-rotating) BH apocenter. The color coding shows the z-averaged gas density, the white and pale gray arrows refer to the BH and disk velocities, respectively. In the insert panel, the trajectories of the gas particles are drawn, as observed in a frame comoving with the orbiting BH. The density wake is in front of the BH trail.

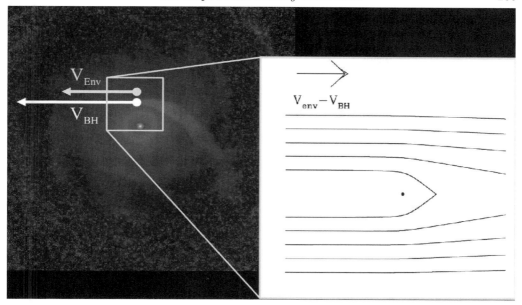

FIGURE 8. Same as Figure 7, but when the BH is at its first pericenter. The density wake is behind the BH trail.

In Figure 8 we plot the gas densities in the disk at the time corresponding to the first passage at the pericenter. The BH has there a velocity higher than the local rotational velocity, so that dynamical friction causes a drag, i.e., a reduction of the BH velocity. Now, a wake of particles lags behind the BH trail. Thus, given this velocity difference, the wake reverses its direction, tangentially decelerating the BH. Now the wake is behind the BH direction of motion. The orbital energy decreases more effectively than angular momentum, since at pericenter the gas density is also higher than at apocenter. The overall effect is a net circularization of the orbit. This seems a generic feature of dynamical friction, regardless the disk composition, and holds as long as the rotational velocity of the gas or/and star particles exceeds the gas sound speed and the stellar velocity dispersion, as in all cases explored. We remark that in spherical backgrounds, dynamical friction tends to increase the eccentricity, both in collisionless (Colpi, Mayer, & Governato 1999; van den Bosch et al. 1999; Arena & Bertin 2007) as well as in gaseous (Sanchez-Salcedo & Brandenburg 2001) environments.

The interaction between the BH and the disk for counter-rotating orbits (run B3) keeps the BH orbit eccentric. When a counter-rotating BH is near apocenter, the gas flows against the BH motion. The over-density that forms is behind the BH trail, and this occurs also near pericenter. Dynamical friction is weaker than for co-rotating orbits given the larger relative velocity between the BH and the gas; thus the corresponding decay timescale is longer (by a factor of $\sim$2).

3.4. *High-resolution run: dynamics*

We run a higher-resolution simulation to study the eccentricity and orbital evolution on scales smaller than one pc. The new initial condition is obtained re-sampling the output of run C1 (for equal mass BHs) with the technique of particle splitting. Re-sampling is performed when the BH separation is $\simeq$14 pc (corresponding to $\simeq$4 Myr after the start of the simulation). Splitting is applied to all particles whose distance from the

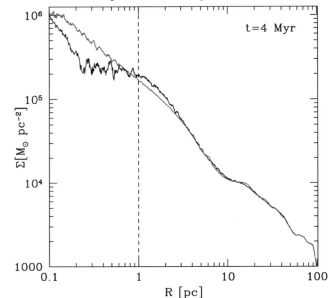

FIGURE 9. Surface density profile of the circumnuclear gaseous disk in run C1 at $t = 4$ Myr. Black line refers to the surface density in the low-resolution simulation, gray line refers to the high-resolution (split) simulation. The dashed vertical line marks the resolution limit in the un-split simulation.

binary center of mass is $\leqslant 42$ pc, so that the total number of particles increases only by a factor of $\simeq 4$, while the local mass resolution in the split region is comparable to that of a standard $\simeq 2 \times 10^6$ particle simulation with uniform resolution. Our choice of the maximum distance for splitting is conservative, since it is aimed at preventing that more massive, unsplit gas particles reach the binary on a timescale shorter than the entire simulation time. In the central split region, the high-mass resolution achieved fulfills the Bate & Burkert (1997) criterion for gravitational softening values down to 0.1 pc.

In Figure 9 we compare the surface density profile of the circumnuclear gaseous disk in run C1 at $t = 4$ Myr, for the low- and high-resolution cases. The two profiles differ only below the scale of the low-resolution limit, $R \lesssim 3$ pc. The decrease of the gravitational softening corresponds to introducing a deeper potential well of the BH within a sphere defined by the former softening radius. Therefore, with the improved resolution, the central surface density increases as the gas reaches a new hydrostatic equilibrium closer to the BH, as shown in Figure 9 (gray line). The lack of noticeable differences in the surface profile at separations $R > 3$ pc confirms the accuracy of the particle splitting technique.

Results of the high resolution run are shown in Figure 10. The separation decays down to 0.1 pc in ~ 10 Myr. In the high resolution run, the dynamical evolution of the BHs is initially identical to the low resolution case. Because of particle splitting, the system granularity is reduced, and therefore the force resolution increases. In the high-resolution run, the binary decreases its eccentricity to ≈ 0 (before the new spatial resolution limit is reached).

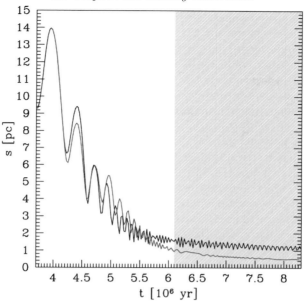

FIGURE 10. Separations s (pc) between the BHs as a function of time. The gray (black) line refers to the (un-)split run C1. The shaded area corresponds to the region where the BH separation is < 1 pc in the split higher resolution.

3.5. *High resolution runs: constraints on accretion processes*

In run G1, a resolution of 0.1 pc allows us to study the properties of the gas bound to each BH. To this purpose, it is useful to divide gaseous particles, bound to each BH, into three subsets, according to their total energy relative to the BH. We then define weakly bound (WB), bound (B), and strongly bound (SB) particles according to the following rule:

$$E < \begin{cases} 0 & \text{(WB)} \\ 0.25\,W & \text{(B)} \\ 0.5\,W & \text{(SB)} \end{cases} \tag{3.4}$$

where E is the sum of the kinetic, internal and gravitational energy (per unit mass), the latter referred to the gravitational potential W of each individual BH. Hereafter WBPs, BPs, and SBPs will denote particles satisfying the WB, B, or SB condition, respectively. Note that, with the above definition, SBPs are a subset of BPs, which in turn are a subset of WBPs.

We find that the mass collected by each BH, relative to WB, B, and SB particles is $M_{\text{WBP}} \approx 0.85 M_{\text{BH}} \approx 3.4 \times 10^6\, M_\odot$, $M_{\text{BP}} \approx 0.41 M_{\text{BH}} \approx 1.6 \times 10^6\, M_\odot$, and $M_{\text{SBP}} \approx 0.02 M_{\text{BH}} \approx 8 \times 10^4\, M_\odot$, respectively (here $M_{\text{BH}} = M_1 = M_2$). These masses are remain constant with time as long as the BH separation is $s \gtrsim 1$ pc. At shorter separations WBPs and BPs are perturbed by the tidal field of the BH companion, and at the end of the simulation M_{WBP} and M_{BP} are reduced of a factor ≈ 0.1. During the same period of time, M_{SBP} associated to the primary (secondary) BH increases by a factor ≈ 4 (≈ 2.5). This result is unaffected by numerical noise since the number of bound particles (associated to each class) is $\gtrsim 1$ SPH kernel ($N_{\text{igh}} = 50$).

The radial density profiles of WBPs, BPs, and SBPs are well resolved during the simulation. Bound particles have a net angular momentum with respect to each BH, and form a pressure-supported spheroid. The half-mass radius is similar for the two BHs: $\simeq 3$ pc, $\simeq 1$ pc, and $\simeq 0.2$ pc for WBPs, BPs, and SBPs, respectively. The disk gas density can be as high as 10^7 cm^{-3}. It is then conceivable that, at these high densities, dissipative processes could be important, possibly reducing the gas internal (turbulent and thermal) energy well below the values adopted in our simulations. If cooling becomes effective, we expect that the bound gas will form a geometrically thin disk with Keplerian angular momentum comparable to what we found in our split simulation. Since $L_z = \sqrt{G\,M_{\rm BH}\,R_{\rm BH,disk}}$, we obtain, for the primary BH, an effective radius $R_{\rm BH,disk} \approx 0.1$ pc (0.03 pc) for WBPs (BPs). The secondary BH is surrounded by particles with a comparatively higher angular momentum with a corresponding effective radius $R_{\rm BH,disd} \approx 1$ (0.13) pc for WBPs (BPs). Finally, for SBPs, both BHs have $R_{\rm BH,disk} \ll 0.01$ pc, which is more than an order of magnitude below our best resolution limit. These simple considerations indicate that a more realistic treatment of gas thermodynamics is necessary to study the details of gas accretion onto the two BHs during the formation of the binary, and the subsequent orbital decay. Nonetheless, our simplified treatment allows us to estimate a lower limit to the accretion timescale, assuming Eddington limited accretion:

$$t_{\rm acc} = \frac{\epsilon}{1-\epsilon}\, t_{\rm Edd}\, \ln\left(1 + \frac{M_{\rm acc}}{M_{\rm BH,0}}\right) \ , \tag{3.5}$$

where ϵ is the radiative efficiency, $t_{\rm Edd} = c\sigma_T/(4\pi G m_{\rm p})$ the Salpeter time, σ_T the Thompson cross section, $M_{\rm BH,0}$ the initial BH mass, and $M_{\rm acc}$ the accreted mass. Assuming $\epsilon = 0.1$, Eddington limited accretion can last for ~ 15 Myr, and only $\lesssim 1$ Myr, if the BHs accrete all the BPs, and SBPs, respectively.

Figure 11 depicts, in a cartoon, the configuration of the two accretion disks surrounding the BHs inside the grand disk. It is expected that the two disks will eventually touch and disrupt tidally, and re-organize to form a circum-binary geometrically thin disk surrounding both BHs; exploring this configuration is the goal of our next series of simulations. Further braking of the BH motion and binary hardening requires energy loss and angular momentum transport through a mechanism that may be reminiscent of planet migration in proto-stellar disks (Gould & Rix 2000): while gravitational torques carry away angular momentum, viscous torques inside the cool Keplerian circum-binary disk sustain the radial motion of the gas toward the BHs, maintaining the binary in near contact with the disk. Equilibrium between the gravitational and viscous torques cause the slow drift of the BHs toward smaller and smaller separations until the gravitational waves guide final inspiral.

4. BH binaries, the $M_{\rm BH}$ versus σ relation and the stalling problem

In the previous section, we followed the orbital decay of a BH binary in a rotationally supported gaseous background, down to a scale of ~ 0.1 pc, and found that the hardening of the BH binary, and thus the corresponding energy and angular momentum loss, results from large-scale density perturbations excited by the BH gravitational. These perturbations have still negligible impact on the overall structure of the Mestel disk: compared with the binding energy of the disk, the energy deposited by the BH binary is still small (approximately an order of magnitude lower). But this may not hold true if the binary continues to harden down to the scale where gravitational wave emission becomes important. Unless gravitational torques are capable of extracting angular momentum with no energy dissipation (making the binary more and more eccentric until $e \to 1$), the energy input from the BH binary may have two main effects: (i) to modify the thermal and

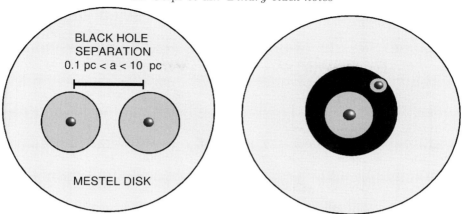

FIGURE 11. To the left is a cartoon depicting the BH binary when the BH separation is $\gtrsim 0.1$ pc, as suggested by our high-resolution simulation. The BHs are surrounded by their own accretion disks. Further orbital decay by gravitational and viscous torques are expected to open a gap leading to the configuration depicted on the right.

density structure of the grand disk; (ii) to halt the BH binary hardening (by "evaporating" the environment) before gravitational wave emission intervenes to guide the BHs toward coalescence. This "negative" feedback on the BH binary fate may indeed cause the "stalling" of the binary, a problem that has been discussed mainly in the context of pure stellar backgrounds (e.g., Merritt 2006a).

In this section we would like to introduce a simple argument based on energy conservation (i.e., assuming that angular momentum transport is accompanied by some energy dissipation) in the attempt to investigate whether the BH binary deposits enough energy to modify its surroundings and influence its fate.

Thus, consider a BH binary of total mass M_{BH}, reduced mass μ_{BH}, and mass ratio q_{BH}. The binary will coalesce in a time t_{GW} under the action of gravitational wave emission if its semi-major axis a is smaller than

$$a_{\mathrm{GW}} = \left(\frac{256}{5} \frac{G^3}{c^5} f(e) M_{\mathrm{BH}}^2 \mu_{\mathrm{BH}} t_{\mathrm{GW}} \right)^{1/4} , \qquad (4.1)$$

where $f(e) = [1 + (73/24)e^2 + (37/96)e^4](1 - e^2)^{-7/2}$ (note that as $e \to 1$, $a_{\mathrm{GW}} \to \infty$). This separation is of only 8×10^{-4} pc if we consider a BH binary of total mass $10^6 \, M_\odot$, with eccentricity $e = 0$, mass ratio $q_{\mathrm{BH}} = 1$, and merging time of $t_{\mathrm{GW}} = 10^9$ yr.

If the BH are initially unbound (as it is the case of a galaxy merger) the total energy ΔE_{BBH} that the BH binary needs to deposit in its environment to reach a_{GW} is

$$\Delta E_{\mathrm{BBH}} = \frac{1}{2} \left(\frac{256}{5} \frac{G^3}{c^5} f(e) \right)^{-1/4} G M_{\mathrm{BH}}^{5/4} \frac{q_{\mathrm{BH}}^{3/4}}{(1 + q_{\mathrm{BH}})^{3/2}} t_{\mathrm{GW}}^{-1/4} . \qquad (4.2)$$

For the binary considered

$$\Delta E_{\mathrm{BBH}} \approx 1.3 \times 10^{55} \left(\frac{M_{\mathrm{BH}}}{10^6 \, M_\odot} \right)^{5/4} \left(\frac{10^9 \, \mathrm{yr}}{t_{\mathrm{GW}}} \right)^{1/4} \mathrm{erg} . \qquad (4.3)$$

If we simply model the binary environment with an isothermal sphere of effective radius R_{e}, mass M_{iso} and 1-D velocity dispersion σ, BH coalescence imposes, as necessary condition

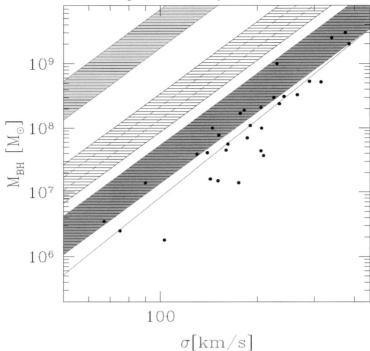

FIGURE 12. M_{BH}–σ plane: thin solid line is the best fit to the data (Tremaine et al. 2002). Black lines and black lines in light-gray-shaded area refer $M_{\mathrm{BH}}^{\max} \equiv \mathcal{M}(t_{\mathrm{GW}}, e, q_{\mathrm{BH}})\,\sigma^{3.7}$ (see eq. [4.8]) computed for $\alpha = 2.63$, $q_{\mathrm{BH}} = 1$, $e = 0$ ($e = 0.99$) and $t_{\mathrm{GW}} = 10^9 - 10^8 - 10^7 - 10^6$ yr (from top to bottom). The dark-gray strip is computed using equation (4.8) for only the gaseous component, assuming $E_{\mathrm{iso,gas}} = 0.1\,E_{\mathrm{iso,bulge}}$.

$$\Delta E_{\mathrm{BBH}} < E_{\mathrm{iso}}\quad, \tag{4.4}$$

where $E_{\mathrm{iso}} = M_{\mathrm{iso}}\sigma^2$ is the binding energy of the isothermal sphere. An energy ΔE_{BBH} of the order of E_{iso} would modify the equilibrium structure of the sphere and halt binary decay. If we identify the sphere with the stellar "bulge" hosting the BH binary, and write $M_{\mathrm{iso,bulge}} = \mathcal{K}\sigma^\alpha$, binary coalescence is possible if

$$\frac{1}{2}\frac{G\mu_{\mathrm{BH}}M_{\mathrm{BH}}}{a_{\mathrm{GW}}} < \mathcal{K}\sigma^{\alpha+2}\quad. \tag{4.5}$$

Then, using the Fundamental Plane relation (Cappellari et al. 2006) and the mass-to-light ratio as in Zibetti et al. (2002), the bulge mass can be expressed as

$$M_{\mathrm{bulge}} \approx 1.54 \times 10^{11}\left(\frac{\sigma}{200\,\mathrm{km\,s^{-1}}}\right)^{2.63} M_\odot\quad, \tag{4.6}$$

and an approximate estimate to the binding energy is

$$E_{\mathrm{iso,bulge}} \approx \times 10^{59}\left(\frac{\sigma}{200\,\mathrm{km\,s^{-1}}}\right)^{4.63}\ \mathrm{erg}\quad. \tag{4.7}$$

Equation (4.5) combined with (4.2) then results in an upper limit for the binary BH mass

$$M_{\mathrm{BH}} < M_{\mathrm{BH}}^{\max} \equiv \mathcal{M}(t_{\mathrm{GW}}, e, q_{\mathrm{BH}})\sigma^{4(\alpha+2)/5}\quad; \tag{4.8}$$

above $M_{\rm BH}^{\rm max}$, the energy deposited by the binary in its hardening would become comparable to the binding energy of the surrounding and coalescence would be halted.

$M_{\rm BH}^{\rm max}$ depends on σ and, for our choice of α, $M_{\rm BH}^{\rm max} \propto \sigma^{3.7}$ while its normalization $\mathcal{M}$ is fixed by the coefficient $\mathcal{K}$ and by the parameters intrinsic to the binary, i.e., $t_{\rm GW}$, e, and mass ratio $q_{\rm BH}$. For $t_{\rm GW}$ between 10^6 yr and 10^9 yr, $e = 0$, and $q_{\rm BH} = 1$, the values of $M_{\rm BH}^{\rm max}$ are inferred from equations (4.6) and (4.8), and plotted in Figure 12 in black colors. In the plane $M_{\rm BH}$-σ we overlaid the BH masses from Tremaine et al. (2002) for a comparison with the observations. If the observed BH masses in the sample of Tremaine are the result of a BH binary merger that also led to the formation of the host elliptical, we then conclude that BH binary coalescences did not affect the equilibrium structure of the bulges. If we consider very eccentric ($e = 0.99$) equal-mass binaries, equations (4.6) and (4.8) lead to upper mass limits (depicted with the light-gray strip) again well above the observed points, thus providing less stringent constraints on the BH binary fate. Unequal-mass binaries move further upwards, on the left side of the diagram.

The decaying BH binary thus has no effect on the overall stellar bulge. However, they can have some influence inside the bulge, in a region much larger than the gravitational influence radius of a "single" BH, defined as

$$r_{\rm BH} = GM_{\rm BH}/\sigma^2 \quad . \tag{4.9}$$

Since the mass within radius r of an isothermal sphere scales as $M_{\rm iso}(r) = 2\,r\,\sigma^2/G$, we can determine the "radius of BH binary gravitational influence," $r_{\rm BBH}$, obtained by equating $\Delta E_{\rm BBH}$ to the energy inside $r_{\rm BBH}$, i.e., $\Delta E_{\rm BBH} = M_{\rm iso}(r_{\rm BBH})\sigma^2$:

$$r_{\rm BBH} = \frac{1}{4}\left(\frac{256}{5}\frac{G^3}{c^5}f(e)\right)^{-1/4}\frac{q_{\rm BH}^{3/4}}{(1+q_{\rm BH})^{3/2}}\,t_{\rm GW}^{-1/4}\left(\frac{G^2 M_{\rm BH}^{5/4}}{\sigma^4}\right) \quad . \tag{4.10}$$

This radius is considerably larger than $r_{\rm BH}$. For a $10^8\,M_\odot$ equal-mass circular-BH binary $r_{\rm BBH}/r_{\rm BH} \approx 25$, assuming $\sigma \sim 200\,{\rm km\,s^{-1}}$. The above relation may thus account for the size of the stellar core seen in the bright elliptical galaxies as extensively discussed by Merritt (2006b).

If the bulge hosts a gaseous nuclear component, the energy released by the binary during its path to coalescence may alter the equilibrium of the surrounding gas and a delicate interplay between heating/cooling of the gas and hardening of the binary may lead to a "self-regulated" evolution of the BH mass and the circum-binary gas. Given the expected widespread values of $E_{\rm iso,gas}$ for a nuclear gaseous component in merging galaxies, due to diversities in the mass gas content, thermodynamics and equilibrium end-states, we simply rescale $E_{\rm iso,gas} = 0.1 E_{\rm iso,bulge}$ and plot the corresponding dark-gray strip in Figure 12, for $e = 0$, $q_{\rm BH} = 1$ and same interval of merging times ($t_{\rm GW} = 10^6$–10^9 yr). The strip now shifts to the right and gets closer to the BHs observed along the $M_{\rm BH}$-σ relation. For $t_{\rm GW}$ less than 10^7 yr, there are uncoalesced BHs and BHs that may have deposited enough energy to affect the surrounding gas. This point will be addressed in future investigations.

A further energy constraint may come from accretion. First notice that the gravitational wave time scale $t_{\rm GW}$ is here treated as a parameter. However, in the nuclear region of the galaxy its value is determined by the mechanisms guiding the inspiral, i.e., material (viscous) and gravitational torques and energy dissipation via shocks and radiation. We know that viscosity is the critical parameter at the heart of BH binary hardening on sub-parsec scales (Armitage & Natarajan 2002), and at the heart of accretion (fixing the magnitude of the mass-transfer rate toward the BHs). As shown in Section 3.4, gas inflows, after orbit circularization, may trigger AGN activity onto the BHs, at least temporarily, and a stringent condition for BH binary coalescence in this context may come from the request that the difference between the energy deposited by the accreting BHs in the gas and the energy radiated away by cooling processes (here denoted as ΔE_{acc}) be less that the binding energy of the gas itself (denoted with $E_{\mathrm{iso,gas}}$, here for simplicity):

$$\Delta E_{\mathrm{acc}} \sim f_{\mathrm{acc}} \left(\frac{L}{L_E} \right) \left(\frac{t}{t_{\mathrm{Edd}}} \right) \epsilon M_{\mathrm{BH}} c^2 < E_{\mathrm{iso,gas}} \quad , \tag{4.11}$$

(with L_E the Eddington luminosity and f_{acc} the fractional energy deposited by the accreting BH in the surrounding gas). Typically

$$\Delta E_{\mathrm{acc}} \approx 2 \times 10^{59} f_{\mathrm{rad}} \left(\frac{L}{L_E} \right) \left(\frac{t}{t_{\mathrm{Edd}}} \right) \left(\frac{\epsilon}{0.1} \right) \left(\frac{M_{\mathrm{BH}}}{10^6 \, M_\odot} \right) \, \mathrm{erg} \quad . \tag{4.12}$$

This energy can be larger than ΔE_{BBH}, and may be comparable to the binding energy of the environment, indicating that a major threat to a BH binary stalling in a gaseous background might come from the radiative and/or mechanical energy emitted by the accreting BHs during their inspiral and hardening. Only short-lived (i.e., $t_{\mathrm{GW}} < t_{\mathrm{Edd}}$), and/or sub-Eddington accretion can guarantee the persistence of a "dense and cool" gaseous structure around the binary. Modeling of realistic nuclear disks will help in quantifying accurately $E_{\mathrm{iso,gas}}$ and the problem of the BH hardening in a gaseous environment (Colpi et al. 2009).

5. Summary

- In gas-rich galaxy–galaxy collisions, the BHs "pair" first under the action of dynamical friction against the dark-matter background. When the merger is sufficiently advanced that tidal forces have perturbed the gaseous/stellar disks of the mother galaxies, each BH is found to be surrounded by a prominent gaseous disk. The BHs now have separations of several kiloparsecs.

- When the merger is completed, a rotationally supported, turbulent, nuclear gaseous disk forms at the center of the remnant galaxy. In this "grand" disk of $\sim 19^9 \, M_\odot$, the BHs excite large-scale density waves: gas-dynamical friction is the main driver of their inspiral down to the parsec-scale. A few million years after the completion of the merger the BHs "couple" to form an eccentric *Keplerian binary*. The binary is embedded in the typical, cool environment of a starburst.

- The binary eccentricity decreases to zero if the BHs bind in the grand disk, along a co-rotating orbit. It remains large ($e > 0$), if the BHs bind along counter-rotating orbits. However, this effect may not be generic: in the presence of a very massive nuclear disk with steep density profile, asymmetric instabilities growing in the innermost self-gravitating region could in principle exert torques on the binary, increasing its eccentricity. The evolution of e is thus sensitive to the dynamical pattern of the gas surrounding the binary.

- If the binary circularizes, a small-scale accretion disk forms around each BH, and on scales less than 10 parsecs, double AGN activity can be sustained for approximately 1 Myr.

- No compelling evidence exists on the actual fate of the BH binary in a gaseous environment: whether it stalls or keeps decaying under the action of gravitational and viscous torques until gravitational wave emission guides the inspiral toward coalescence. This may depend sensitively on the BH mass, mass ratio, eccentricity, and thermodynamical response of the gaseous environment to the BH perturbation. We note here that a possible threat may come from energy injection by the BHs, should they accrete, and/or (to a lesser extent) from the energy extracted from the orbit and deposited in the surroundings.

REFERENCES

ARENA, S. E. & BERTIN, G. 2007 *A&A* **463**, 921.

ARMITAGE, P. J & NATARAJAN, P. 2002 *ApJ* **567**, 9.

BALLO, L., BRAITO, V., DELLA CECA, R., MARASCHI, L., TAVECCHIO, F., & DADINA, M. 2004 *ApJ* **600**, 634.

BATE, M. R. & BURKERT, A. 1997 *MNRAS* **288**, 106.

BEGELMAN, M. C., BLANDFORD, R. D., & REES, M. J. 1980 *Nature* **287**, 307.

BENDER, P., ET AL. 1994 *Laser Interferometer Space Antenna for gravitational waves measurements: ESA assessment study report.* ESA.

BERCZIK, P., MERRITT, D., & SPURZEM, R. 2005 *ApJ* **633**, 680.

CAPPELLARI, M., ET AL. 2006 *MNRAS* **366**, 1126.

COLPI, M., CALLEGARI, S., DOTTI, M., KAZANTZIDIS, S., & MAYER, L. 2007. In *The Multicolored Landscape of Compact Objects and Their Explosive Origins.* AIP Conf. Proc. Vol. 924, p. 705. AIP.

COLPI, M., CALLEGARI, S., DOTTI, M., & MAYER, L. 2009 *Class. Quantum Grav.* **26**, 094029.

COLPI, M., MAYER, L., & GOVERNATO, F. 1999 *ApJ* **525**, 720.

DELLA CECA, R., ET AL. 2002 *ApJ* **581**, 9.

DI MATTEO, T., COLBERG, J., SPRINGEL, V., HERNQUIST, L., & SIJACKI, D. 2008 *ApJ* **676**, 33.

DI MATTEO, T., SPRINGEL, V., & HERNQUIST, L. 2005 *Nature* **433**, 604.

DOTTI, M., COLPI, M., & HAARDT, F. 2006 *MNRAS* **367**, 103.

DOTTI, M., COLPI, M., HAARDT, F., & MAYER, L. 2007 *MNRAS* **379**, 956.

DOWNES, D. & SOLOMON, P. M. 1998 *ApJ* **507**, 615.

ESCALA, A., LARSON, R. B., COPPI, P. S., & MARADONES, D. 2004 *ApJ* **607**, 765.

ESCALA, A., LARSON, R. B., COPPI, P. S., & MARADONES, D. 2005 *ApJ* **630**, 152.

FERRARESE, L. & MERRITT, D. 2000 *ApJ* **539**, 9.

GEBHARDT, K., ET AL. 2000 *ApJ* **543**, 5.

GOULD, A. & RIX, H. 2000 *ApJ* **532**, 29.

GOVERNATO, F., COLPI, M., & MARASCHI, L. 1994 *MNRAS* **271**, 317.

GRAHAM, A. W. & DRIVER, S. P. 2007 *ApJ* **655**, 77.

GRANATO, G. L., DE ZOTTI, G., SILVA, L., BRESSAN, A., & DANESE, L. 2004 *ApJ* **600**, 580.

HAEHNELT, M. G. 1994 *MNRAS* **269**, 199.

HERNQUIST, L. 1993 *ApJS* **86**, 389.

JAFFE, A. H. & BACKER, D. C. 2003 *ApJ* **583**, 616.

KATZ, N. 1992 *ApJ* **391**, 502.

KAUFMANN, T., MAYER, L., WADSLEY, J., STADEL, J., & MOORE, B. 2006 *MNRAS* **370**, 1612.

KAZANTZIDIS, S., ET AL. 2005 *ApJ* **623**, 67.

KHOCHFAR, S. & BURKERT, A. 2006 *A&A* **445**, 403.

KING, A. R. 2003 *ApJ* **596**, 27.

KITSIONAS, S. & WHITWORTH, A. P. 2002 *MNRAS* **330**, 129.

KLESSEN, R. S., SPAANS, M., & JAPPSEN, A. 2007 *MNRAS* **374**, 29.

KLYPIN, A., ZHAO, H., & SOMERVILLE, R. S. 2002 *ApJ* **573**, 597.

KOMOSSA, S. 2006 *Memorie della Societa Astronomica Italiana* **77**, 733.

KOMOSSA, S., BURWITZ, V., HASINGER, G., PREDEHL, P., KAASTRA, J. S., & IKEBE, Y. 2003 *ApJ* **582**, 15.

KORMENDY, J. & RICHSTONE, D. 1995 *ARA&A* **33**, 581.

MAGORRIAN, J., ET AL. 1998 *AJ* **115**, 2285.

MAKINO, J. & EBISUZAKI, T. 1996 *ApJ* **465**, 527.

MAKINO, J. & FUNATO, Y. 2004 *ApJ* **602**, 93.

MAX, C. E., CANALIZO, G., & DE VRIES, W. H. 2007 *Science* **316**, 1877.

MAYER, L., KAZANTZIDIS, S., MADAU, P., COLPI, M., QUINN, T., & WADSLEY, J. 2007 *Science* **316**, 1874.

MERRITT, D. 2006a *Rept. Prog. Phys.* **69**, 2513.

MERRITT, D. 2006b *ApJ* **648**, 976.

MILOSAVLJEVIĆ, M. & MERRITT, D. 2001 *ApJ* **563**, 34.

NAVARRO, J. F., FRENK, C. S., & WHITE, S. D. M. 1996 *ApJ* **462**, 563.

OSTRIKER, E. C. 1999 *ApJ* **513**, 252.

RICHSTONE, D. 1998 *Nature* **395**, 14.

RODRIGUEZ, C., TAYLOR, G. B., ZAVALA, R. T., PECK, A. B., POLLACK, L. K., & ROMANI, R. W. 2006 *ApJ* **646**, 623.

SANCHEZ-SALCEDO, F. J. & BRANDENBURG, A. 2001 *MNRAS* **322**, 67.

SESANA, A., HAARDT, F., & MADAU, P., 2007 *ApJ* **660**, 546.

SESANA, A., HAARDT, F., MADAU, P., & VOLONTERI, M. 2005 *ApJ* **623**, 23.

SILK, J. & REES, M. J. 1998 *A&A* **331**, 1.

SPAANS, M. & SILK, J. 2000 *ApJ* **538**, 115.

SPRINGEL, V., DI MATTEO, T., & HERNQUIST, L. 2005a *MNRAS* **361**, 776.

SPRINGEL, V., DI MATTEO, T., & HERNQUIST, L. 2005b *ApJ* **620**, 79.

SPRINGEL, V., FRENK, C. S., & WHITE, S. D. M. 2006 *Nature* **440**, 1137.

SPRINGEL, V. & HERNQUIST, L. 2003 *MNRAS* **339**, 289.

SPRINGEL, V., YOSHIDA, N., & WHITE S. D. M. 2001 *New Astron.* **6**, 79.

TREMAINE, S., ET AL. 2002 *ApJ* **574**, 740.

VAN DEN BOSCH, F. C., LEWIS, G. F., LAKE, G., & STADEL, J. 1999 *ApJ* **515**, 50.

VECCHIO, A. 2004 *Phys. Rev. Letts.* **70**, 2001.

VITALE, S., ET AL. 2002 *Nuclear Physics B; Proceeding Supplements* **110**, 209.

VOLONTERI, M., HAARDT, F., & MADAU, P. 2003 *ApJ* **582**, 559.

WADSLEY, J., STADEL, J., & QUINN, T. 2004 *New Astr.* **9**, 137.

YU, Q. 2002 *MNRAS* **331**, 935.

ZIBETTI, S., GAVAZZI, G., SCODEGGIO, M., FRANZETTI, P., & BOSELLI, A. 2002 *ApJ* **579**, 261.

Dynamics around supermassive black holes

By ALESSIA GUALANDRIS AND DAVID MERRITT

Department of Physics and Center for Computational Relativity and Gravitation,
Rochester Institute of Technology, Rochester, NY 14623, USA

The dynamics of galactic nuclei reflect the presence of supermassive black holes (SBHs) in many ways. Single SBHs act as sinks, destroying a mass in stars equal to their own mass in roughly one relaxation time and forcing nuclei to expand. Formation of binary SBHs displaces a mass in stars roughly equal to the binary mass, creating low-density cores and ejecting hyper-velocity stars. Gravitational radiation recoil can eject coalescing binary SBHs from nuclei, resulting in offset SBHs and lopsided cores. We review recent work on these mechanisms and discuss the observable consequences.

1. Characteristic scales

Supermassive black holes (SBHs) are ubiquitous components of bright galaxies and many have been present with roughly their current masses ($\sim 10^9 \, M_\odot$) since very early times, as early as ~ 1 Gyr after the Big Bang (Fan et al. 2003; Marconi et al. 2004). An SBH strongly influences the motion of stars within a distance r_h, the gravitational influence radius, where

$$r_h = \frac{GM_\bullet}{\sigma_c^2} \; ; \tag{1.1}$$

$M_\bullet$ is the SBH mass and σ_c is the stellar (1d) velocity dispersion in the core. Using the tight empirical correlation between $M_\bullet$ and σ_c:

$$\left(\frac{M_\bullet}{10^8 \, M_\odot} \right) = (1.66 \pm 0.24) \left(\frac{\sigma_c}{200 \text{ km s}^{-1}} \right)^\alpha , \quad \alpha = 4.86 \pm 0.4 , \tag{1.2}$$

(Ferrarese & Ford 2005), this can be written

$$r_h \approx 18 \text{ pc} \left(\frac{\sigma_c}{200 \text{ km s}^{-1}} \right)^{2.86} \approx 13 \text{ pc} \left(\frac{M_\bullet}{10^8 \, M_\odot} \right)^{0.59} . \tag{1.3}$$

While the velocities of stars must increase—by definition—inside r_h, this radius is not necessarily associated with any other observational marker. Such is the case at the Galactic center, for instance, where the stellar density exhibits no obvious feature at $r_h \approx 3$ pc. However the most luminous elliptical galaxies always have cores, regions near the center where the stellar density is relatively low. Core radii are of order r_h in these galaxies, and the stellar mass that was (apparently) removed in creating the core is of order $M_\bullet$. These facts suggest a connection between the cores and the SBHs, and this idea has motivated much recent work, reviewed here, on binary SBHs and on the consequences of displacing SBHs temporarily or permanently from their central locations in galaxies.

An important time scale associated with galactic nuclei (not just those containing SBHs) is the relaxation time, defined as the time for gravitational encounters between stars to establish a locally Maxwellian velocity distribution. The nuclear relaxation time is (Spitzer 1987)

$$T_R \approx \frac{0.34 \sigma_c^3}{G^2 \rho_c m_\star \ln \Lambda} \tag{1.4a}$$

$$\approx 1.2 \times 10^{10} \text{ yr} \left(\frac{\sigma_c}{100 \text{ km s}^{-1}} \right)^3 \left(\frac{\rho_c}{10^5 \, M_\odot \text{ pc}^{-3}} \right)^{-1} \left(\frac{m_\star}{M_\odot} \right)^{-1} \left(\frac{\ln \Lambda}{15} \right)^{-1} , \tag{1.4b}$$

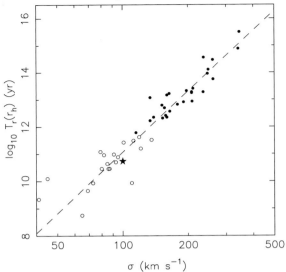

FIGURE 1. Relaxation time, measured at the SBH influence radius, in a sample of early-type galaxies (Côté et al. 2004), vs. the central stellar velocity dispersion. Filled symbols are galaxies in which the SBH's influence radius is resolved; the star is the Milky Way bulge (from Merritt, Mikkola, & Szell 2007).

with ρ_c the nuclear density and $\ln \Lambda$ the Coulomb logarithm. Figure 1 shows estimates of T_R measured at r_h in a sample of early-type galaxies, assuming $m_\star = 1\,M_\odot$. A least-squares fit to the points (shown as the dashed line in the figure) gives

$$T_R(r_h) \approx 2.5 \times 10^{13} \ \mathrm{yr} \left(\frac{\sigma}{200 \ \mathrm{km \ s^{-1}}} \right)^{7.47} \approx 9.6 \times 10^{12} \ \mathrm{yr} \left(\frac{M_\bullet}{10^8 \ M_\odot} \right)^{1.54} . \qquad (1.5)$$

"Collisional" nuclei can be defined as those with $T_R(r_h) \lesssim 10$ Gyr; Figure 1 shows that such nuclei are uniquely associated with galaxies that are relatively faint—as faint as or fainter than the Milky Way bulge, which has $T_R(r_h) \approx 4 \times 10^{10}$ yr. Furthermore, relaxation-driven changes in the stellar distribution around an SBH are generally confined to radii $\lesssim 10^{-1}\,r_h$, making them all but unobservable in galaxies beyond the Local Group (T. Alexander, these proceedings). But the relaxation time also fixes the rate of gravitational scattering of stars into the central "sink"—either a single or a binary SBH—and this fact has important consequences for nuclear evolution in low-luminosity galaxies, as discussed below.

2. Core structure

The "core" of a galaxy can loosely be defined as the region near the center where the density of starlight drops significantly below what is expected based on an inward extrapolation of the overall luminosity profile. At large radii, the surface brightness profiles of early-type galaxies are well fit by the Sérsic (1968) model,

$$\ln I(R) = \ln I_e - b(n) \left[(R/R_e)^{1/n} - 1 \right] . \qquad (2.1)$$

The quantity b is normally chosen such that R_e is the projected radius containing one-half of the total light. The shape of the profile is then determined by n; $n = 4$ is the de Vaucouleurs (1948) model, which is a good representation of bright elliptical (E) galaxies (Kormendy & Djorgovski 1989), while $n = 1$ is the exponential model, which approximates the luminosity profiles of dwarf elliptical (dE) galaxies (Binggeli, Sandage, & Tarenghi 1984). An alternative way to write Eq. (2.1) is

$$\frac{d\ln I}{d\ln R} = -\frac{b}{n}\left(\frac{R}{R_e}\right)^{1/n} \quad , \tag{2.2}$$

i.e., the logarithmic slope varies as a power of the projected radius. While there is no consensus on why the Sérsic model is such a good representation of stellar spheroids, a possible hint comes from the dark-matter halos produced in N-body simulations of hierarchical structure formation, which are also well described by Eq. (2.2) (Navarro et al. 2004), suggesting that Sérsic's model applies generally to systems that form via dissipationless clustering (Merritt et al. 2005).

Sérsic's model is known to accurately reproduce the luminosity profiles in some galaxies over at least three decades in radius (e.g., Graham et al. 2003), but deviations often appear near the center. Galaxies fainter than absolute magnitude $M_B \approx -19$ tend to have *higher* central surface brightness than predicted by Sérsic's model; the structure of the central excess is typically unresolved, but its properties are consistent with those of a compact, intermediate-age star cluster (Carollo, Stiavelli, & Mack 1998; Côté et al. 2006; Balcells et al. 2007). Galaxies brighter than $M_B \approx -20$ have long been known to exhibit central *deficits* (e.g., Kormendy 1985a); these have traditionally been called simply "cores," perhaps because a flat central density profile was considered *a priori* most natural (Tremaine 1997).

For about two decades, it was widely believed that dE galaxies were distinct objects from the more luminous E galaxies. The dividing line between the two classes was put at absolute magnitude $M_B \approx -18$, based partly on the presence of cores in bright galaxies, and also on the relation between total luminosity and mean surface brightness (Kormendy 1985b). This view was challenged by Jerjen and Binggeli (1997), and in a compelling series of papers, A. Graham and collaborators showed that—aside from the cores—early-type galaxies display a remarkable continuity of structural properties, from $M_B \approx -13$ to $M_B \approx -22$ (Graham & Guzman 2003; Graham et al. 2003; Trujillo et al. 2004). This continuity of properties has recently been extended to the centers: Côté et al. (2007) have shown that the net deviation from a Sérsic profile is a smooth function of galaxy luminosity, changing from negative (core) to positive (nuclear cluster) at $M_B \approx -19.5$.

The connection between nuclear star clusters and SBHs, if any, is unclear; in fact it has been suggested that the two are mutually exclusive (Ferrarese et al. 2006; Wehner & Harris 2006).

Here we focus on the cores. The cores extend outward to a break radius r_b that is roughly a few times r_h, or from ~ 0.01 to ~ 0.05 times R_e. A more robust way of quantifying the cores is in terms of their mass (i.e., light): the "mass deficit" (Milosavljević et al. 2002) is defined as the difference in integrated mass between the observed density profile $\rho(r)$ and an inward extrapolation of the outer profile, $\rho_{\rm out}(r)$, typically modeled as a Sérsic profile (Figure 2):

$$M_{\rm def} \equiv 4\pi \int_0^{r_b} \left[\rho_{\rm out}(r) - \rho(r)\right] r^2 dr \quad . \tag{2.3}$$

Figure 2 shows mass deficits for a sample of "core" galaxies, expressed in units of the SBH mass. There is a clear peak at $M_{\rm def} \approx 1\,M_\bullet$, although some galaxies have much larger cores.

The fact that core and SBH masses are often so similar suggests a connection between the two. Ejection of stars by binary SBHs during galaxy mergers is a natural model

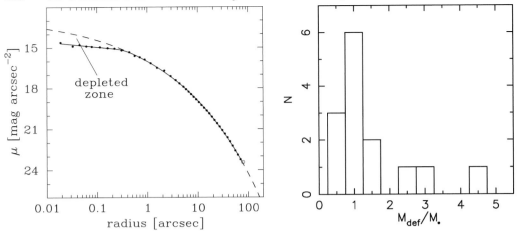

FIGURE 2. Left: Surface brightness profile in the R band of NGC 3348, a "core" galaxy. The dashed line is the best-fitting Sérsic model; the observed profile (points, and solid line) fall below this inside of a break radius $r_b \approx 0\rlap{.}''35$ (from Graham 2004). Right: Histogram of observed mass deficits for the sample of core galaxies in Graham (2004) and Ferrarese et al. (2006; adapted from Merritt 2006a).

(Begelman, Blandford, & Rees 1980); the non-existence of cores in fainter galaxies could then be due to regeneration of a steeper density profile by star formation (e.g., McLaughlin et al. 2006) or by dynamical evolution associated with the (relatively) short relaxation times in faint galaxies (e.g., Merritt & Szell 2006). However, the largest cores are difficult to explain via the binary model (Milosavljević & Merritt 2001).

3. Massive binaries

A typical mass ratio for galaxy mergers in the local Universe is $\sim 10:1$ (e.g., Sesana et al. 2004). To a good approximation, the initial approach of the two SBHs can therefore be modelled by assuming that the galaxy hosting the smaller BH spirals inward under the influence of dynamical friction from the fixed distribution of stars in the larger galaxy. Modelling both galaxies as singular isothermal spheres ($\rho \sim r^{-2}$) and assuming that the smaller galaxy spirals in on a circular orbit, its tidally truncated mass is $m_2 \approx \sigma_2^2 r/2G\sigma$, where σ_2 and σ are the velocity dispersion of the small and large galaxy, respectively (Merritt 1984). Chandrasekhar's (1943) formula then gives for the orbital decay rate and infall time

$$\frac{dr}{dt} = -0.30 \frac{Gm_2}{\sigma r} \ln \Lambda, \qquad t_{\text{infall}} \approx 3.3 \frac{r(0)\sigma^2}{\sigma_2^3}, \tag{3.1}$$

where $\ln \Lambda$ has been set to 2. Using Eq. (1.2) to relate σ and σ_2 to the respective SBH masses M_1 and M_2, this becomes

$$t_{\text{infall}} \approx 3.3 \frac{r(0)}{\sigma} \left(\frac{M_2}{M_1}\right)^{-0.62}, \tag{3.2}$$

i.e., t_{infall} exceeds the crossing time of the larger galaxy by a factor $\sim q^{-0.6}$, $q \equiv M_2/M_1 \leqslant 1$. Thus for mass ratios $q \gtrsim 10^{-3}$, infall requires less than $\sim 10^2 T_{\text{cr}} \approx 10^{10}$ yr. This mass ratio is roughly the ratio between the masses of the largest ($\sim 10^{9.5} M_\odot$) and smallest ($\sim 10^{6.5} M_\odot$) known SBHs and so it is reasonable to assume that galaxy mergers will almost always lead to formation of a binary SBH in a time less than 10 Gyr. This conclusion is strengthened if the effects of gas are taken into account (e.g., Mayer et al. 2007).

Equation (3.1) begins to break down when the two SBHs approach more closely than $\sim r_h$, the influence radius of the larger hole, since the orbital energy of M_2 is absorbed by the stars, lowering their density and reducing the frictional force. In spite of this slowdown, N-body integrations (Merritt & Cruz 2001; Milosavljević & Merritt 2001; Makino & Funato 2004; Berczik et al. 2005) show that the separation between the two SBHs continues to drop rapidly until the binary semi-major axis is $a \approx a_h$, where

$$a_h \equiv \frac{G\mu}{4\sigma^2} \approx \frac{1}{4}\frac{q}{(1+q)^2}r_h \tag{3.3a}$$

$$\approx 3.3\mathrm{pc}\frac{q}{(1+q)^2}\left(\frac{M_1+M_2}{10^8 M_\odot}\right)^{0.59}, \tag{3.3b}$$

and $\mu \equiv M_1 M_2/(M_1 + M_2)$ is the binary reduced mass. At this separation—the "hard binary" separation—the binary's binding energy per unit mass is $\sim\sigma^2$, and it ejects stars that pass within a distance $\sim a$ with velocities large enough to remove them from the nucleus (Mikkola & Valtonen 1992; Quinlan 1996).

What happens next depends on the density and geometry of the nucleus. In a spherical or axisymmetric galaxy, the mass in stars on orbits that intersect the binary is small, $\lesssim M_1 + M_2$, and the binary rapidly ejects these stars; no stars then remain to interact with the binary and its evolution stalls (Figure 3). In non-axisymmetric (e.g., triaxial) nuclei, the mass in stars on centrophilic orbits can be much larger, allowing the binary to continue shrinking past a_h. And in collisional nuclei of any geometry, gravitational scattering of stars can repopulate depleted orbits. These different possibilities are discussed in more detail below.

If the binary does stall at $a \approx a_h$, it will have given up an energy

$$\Delta E \approx -\frac{GM_1 M_2}{2r_h} + \frac{GM_1 M_2}{2a_h} \tag{3.4a}$$

$$\approx -\frac{1}{2}M_2\sigma^2 + 2(M_1 + M_2)\sigma^2 \tag{3.4b}$$

$$\approx 2(M_1 + M_2)\sigma^2 \tag{3.4c}$$

to the stars in the nucleus, i.e., the energy transferred from the binary to the stars is roughly proportional to the *combined* mass of the two SBHs. The reason for this counter-intuitive result is the $a_h \propto M_2$ dependence of the stalling radius (Eq. [3.3]): smaller infalling BHs form tighter binaries. Detailed N-body simulations (Merritt 2006a) verify that the mass deficit generated by the binary in evolving from $\sim r_h$ to $\sim a_h$ is a weak function of the mass ratio

$$M_{\mathrm{def,h}} \approx 0.4(M_1 + M_2)\left(\frac{q}{0.1}\right)^{0.2}, \tag{3.5}$$

for $0.025 \lesssim q \lesssim 0.5$. A mass deficit of $\sim 0.5~M_\bullet$ is still a factor ~ 2 too small to explain the observed peak in the $M_{\mathrm{def}}/M_\bullet$ histogram (Figure 2). On the other hand, bright elliptical galaxies have probably undergone numerous mergers, and the proportionality between M_{def} and $M_1 + M_2$ (rather than, say, M_2) implies that the mass deficit following $\mathcal{N}$ mergers is $\sim 0.5\mathcal{N}$ times the *accumulated* BH mass. Mass deficits in the range $0.5 \lesssim M_{\mathrm{def}}/M_\bullet \lesssim 1.5$ therefore imply $1 \lesssim \mathcal{N} \lesssim 3$ mergers, consistent with the number of major mergers expected for bright galaxies since the epoch at which most of the gas was depleted (e.g., Haehnelt & Kauffmann 2002). Hierarchical growth of cores tends to saturate after a few mergers however making it difficult to explain mass deficits greater

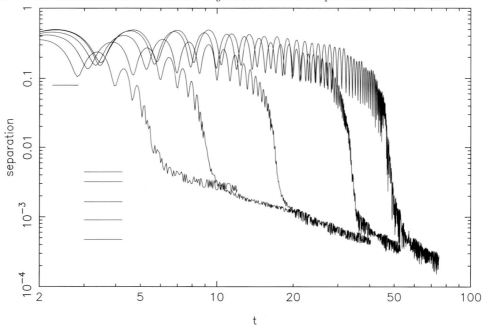

FIGURE 3. Early evolution of binary SBHs in N-body galaxies, for different values of the binary mass ratio, $M_2/M_2 = 0.5, 0.25, 0.1, 0.05, 0.025$ (left to right). The upper horizontal line indicates r_h, the influence radius of the more massive hole. Lower horizontal lines show a_h (Eq. [3.3]), the "hard-binary" separation. The evolution of the binary slows drastically when $a \approx a_h$; in a real (spherical) galaxy with much larger N, evolution would stall at this separation. The smaller the infalling BH, the farther it spirals in before stalling (adapted from Merritt 2006a).

than $\sim 2\, M_\bullet$ in this way. An effective way to enlarge cores still more is to kick the SBH out, at least temporarily, as discussed in Section 4.

The first convincing evidence for a true, binary SBH was recently presented by Rodriguez et al. (2006), who discovered two compact, flat-spectrum AGN at the center of a single elliptical galaxy, with a projected separation of ~ 7 pc. This is consistent with the radius at formation of a $\sim 10^{7.5}\, M_\odot$ binary (Eq. [1.3]), or the stalling radius for a binary of at least $\sim 10^{9.3}\, M_\odot$ (Eq. [3.3]). Rodriguez et al. estimate a binary mass of $\sim 10^8\, M_\odot$, but with considerable uncertainty. All other examples of "binary" SBHs in single galaxies have separations $\gg r_h$ (Komossa 2006).

Even in a spherical galaxy, the stalling that occurs at $a \approx a_h$ can be avoided if stars continue to be scattered onto orbits that intersect the binary (Valtonen 1996; Yu 2002; Milosavljević & Merritt 2003). Such "collisional loss-cone repopulation" requires that the two-body (star-star) relaxation time at $r \approx r_h$ be less then $\sim 10^{10}$ yr; according to Eq. (1.5), this is the case in galaxies with $M_\bullet \lesssim 10^6\, M_\odot$, i.e., at the extreme low end of the SBH mass distribution. Collisional loss cone repopulation is therefore irrelevant to the luminous galaxies that are observed to have cores but may be important in the mass range ($M_\bullet \lesssim 10^7\, M_\odot$) of most interest to space-based gravitational wave interferometers like *LISA* (Hughes 2006).

N-body simulation would seem well suited to this problem (e.g., Governato, Colpi, & Maraschi 1994; Makino 1997; Milosavljević & Merritt 2001; Makino & Funato 2004). The difficulty is noise—or more precisely, getting the level of noise just right. In a real galaxy there is a clear separation of time scales between an orbital period, and the

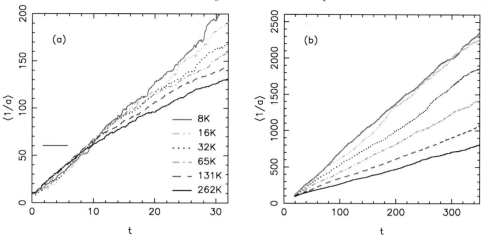

FIGURE 4. Short-term (a) and long-term (b) evolution of a massive binary in a series of N-body integrations. Vertical axis is the inverse semi-major axis (i.e., energy) of the binary, computed by averaging several independent N-body runs; different curves correspond to different values of N, the number of "star" particles. The evolution of the binary is independent of N until $a \approx a_h$ (horizontal line); thereafter the evolution rate is limited by how quickly stars are scattered onto orbits that intersect the binary, and decreases with increasing N (from Merritt, Mikkola, & Szell 2007).

time for stars to be scattered onto depleted orbits: the first is typically much shorter than the second, which means that orbits intersecting the binary will remain empty for many periods before a new star is scattered in. This is called the "empty loss cone" regime and it implies that supply of stars to the binary will take place diffusively. In an N-body simulation, however, N is much smaller than its value in real galaxies and orbits are repopulated too quickly. This is the reason that binary evolution rates in N-body simulations typically scale as $N^{-\alpha}, \alpha < 1$ rather than the $\sim T_R^{-1} \sim N^{-1}$ dependence expected if stars diffused gradually into an empty loss cone (Milosavljević & Merritt 2003). Figure 4 provides an illustration: early evolution of the binary, until $a \approx a_h$, is N-independent; formation of a hard binary then depletes the loss cone and continued hardening occurs at a rate that is a decreasing function of N, though less steep than N^{-1}.

An alternative approach is based on the Fokker-Planck equation. Both single and binary SBHs can be modeled as "sinks" located at the centers of galaxies (Yu 2002). The main differences are the larger physical extent of the binary ($\sim G(M_1 + M_2)/\sigma^2$ vs. $GM_\bullet/c^2$) and the fact that the binary gives stars a finite kick rather than disrupting or consuming them completely. However, the diffusion rate of stars into a central sink varies only logarithmically with the size of the sink (Lightman & Shapiro 1978), and a hard binary ejects most stars well out of the core with $V \gg \sigma$, so the analogy is fairly good. The Fokker-Planck equation describing nuclei with sinks is (Bahcall & Wolf 1977)

$$\frac{\partial N}{\partial t} = 4\pi^2 p(E) \frac{\partial f}{\partial t} = -\frac{\partial F_E}{\partial E} - \mathcal{F}(E, t) \quad . \tag{3.6}$$

Here $N(E, t) = 4\pi p(E) f(E, t)$ is the distribution of stellar energies, $f(E, t)$ is the phase space density and $p(E)$ is a phase-space volume element. The first term on the RHS of Eq. (3.6) describes the response of f to the flux F_E of stars in energy space due to encounters. The second term, $-\mathcal{F}$, is the flux of stars into the sink, which is dominated by scattering in angular momentum (Frank & Rees 1976). A proper treatment of the latter term requires a 2d (energy, angular momentum) analysis, but a good approximation to $\mathcal{F}$ can be derived by assuming that the distribution of stars has reached a quasi steady-state near the loss cone boundary in phase space (Cohn & Kulsrud 1978). If the sink is a binary SBH, a second equation is needed that relates the flux of stars into the loss cone to the rate of change of the binary's semi-major axis:

$$\frac{d}{dt}\left(\frac{1}{a}\right) = \frac{2\langle C \rangle}{a(M_1 + M_2)} \int \mathcal{F}(E,t) dE \quad , \tag{3.7}$$

with $\langle C \rangle \approx 1.25$ a dimensionless mean energy change for stars that interact with the binary.

Both terms on the RHS of Eq. (3.6) imply changes in a time $\sim T_R$. The first term on its own implies evolution toward the Bahcall-Wolf (1976) "zero flux" solution, $\rho \sim r^{-7/4}$. The second term implies that a mass of order $\sim M_\bullet$ will be scattered into the sink in a time of $T_R(r_h)$. When the sink is a binary SBH, the binary responds by ejecting the incoming stars and shrinking, according to Eq. (3.7). As a result, changes in the structure of the nucleus on a relaxation time scale (e.g., growth of a core) are directly connected to changes in the binary semi-major axis.

Numerical solutions to Eqs. (3.6), (3.7)—including also the effects of a changing gravitational potential—have been presented by Merritt et al. (2007). The solutions are well fit by

$$\ln\left(\frac{a_h}{a}\right) = -\frac{B}{A} + \sqrt{\frac{B^2}{A^2} + \frac{2}{A}\frac{t}{T_R(r_h)}} \quad , \tag{3.8}$$

where t is defined as the time since the binary first became hard ($a = a_h$), and the coefficients $A \approx 0.016$, $B \approx 0.08$ depend weakly on the binary mass ratio. Including the effect of energy lost to gravitational radiation:

$$\frac{d}{dt}\left(\frac{1}{a}\right) = \frac{d}{dt}\left(\frac{1}{a}\right)_{\text{stars}} + \frac{d}{dt}\left(\frac{1}{a}\right)_{\text{GR}} \quad , \tag{3.9}$$

allows one to compute the time to full coalescence, T_{coal}. Figure 5 shows T_{coal} (right panel), and the time spent by the binary in unit intervals of $\ln a$ prior to coalescence (left panel), as functions of binary mass. The time to coalescence is well fit by

$$Y = C_1 + C_2 X + C_3 X^2 \quad , \tag{3.10a}$$

$$Y \equiv \log_{10}\left(\frac{T_{\text{coal}}}{10^{10} \text{ yr}}\right) \quad , \tag{3.10b}$$

$$X \equiv \log_{10}\left(\frac{M_1 + M_2}{10^6 M_\odot}\right) \quad . \tag{3.10c}$$

with

$$M_2/M_1 = 1 : \ C_1 = -0.372, \ C_2 = 1.384, \ C_3 = -0.025 \tag{3.11a}$$

$$M_2/M_1 = 0.1 : \ C_1 = -0.478, \ C_2 = 1.357, \ C_3 = -0.041 \quad . \tag{3.11b}$$

Based on the figure, binary SBHs would be able to coalesce via interaction with stars alone in galaxies with $M_\bullet \lesssim 2 \times 10^6 M_\odot$. For $M_1 + M_2 \gtrsim 10^7 M_\odot$, evolution for 10 Gyr only brings the binary separation slightly below a_h; in such galaxies the most likely separation to find a massive binary (in the absence of other sources of energy loss) would be near a_h.

The core continues to grow as the binary shrinks, but the mass deficit is not related in a simple way to the mass in stars "ejected" by the binary (e.g., Quinlan 1996). Rather it

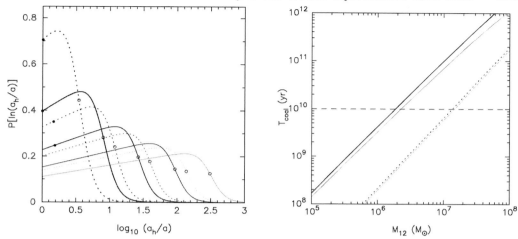

FIGURE 5. Evolution of a binary SBH in a collisional nucleus, based on a Fokker-Planck model that allows for evolution of the stellar distribution (Merritt, Mikkola, & Szell 2007). Left: Probability of finding the binary in a unit interval of $\ln a$. From left to right, curves are for $M_1 + M_2 = (0.1, 1, 10, 100) \times 10^6 \, M_\odot$. Solid (dashed) curves are for $M_2/M_1 = 1(0.1)$. Open circles indicate when the rate of energy loss to stars equals the loss rate to gravitational waves; filled circles correspond to an elapsed time since $a = a_h$ of 10^{10} yr. For the two smallest values of $M_\bullet$, the latter time occurs off the graph to the right. Right: Total time for a binary to evolve from $a = a_h$ to gravitational wave coalescence as a function of binary mass. The thick curve is for $M_2/M_1 = 1$ and the thin curve is for $M_2/M_1 = 0.1$. Dotted curves show the time spent in the gravitational radiation regime only.

results from a competition between loss of stars to the binary, represented by $-\mathcal{F}(E, t)$, and the change in $N(E, t)$ due to diffusion of stars in energy, represented by $-\partial F_E/\partial E$. As the mass deficit increases, so do gradients in f, which increases the flux of stars toward the center and counteracts the drop in density. In principle the two terms could balance, but at some distance from the center the relaxation time is so long that local $F_E(E)$ must drop below the integrated loss term $\int_E^\infty \mathcal{F}(E)dE$—stars can not diffuse in fast enough to replace those being lost to the binary and the density drops. The Fokker-Planck solutions show that the mass deficit increases with binary binding energy as

$$M_{\mathrm{def,c}} \approx 1.7 \, (M_1 + M_2) \log_{10} (a_h/a) \quad , \tag{3.12}$$

again with a weak dependence on M_2/M_1. The mass deficit at the onset of the gravitational radiation regime is found to be

$$M_{\mathrm{def,c}} \approx (4.5, 3.5, 2.6, 1.6)(M_1 + M_2) \quad (M_2/M_1 = 1) \tag{3.13a}$$
$$\approx (3.4, 2.6, 1.7, 0.9)(M_1 + M_2) \quad (M_2/M_1 = 0.1) \quad , \tag{3.13b}$$

for $M_1 + M_2 = (10^5, 10^6, 10^7, 10^8) \, M_\odot$. These values should be added to the mass deficits Eq. (3.5) generated during formation of the binary when predicting core sizes in real galaxies.

Are such mass deficits observed? Only a handful of galaxies in the relevant mass range ($M_{\mathrm{gal}} \lesssim 10^{10} \, M_\odot$) are near enough that their cores could be resolved, even if present; of these, neither the Milky Way nor M32 exhibit cores. Also, as noted above, many low-luminosity spheroids have compact central excesses rather than cores. These facts do not rule out the past existence of massive binaries in these galaxies however.

(1) Binary evolution might have been driven more by gas dynamical torques than by ejection of stars; gas content during the most recent major merger is believed to be a steep inverse function of galaxy luminosity (Kauffmann & Haehnelt 2000).

(2) Star formation can create a dense core after the two SBHs have coalesced (Mihos & Hernquist 1994).

(3) A two-body relaxation time short enough to bring the two SBHs together would also allow a Bahcall-Wolf cusp to be regenerated in a comparable time after the two SBHs combine, tending to erase the core (Merritt & Szell 2006).

From the point of view of physicists hoping to detect gravitational waves, it is disappointing that this model only guarantees coalescence at the extreme low end of the SBH mass distribution. (Astronomers hoping to detect binary SBHs may take the opposite point of view.) Fortunately, there is no dearth of ideas for overcoming the "final parsec problem" and allowing binary SBHs to merge efficiently, even in massive galaxies.

3.1. *Non-axisymmetric geometries*

Real galaxies are not spherical nor even axisymmetric; parsec-scale bars are relatively common and departures from axisymmetry are often invoked to enhance fueling of AGN (e.g., Shlosman et al. 1990). Orbits in a triaxial nucleus can be "centrophilic," passing arbitrarily close to the center after a sufficiently long time (Poon & Merritt 2001, 2004). This implies feeding rates for a central binary that can approach the "full loss cone" rate in spherical geometries, or

$$\frac{d}{dt}\left(\frac{1}{a}\right) \approx 2.5 F_{\rm c} \frac{\sigma}{r_h^2} \tag{3.14}$$

if the fraction $F_{\rm c}$ of centrophilic orbits is large (Merritt & Poon 2004). While $F_{\rm c}$ is impossible to know in any particular galaxy, even small values imply much larger feeding rates than in a diffusively repopulated loss cone. Figure 6 shows results from N-body simulations that support this idea.

3.2. *Secondary slingshot*

Stars ejected by a massive binary can interact with it again if they return to the nucleus on nearly-radial orbits. The total energy extracted from the binary via this "secondary slingshot" will be the sum of the discreet energy changes during the interactions. Milosavljević and Merritt (2003) showed that a mass $\mathcal{M}_\star$ of stars initially in the binary's loss cone causes the binary to evolve as

$$\frac{1}{a} \approx \frac{1}{a_h} + \frac{4}{r_h} \ln\left(1 + \frac{t}{t_0}\right), \quad t_0 = \frac{2\mu\sigma^2}{\mathcal{M}_\star \langle \Delta E \rangle} P(E) \ , \tag{3.15}$$

in the absence of diffusive loss cone repopulation, where $\langle \Delta E \rangle$ is the specific energy change after one interaction with the binary, E is the initial energy and $P(E)$ is the orbital period. The secondary slingshot runs its course after a few orbital periods. Sesana et al. (2007) sharpened this analysis by carrying out detailed three-body scattering experiments and recording the precise changes in energy of stars as they underwent repeated interactions with the binary. They inferred modest ($\sim \times 2$) changes in $1/a$ due to the secondary slingshot, but their assumption of a $\rho \sim r^{-2}$ density profile around the binary was probably over-optimistic; such steep density profiles are never observed and even if present initially would be rapidly destroyed when the binary first formed.

3.3. *Bound subsystems*

As noted above, recent observational studies have greatly increased the number of galaxies believed to harbor compact nuclear star clusters; inferred masses for the clusters are

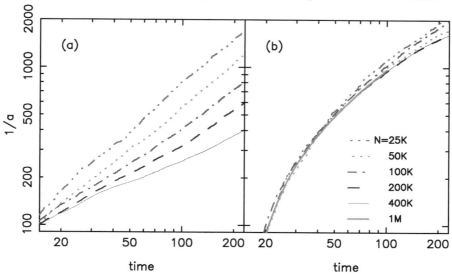

FIGURE 6. Efficient merger of binary SBHs in barred galaxies. Plots are based on N-body simulations (no gas) of equal-mass binaries at the centers of galaxy models, with and without rotation. (a) Spherical models. The binary hardening rate declines with increasing N, as in Figure 4, implying that the evolution would stall in the large-N limit. (b) Binary evolution in a flattened, rotating version of the same galaxy model. At $t \approx 10$, the rotating model forms a triaxial bar. Binary hardening rates in this model are essentially independent of N, indicating that the supply of stars to the binary is not limited by collisional loss-cone refilling as in the spherical models. This is currently the only simulation that follows two SBHs from kiloparsec to sub-parsec separations and that can be robustly scaled to real galaxies (from Berczik et al. 2006).

comparable with the mass that would normally be associated with a SBH. It is not yet clear whether these subsystems co-exist with SBHs, but if they do, they could provide an extra source of stars to interact with a massive binary. Zier (2006, 2007) explored this idea, assuming a steeply rising density profile around the binary, $\rho \propto r^{-\gamma}$, at the time that its separation first reached $\sim a_h$. Zier concluded that a cluster having total mass $\sim M_1 + M_2$, distributed as a steep power law, $\gamma \gtrsim 2.5$, could extract enough energy from the binary to allow gravitational wave coalescence in less than 10 Gyr. N-body tests of this hypothesis are sorely needed; as in the Sesana (2007) study, Zier's approach did not allow him to self-consistently follow the effect of formation of the binary on the surrounding mass distribution.

3.4. *Massive perturbers*

In a nucleus containing a spectrum of masses, the gravitational scattering rate is proportional to

$$\tilde{m} = \frac{\int n(m) m^2 dm}{\int n(m) m \, dm} \quad , \tag{3.16}$$

(e.g., Merritt 2004). Perets et al. (2007) argued that "massive perturbers" near the center of the Milky Way—massive stars, star clusters, giant molecular clouds—are sufficiently numerous to dominate $\tilde{m}$, implying potentially much higher rates of gravitational scattering into a central sink than in the case of solar-mass perturbers. Perets and Alexander (2008) extended this argument to galaxies in general, emphasizing in particular the early stages following a galactic merger, and concluded that collisional loss cone repopulation would be sufficient to guarantee coalescence of binary SBHs in less than 10 Gyr for all but the most massive binaries. As in the studies of Sesana et al. (2007) and Zier (2007), Perets and Alexander (2008) optimistically assumed a steep ($\rho \propto r^{-2}$) density profile around the binary, in spite of N-body studies showing rapid destruction of the cusps. Their arguments for massive perturbers in giant E galaxies are also rather speculative.

3.5. *Multiple SBHs*

An extreme case of a "massive perturber" is a third SBH, which might scatter stars into a central binary (Zhao et al. 2002), or perturb the binary directly, driving the two SBHs into an eccentric orbit and shortening the time scale for gravitational wave losses (Valtonen et al. 1994; Makino & Ebisuzaki 1994; Blaes et al. 2002; Volonteri et al. 2003; Iwasawa et al. 2006; Hoffman & Loeb 2007). The likelihood of multiple-SBH systems forming is probably highest in the brightest E galaxies since massive binaries are most likely to stall (low stellar density, little gas), and since large galaxies experience the most frequent mergers. Here again, more N-body simulations, including post-Newtonian terms, are needed; among other dynamical effects that could then be self-consistently included are changes in core structure, and BH-core oscillations like those described in the next section.

3.6. *Gas*

The same galaxy mergers that create binary SBHs can also drive gas into the nucleus, and there is abundant observational evidence for cold (e.g., Jackson et al. 1993; Gallimore et al. 2001; Greenhill et al. 2003) and hot (e.g., Baganoff et al. 2003) gas near the centers of at least some galaxies. Dense concentrations of gas can substantially accelerate the evolution of a massive binary by increasing the drag on the individual BHs (Escala et al. 2004, 2005; Dotti et al. 2007). However, the plausibility of such dense accumulations of gas, with mass comparable to the mass of the SBHs, is unclear (e.g., Sakamoto et al. 1999; Christopher et al. 2005). Large-scale galaxy merger simulations (Kazantzidis et al. 2005; Mayer et al. 2007) show that the presence of gas leads to more rapid formation of the massive binary, but these simulations still lack the resolution to follow the binary past $a \approx r_h$, and so have nothing relevant to say (yet) concerning the final parsec problem.

As this summary indicates, many possible solutions to the "final parsec problem" exist, but none is guaranteed to be effective in all, or even most, galaxies. The safest bet is that both coalesced and uncoalesced binary SBHs exist, but with what relative frequency is still anyone's guess.

4. SBH/IBH binaries

Secure dynamical evidence exists for SBHs in the mass range $10^{6.5} \lesssim M_\bullet/M_\odot \lesssim 10^{9.5}$ (Ferrarese & Ford 2005) and compelling arguments have been made for BHs with masses $10^5 \lesssim M_\bullet/M_\odot \lesssim 10^7$ in active nuclei (Greene & Ho 2004). Binary mass ratios as extreme as 1000:1, and possibly greater, are therefore to be expected. This possibility has received most attention in the context of intermediate-mass black holes (IBHs) in the Milky Way, where they could form in dense star clusters like the Arches or Quintuplet before spiraling into the center and forming a tight binary with the $\sim 3.5 \times 10^6\ M_\odot$ SBH (Portegies Zwart & McMillan 2002; Hansen & Milosavljević 2003).

The predicted hard-binary separation for a SBH/IBH pair is Eq. (3.3)

$$a_h \approx 0.5\ \text{mpc} \left(\frac{q}{10^{-3}} \right) \left(\frac{M_\bullet}{3.5 \times 10^6\ M_\odot} \right)^{0.59}, \quad q \equiv \frac{M_{\text{IBH}}}{M_\bullet}\ . \qquad (4.1)$$

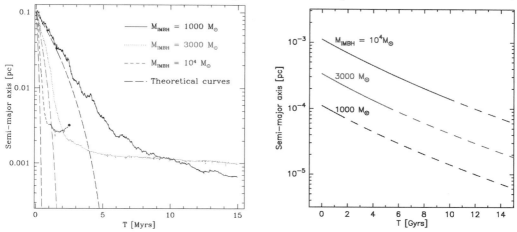

FIGURE 7. Left: N-body simulations of the inspiral of IBHs into the center of the Milky Way. Solid lines show the separation between the two BHs and dashed lines are theoretical predictions that ignore loss-cone depletion or changes in the structure of the core. Smaller IBHs spiral in farther before "stalling"; the $M_{\rm IBH} = 10^3 \, M_\odot$ simulation ends before the stalling radius is reached (from Baumgardt, Gualandris, & Portegies Zwart 2006). Right: Evolution beyond $a = a_h$, based on the Fokker-Planck model of Merritt, Mikkola, and Szell (2007). Dashed lines indicate when the evolution time due to gravitational radiation losses is less than 10 Gyr.

This separation—$\sim 10^2$ AU—is comparable to the periastron distances of the famous "S" stars (Eckart et al. 2002; Ghez et al. 2005). Dynamical constraints on the existence of an IBH at this distance from the SBH are currently weak (Yu & Tremaine 2003; Hansen & Milosavljević 2003; Reid & Brunthaler 2004). Figure 7a shows N-body simulations designed to mimic inspiral of IBHs into the Galactic center. The figure confirms the expected slowdown in the inspiral rate at a separation $\sim a_h$. Figure 7b plots evolutionary tracks for the same three IBH masses as in the left panel, based on the Fokker-Planck model of Merritt et al. (2007). For $M_{\rm IBH} \lesssim 10^3 \, M_\odot$, evolution of the binary is dominated by gravitational wave losses already at $a = a_h$.

The same inward flux of stars that allows the binary to shrink also implies an outward flux of stars ejected by the binary. The latter are a potential source of "hyper-velocity stars" (HVSs), stars moving in the halo with greater than Galactic escape velocity (Hills 1988). The relation between the stellar ejection rate and the binary hardening rate, when $a \leqslant a_h$, is given by Eq. (3.7) after rewriting it as

$$T_{\rm hard} \equiv a \frac{d}{dt}\left(\frac{1}{a}\right) \approx \frac{2.5}{M_\bullet} \times \text{flux} \quad ; \tag{4.2}$$

here "flux" is the total mass in stars per unit time, from all energies, that are scattered into (and ejected by) the binary. Combining Eq. (4.2) with Eq. (3.8), the flux is

$$\sim 5.0 \frac{M_\bullet}{T_R(r_h)} \left[1 + \frac{5t}{T_R(r_h)}\right]^{-1/2} \tag{4.3a}$$

$$\lesssim 350 \, M_\odot \, {\rm yr}^{-1} \quad . \tag{4.3b}$$

where the second line uses values appropriate to the Galactic center. Relating the total ejected flux to the number of HVSs that would be observed is not straightforward; for instance, only a fraction ($\lesssim 10\%$) would be ejected with high enough velocity to still be moving faster than ~ 500 km s^{-1} after climbing through the Galactic potential

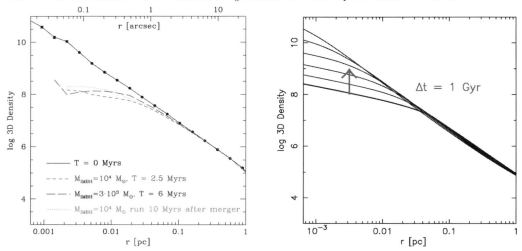

FIGURE 8. Left: Creation of a core by inspiral of IBHs into the Galactic center. The initial density profile is shown by the solid line. Dotted line shows the core 10 Myr after the $10^4\,M_\odot$ IBH has merged with the SBH; almost no change occurs during this time (from Baumgardt, Gualandris, & Portegies Zwart 2006). Right: Fokker-Planck model showing how the cusp regenerates due to two-body scattering, on Gyr timescales (adapted from Merritt, Mikkola, & Szell 2007).

(Gualandris et al. 2005; Baumgardt et al. 2006), and targeted searches for HVSs only detect certain stellar types so that knowledge of the stellar mass function is also required (Brown et al. 2006).

Inspiral of the IBH creates a core of radius ~ 0.05 pc $\approx 1''$ (Figure 8a). Such a core might barely be detectable at the center of the Milky Way from star counts. There is no clear indication of a core (Schoedel et al. 2007), but if the inspiral occurred more than a few Gyr ago, star-star gravitational scattering would have gone some way toward "refilling" the region depleted by the binary (Merritt & Wang 2005; Merritt & Szell 2006; Figure 7b). In this case, however, the ejected stars would almost all have moved beyond the range of HVS surveys by now.

The angular distribution of the ejected stars has been proposed as a test for their origin; unlike other possible sources of HVSs, a SBH/IBH binary tends to eject stars parallel to the orbital plane or, if the orbit is eccentric, in a particular direction (Levin 2006; Sesana et al. 2006). In two N-body simulations of IBH inspiral, however (Baumgardt, Gualandris, & Portegies Zwart 2006; Matsubayashi et al. 2007), the orientation of the binary began to change appreciably, in the manner of a random walk, after it became hard. This was due to "rotational Brownian motion" (Merritt 2002): torques from passing stars—the same stars that extract energy and angular momentum from the binary—also change the direction of the binary's orbital angular momentum vector. In one hardening time $|a/\dot{a}|$ of the binary, its orientation changes by

$$\Delta\theta \approx q^{-1/2}\left(\frac{m_\star}{M_\bullet}\right)^{1/2}\left(1-e^2\right)^{-1/2} \tag{4.4a}$$

$$\approx 9.0^\circ \left(\frac{q}{10^{-3}}\right)^{-1/2}\left(\frac{M_\bullet}{10^6 m_\star}\right)^{-1/2}\left[\frac{\left(1-e^2\right)^{-1/2}}{5}\right] . \tag{4.4b}$$

(The eccentricity dependence in Eq. [4.4b] is approximate; the numerical coefficient in this equation has only been confirmed by detailed scattering experiments for $e = 0$.) In both of the cited N-body studies, the binary eccentricity evolved appreciably away from zero before the orientation changes became significant. However, rotational Brownian motion might not act quickly enough to randomize the orientation of a SBH/IBH binary in a time of $\sim 10^8$ yr, the flight time from the Galactic center to the halo, unless perturbers more massive than Solar-mass stars are present near the binary (Merritt 2002; Perets & Alexander 2008).

5. Kicks and cores

After seeming to languish for several decades, the field of numerical relativity has recently experienced exciting progress. Following the breakthrough papers of Pretorius (2005), Campanelli et al. (2006), and Baker et al. (2006a,b), several groups have now successfully simulated the evolution of binary BHs all the way to coalescence. The final inspiral is driven by emission of gravitational waves, and in typical (asymmetric) inspirals, a net impulse—a "kick"—is imparted to the system due to anisotropic emission of the waves (Bekenstein 1973; Fitchett 1983; Favata et al. 2004). Early arguments that the magnitude of the recoil velocity would be modest for non-spinning BHs were confirmed by the simulations, which found $V_{\mathrm{kick}} \lesssim 200$ km s^{-1} in the absence of spins (Baker et al. 2006a,b; González et al. 2007b; Herrmann et al. 2007). The situation changed dramatically following the first (Campanelli et al. 2007a) simulations of "generic" binaries, in which the individual BHs were spinning and tilted with respect to the orbital angular momentum vector. Kicks as large as ~ 2000 km s^{-1} have now been confirmed (Campanelli et al. 2007b; González et al. 2007a; Tichy & Marronetti 2007), and scaling arguments based on the post-Newtonian approximation suggest that the maximum kick velocity would probably increase to ~ 4000 km s^{-1} in the case of maximally spinning holes (Campanelli et al. 2007b). The most propitious configuration for the kicks appears to be an equal-mass binary in which the individual spin vectors are oppositely aligned and oriented parallel to the orbital plane. For unequal-mass binaries, the maximum kick is

$$V_{\mathrm{max}} \approx 6 \times 10^4 \text{ km s}^{-1} \frac{q^2}{(1+q)^4} \quad , \tag{5.1}$$

where $q \equiv M_2/M_1 \leqslant 1$ is the binary mass ratio and maximal spins have been assumed (Campanelli et al. 2007c). Orienting the BHs with their spins perpendicular to the orbital angular momentum may seem odd (Bogdanović, Reynolds, & Miller 2007), but there is considerable evidence that SBH spins bear no relation to the orientations of the gas disks that surround them (e.g., Kinney et al. 2000; Gallimore et al. 2006) and this is presumably even more true with respect to the directions of infalling BHs. Galaxy escape velocities are $\lesssim 3000$ km s^{-1} (Merritt et al. 2004), so gravitational wave recoil can in principle eject coalescing SBHs completely from galaxies.

Detailed N-body simulations show that the motion of an SBH that has been kicked with enough velocity to eject it out of the core, but not fast enough to escape the galaxy entirely, exhibits three distinct phases (Gualandris & Merritt 2008):

• *Phase I*: The SBH oscillates with decreasing amplitude, losing energy via dynamical friction each time it passes through the core. Chandrasekhar's theory accurately reproduces the motion of the SBH in this regime for values $2 \lesssim \ln \Lambda \lesssim 3$ of the Coulomb logarithm, if the gradually-decreasing core density is taken into account.

• *Phase II*: When the amplitude of the motion has decayed to roughly the core radius, the SBH and core begin to exhibit oscillations about their common center of mass (Figure 9). These oscillations decay exponentially (Figure 10), but with a time constant that

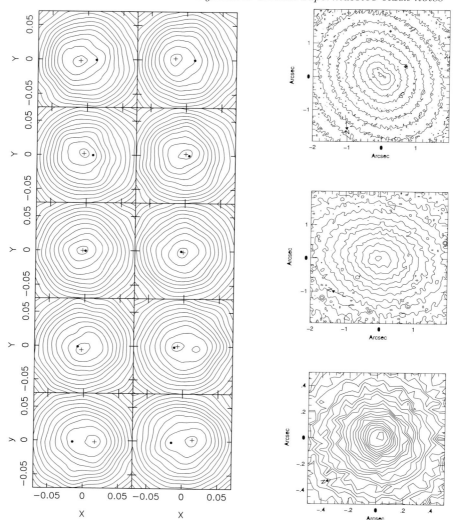

FIGURE 9. Left: Core oscillations in an *N*-body simulation of ejection of an SBH from the center of a galaxy; the kick velocity was 60% of the escape velocity. Contour plots show the stellar density at equally spaced times, spanning ~1/2 of the SBH's orbital period. Filled circles are the SBH and crosses indicate the location of the (projected) density maxima (from Gualandris & Merritt 2008). Right: Surface brightness contours of three "core" galaxies with double or offset nuclei, from Lauer et al. (2005). Top: NGC 4382; middle: NGC 507; bottom: NGC 1374.

is 10–20 times longer than would be predicted by a naive application of the dynamical friction formula.

• *Phase III*: Eventually the SBH's kinetic energy drops to an average value

$$\frac{1}{2}M_{\bullet}V_{\bullet}^2 \approx \frac{1}{2}m_{\star}\mathrm{v}_{\star}^2 \tag{5.2}$$

i.e., to the kinetic energy of a single star. This is the regime of gravitational Brownian motion (Bahcall & Wolf 1976; Young 1977; Merritt et al. 2007).

A natural definition of the "return time" of a kicked SBH is the time to reach the Brownian regime. Unless the kick is very close to the escape velocity, the return time is

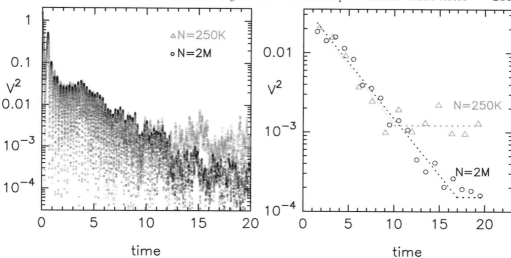

FIGURE 10. Evolution of the SBH kinetic energy following a kick of 60% the central escape velocity, in two N-body simulations of a galaxy represented by N stars, with $N = (2.5 \times 10^5, 2 \times 10^6)$. The mass of the SBH and the total mass of the galaxy are the same in the two simulations; all that varies is the mass of the "star" particles. The right-hand panel shows binned values of V^2. Most of the elapsed time is spent in SBH/core oscillations like those illustrated in Figure 9. Eventually, the SBH's kinetic energy decays to the Brownian value, shown as the horizontal dashed lines in the right panel. The Brownian velocity scales as $m_\star^{1/2}$ and so is smaller for larger N. Scaled to a $3 \times 10^{10}\ M_\odot$ galaxy, the time to reach the Brownian regime would be $\sim 10^8$ yr (adapted from Gualandris & Merritt 2008).

dominated by the time spent in "Phase II"; during this time, the SBH's energy decays roughly as

$$E \approx \Phi(0) + \Phi(r_c)e^{-(t-t_c)/\tau} \qquad (5.3)$$

(Gualandris & Merritt 2008); t_c is the time when the SBH re-enters the core whose radius is r_c. The damping time in the N-body simulations, τ, is

$$\tau \approx 15\frac{\sigma_c^3}{G^2\rho_c M_\bullet} \qquad (5.4a)$$

$$\approx 1.2 \times 10^7\ \text{yr}\left(\frac{\sigma_c}{250\ \text{km s}^{-1}}\right)^3\left(\frac{\rho_c}{10^3\ M_\odot\ \text{pc}^{-3}}\right)^{-1}\left(\frac{M_\bullet}{10^9\ M_\odot}\right)^{-1}, \qquad (5.4b)$$

with ρ_c and σ_c the core density and velocity dispersion respectively. The number of decay times required for the SBH's energy to reach the Brownian level is $\sim \ln(M_\bullet/m_\star) \approx 20$, implying that a kicked SBH will remain significantly off-center for a long time, as long as ~ 1 Gyr in a bright galaxy with a low-density core.

In fact, asymmetric cores are rather common. These include off-center nuclei (Binggeli et al. 2000; Lauer et al. 2005); double nuclei (Lauer et al. 1996); and cores with a central minimum in the surface brightness (Lauer et al. 2002). Three examples from Lauer et al. (2005) are reproduced here on the right side of Figure 9; all are luminous "core" galaxies, and each strikingly resembles at least one frame from the N-body montage on the left. The longevity of the "Phase II" oscillations makes the kicks a plausible model for the observed asymmetries. This explanation is probably not appropriate for the famous double nucleus of M31, since M31 is not a "core" galaxy, and since one of the brightness peaks in M31 (the one associated with the SBH) lies essentially at the galaxy photocenter;

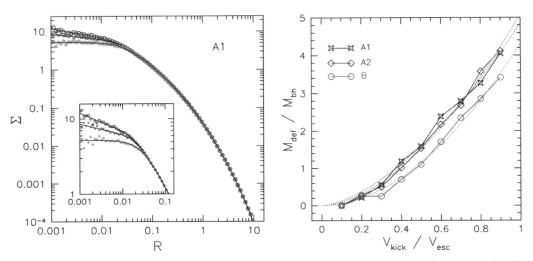

FIGURE 11. Left: Points show projected density profiles computed from N-body models after a kicked SBH has returned to the center, for three different values of the kick velocity ($V_{kick} = (0.2, 0.4, 0.8) \times V_{esc}$). Each set of points is compared with the best-fitting core-Sérsic model (lines). The insert shows a zoom into the central region. Right: Mass deficits generated by kicked SBHs. The different sets of points correspond to different galaxy models (adapted from Gualandris & Merritt 2008).

Figure 9 suggests that an oscillating SBH would typically (though not always) be found on the opposite side of the galaxy from the point of peak brightness. The M31 double nucleus has been successfully modeled as a clump of stars on eccentric orbits which maintain their lopsidedness by virtue of moving deep within the Keplerian potential of the SBH (Tremaine 1995).

The kicks are quite effective at inflating cores (Merritt et al. 2004; Boylan-Kinchin et al. 2004). Figure 11, from Gualandris and Merritt (2008), illustrates this: the mass deficit generated by the kick is approximately

$$M_{def,k} \approx 5M_\bullet \left(V_{kick}/V_{esc}\right)^{1.75} \quad . \tag{5.5}$$

Even modest kicks can generate cores substantially larger than those produced during the formation of a massive binary ($\sim 1\ M_\bullet$; Eq. [3.5]). Furthermore, this mechanism is potentially effective in even the most luminous galaxies, unlike the relaxation-driven model for core growth discussed above (Eq. [3.12]), which only applies to nuclei with short relaxation times. Gravitational radiation recoil is therefore a tenable explanation for the subset of luminous E galaxies with large mass deficits (Figure 2). An alternative explanation for the over-sized cores (Lauer et al. 2007) postulates that the SBHs in these galaxies are "hypermassive," $M_\bullet \gtrsim 10^{10}\ M_\odot$ and that the cores are a consequence of slingshot ejection by a massive binary.

The kicks have a number of other potentially observable consequences, including spatially and/or kinematically offset AGN (Madau & Quataert 2004; Haehnelt et al. 2006; Merritt et al. 2006b; Bonning et al. 2007) and distorted or wiggling radio jets (Gualandris & Merritt 2008). Many of these manifestations were first discussed by R. Kapoor in a remarkably prescient series of papers (Kapoor 1976, 1983a,b, 1985).

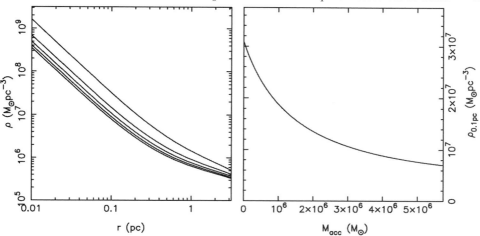

FIGURE 12. Black-hole-driven expansion of a nucleus; this Fokker-Planck model was given parameters such that the density at the end is similar to what is currently observed in the nucleus of M32, with a final influence radius $r_h \approx 3$ pc. The left panel shows density profiles at constant time intervals after a Bahcall-Wolf cusp has been established; the right panel shows the evolution of the density at 0.1 pc as a function of $M_{\rm acc}$, the accumulated mass in tidally disrupted stars. As scaled to M32, the final time is roughly 2×10^{10} yr. This plot suggests that the densities of collisional nuclei like those of M32 and the Milky Way were once higher, by factors of $\sim$a few, than at present (from Merritt 2006b).

6. Black-hole-driven expansion

The growth of a core around a shrinking, binary SBH was discussed above: beyond a certain radius, the relaxation time becomes so long that the encounter-driven flux of stars toward the center cannot compensate for losses to the binary, forcing the density to drop. A similar process takes place around a single SBH (Shapiro 1977; Dokuchaev 1989): stars coming too near are consumed, or disrupted, and the density drops. This effect is absent from the classical equilibrium models for stars around a BH (e.g., Bahcall & Wolf 1976; Cohn & Kulsrud 1978) since these solutions fix the phase-space density far from the BH, enforcing an inward flux of stars precisely large enough to replace the stars being consumed by the sink. In reality, the BH acts as a heat source, in much the same way that hard binary stars inject energy into a post-core-collapse globular cluster and cause it to re-expand.

A simple model that produces self-similar expansion of a nucleus containing a SBH can be constructed by simply changing the outer boundary condition in the Bahcall and Wolf (1976) problem from $f(0) = f_0$ to $f(0) = 0$. One finds that the evolution after $\sim$one relaxation time can be described as $\rho(r, t) = \rho_c(t)\rho^*(r)$, with $\rho^*(r)$ slightly steeper than the $\rho \sim r^{-7/4}$ Bahcall-Wolf form; the normalization drops off as $\rho_c \propto t^{-1}$ at late times. Figure 12 shows the results of a slightly more realistic calculation in a model designed to mimic the nearby dE galaxy M32. After reaching approximately the Bahcall-Wolf form, the density drops in amplitude with roughly fixed slope for $r \lesssim r_h$. This example suggests that the nuclei of galaxies like M32 or the Milky Way might have been $\sim$a few times denser in the past than they are now, with correspondingly higher rates of stellar tidal disruption and stellar collisions.

Expansion due to a central BH has been observed in a handful of studies based on fluid (Amaro-Seoane et al. 2004), Monte-Carlo (Shapiro & Marchant 1978; Marchant & Shapiro 1980; Freitag et al. 2006), Fokker-Planck (Murphy et al. 1991), and N-body

(Baumgardt et al. 2004) algorithms. All of these studies allowed stars to be lost into or destroyed by the BH; however most adopted parameters more suited to globular clusters than to nuclei, e.g., a constant-density core. Murphy et al. (1991) applied the isotropic, multi-mass Fokker-Planck equation to the evolution of nuclei containing SBHs, including an approximate loss term in the form of Eq. (3.6) to model the scattering of low-angular-momentum stars into the SBH. Most of their models had what would now be considered uncharacteristically high densities and the evolution was dominated by physical collisions between stars. However in two models with lower densities, they reported observing significant expansion over 10^{10} yr; these models had initial central relaxation times of $T_r \lesssim 10^9$ yr when scaled to real galaxies, similar to the relaxation times near the centers of M32 and the Milky Way. The $\rho \sim r^{-7/4}$ form of the density profile near the SBH was observed to be approximately conserved during the expansion. Freitag et al. (2006) carried out Monte-Carlo evolutionary calculations of a suite of models containing a mass spectrum, some of which were designed to mimic the Galactic center star cluster. After the stellar-mass BHs in their models had segregated to the center, they observed a roughly self-similar expansion. Baumgardt et al. (2004) followed core collapse in N-body models with and without a massive central particle; "tidal destruction" was modeled by simply removing stars that came within a certain distance of the massive particle. When the "black hole" was present, the cluster expanded almost from the start and in an approximately self-similar way. These important studies notwithstanding, there is a crucial need for more work on this problem in order to understand how the rates of processes like stellar tidal disruption vary over cosmological times (e.g., Milosavljević et al. 2006).

We thank H. Baumgardt, A. Graham and T. Lauer for permission to reproduce figures from their published work. We acknowledge support from the National Science Foundation under grants AST-0420920 and AST-0437519 and from the National Aeronautics and Space Administration under grants NNG04GJ48G and NNX07AH15G.

REFERENCES

Amaro-Seoane, P., Freitag, M., & Spurzem, R. 2004 *MNRAS* **352**, 655.

Baganoff, F. K., et al. 2003 *ApJ* **591**, 891.

Bahcall, J. N. & Wolf, R. A. 1976 *ApJ* **209**, 214.

Bahcall, J. N. & Wolf, R. A. 1977 *ApJ* **216**, 883.

Baker, J. G., Centrella, J., Choi, D.-I., Koppitz, M., & van Meter, J. 2006a *Phys. Rev. Letts.* **96**, 111102.

Baker, J. G., Centrella, J., Choi, D.-I., Koppitz, M., van Meter, J. R., & Miller, M. C. 2006b *ApJ* **653**, L93.

Balcells, M., Graham, A. W., & Peletier, R. F. 2007 *ApJ* **665**, 1084.

Baumgardt, H., Gualandris, A., & Portegies Zwart, S. 2006 *MNRAS* **372**, 174.

Baumgardt, H., Makino, J., & Ebisuzaki, T. 2004 *ApJ* **613**, 1133.

Begelman, M. C., Blandford, R. D., & Rees, M. J. 1980 *Nature* **287**, 307.

Bekenstein, J. D. 1973 *ApJ* **183**, L657.

Berczik, P., Merritt, D., & Spurzem, R. 2005 *ApJ* **633**, 680.

Berczik, P., Merritt, D., Spurzem, R., & Bischof, H.-P. 2006 *ApJ* **642**, L21.

Binggeli, B., Barazza, F., & Jerjen, H. 2000 *A&A* **359**, 447.

Binggeli, B., Sandage, A., & Tarenghi, M. 1984 *AJ* **89**, 64.

Blaes, O., Lee, M. H., & Socrates, A. 2002 *ApJ* **578**, 775.

Bogdanović, T., Reynolds, C. S., & Miller, M. C. 2007 *ApJ* **661**, L147.

Bonning, E. W., Shields, G. A., & Salviander, S. 2007 *ApJ* **666**, L13.

Boylan-Kolchin, M., Ma, C.-P., & Quataert, E. 2004 *ApJ* **613**, L37.

BROWN, W. R., GELLER, M. J., KENYON, S. J., & KURTZ, M. J. 2006 *ApJ* **640**, L35.

CAMPANELLI, M., LOUSTO, C. O., MARRONETTI, P., & ZLOCHOWER, Y. 2006 *Phys. Rev. Letts.* **96**, 111101.

CAMPANELLI, M., LOUSTO, C. O., ZLOCHOWER, Y., KRISHNAN, B., & MERRITT, D. 2007a *Phys. Rev. D* **75**, 064030.

CAMPANELLI, M., LOUSTO, C., ZLOCHOWER, Y., & MERRITT, D. 2007b *ApJ* **659**, L5.

CAMPANELLI, M., LOUSTO, C. O., ZLOCHOWER, Y., & MERRITT, D. 2007c *Phys. Rev. Letts.* **98**, 231102.

CAROLLO, C. M., STIAVELLI, M., & MACK, J. 1998 *AJ* **116**, 68.

CHANDRASEKHAR, S. 1943 *ApJ* **97**, 255.

CHRISTOPHER, M. H., SCOVILLE, N. Z., STOLOVY, S. R., & YUN, M. S. 2005 *ApJ* **622**, 346.

COHN, H. & KULSRUD, R. M. 1978 *ApJ* **226**, 1087.

CÔTÉ, P., ET AL. 2004 *ApJS* **153**, 223.

CÔTÉ, P., ET AL. 2006 *ApJS* **165**, 57.

CÔTÉ, P., ET AL. 2007 *ApJ* **671**, 1456.

DE VAUCOULEURS, G. 1948 *Annales d'Astrophysique* **11**, 247.

DOKUCHAEV, V. I. 1989 *Sov. Astron. Letts.* **15**, 167.

DOTTI, M., COLPI, M., HAARDT, F., & MAYER, L. 2007 *MNRAS* **582**, 956.

ECKART, A., GENZEL, R., OTT, T., & SCHÖDEL, R. 2002 *MNRAS* **331**, 917.

ESCALA, A., LARSON, R. B., COPPI, P. S., & MARDONES, D. 2004 *ApJ* **607**, 765.

ESCALA, A., LARSON, R. B., COPPI, P. S., & MARDONES, D. 2005 *ApJ* **630**, 152.

FAN, X., ET AL. 2003 *AJ* **125**, 1649.

FAVATA, M., HUGHES, S. A., & HOLZ, D. E. 2004 *ApJ* **607**, L5.

FERRARESE, L., ET AL. 2006 *ApJ* **644**, L21.

FERRARESE, L. & FORD, H. 2005 *Sp. Sci. Rev.* **116**, 523.

FITCHETT, M. J.1983 *MNRAS* **203**, 1049.

FRANK, J. & REES, M. J. 1976 *MNRAS* **176**, 633.

FREITAG, M., AMARO-SEOANE, P., & KALOGERA, V. 2006 *ApJ* **649**, 91.

GALLIMORE, J. F., ET AL. 2001 *ApJ* **556**, 694.

GALLIMORE, J. F., AXON, D. J., O'DEA, C. P., BAUM, S. A., & PEDLAR, A. 2006 *AJ* **132**, 546.

GHEZ, A. M. 2005 *ApJ* **620**, 744.

GONZÁLEZ, J. A., HANNAM, M., SPERHAKE, U., BRÜGMANN, B., & HUSA, S. 2007a *Phys. Rev. Letts.* **98**, 231101.

GONZÁLEZ, J. A., SPERHAKE, U., BRÜGMANN, B., HANNAM, M., & HUSA, S. 2007b *Phys. Rev. Letts.* **98**, 091101.

GOVERNATO, F., COLPI, M., & MARASCHI, L. 1994 *MNRAS* **271**, 317.

GRAHAM, A. W. 2004 *ApJ* **613**, L33.

GRAHAM, A. W., ERWIN, P., TRUJILLO, I., & ASENSIO RAMOS, A. 2003 *AJ* **125**, 2951.

GRAHAM, A. W. & GUZMÁN, R. 2003 *AJ* **125**, 2936.

GREENE, J. E. & HO, L. C. 2004 *ApJ* **610**, 722.

GREENHILL, L. J., ET AL. 2003 *ApJ* **590**, 162.

GUALANDRIS, A. & MERRITT, D. 2008 *ApJ* **678**, 780.

GUALANDRIS, A., PORTEGIES ZWART, S., & SIPIOR, M. S. 2005 *MNRAS* **363**, 223.

HAEHNELT, M. G., DAVIES, M. B., & REES, M. J. 2006 *MNRAS* **366**, L22.

HAEHNELT, M. G. & KAUFFMANN, G. 2002 *MNRAS* **336**, L61.

HANSEN, B. M. S. & MILOSAVLJEVIĆ, M. 2003 *ApJ* **593**, L77.

HERRMANN, F., HINDER, I., SHOEMAKER, D., & LAGUNA, P. 2007 *Class. Quantum Gravity* **24**, 33.

HILLS, J. G. 1988 *Nature* **331**, 687.

HOFFMAN, L. & LOEB, A. 2007 *MNRAS* **377**, 957.

HUGHES, S. A. 2006. In *Laser Interferometer Space Antenna: 6th International LISA Symposium*, AIP Conference Proceedings, Vol. 873, p. 13. AIP.

IWASAWA, M., FUNATO, Y., & MAKINO, J. 2006 *ApJ* **651**, 1059.

JACKSON, J. M., ET AL. 1993 *ApJ* **402**, 173.

JERJEN, H. & BINGGELI, B. 1997. In *The Nature of Elliptical Galaxies; 2nd Stromlo Symposium* (eds. M. Arnaboldi, G. S. Da Costa, & P. Saha) ASP Conference Series, Vol. 116, p. 239. ASP.

KAPOOR, R. C. 1976 *Pramana* **7**, 334.

KAPOOR, R. C. 1983a *Ap&SS* **93**, 79.

KAPOOR, R. C. 1983b *Ap&SS* **95**, 425.

KAPOOR, R. C. 1985 *Ap&SS* **112**, 347.

KAUFFMANN, G. & HAEHNELT, M. 2000 *MNRAS* **311**, 576.

KAZANTZIDIS, S., ET AL. 2005 *ApJ* **623**, L67.

KINNEY, A. L., SCHMITT, H. R., CLARKE, C. J., PRINGLE, J. E., ULVESTAD, J. S., & ANTONUCCI, R. R. J. 2000 *ApJ* **537**, 152.

KOMOSSA, S. 2006 *Memorie della Societa Astronomica Italiana* **77**, 733.

KORMENDY, J. 1985a *ApJ* **292**, L9.

KORMENDY, J. 1985b *ApJ* **295**, 73.

KORMENDY, J. & DJORGOVSKI, S. 1989 *ARAA* **27**, 235.

LAUER, T. R., ET AL. 1996 *ApJ* **471**, L79.

LAUER, T. R., ET AL. 2002 *AJ* **124**, 1975.

LAUER, T. R., ET AL. 2005 *AJ* **129**, 2138.

LAUER, T. R., ET AL. 2007 *AJ* **662**, 808.

LEVIN, Y. 2006 *ApJ* **653**, 1203.

LIGHTMAN, A. P./ & SHAPIRO, S. L. 1978 *Rev. Mod. Phys.* **50**, 437.

MADAU, P. & QUATAERT, E. 2004 *ApJ* **606**, L17.

MAKINO, J. 1997 *ApJ* **478**, 58.

MAKINO, J. & EBISUZAKI, T. 1994 *ApJ* **446**, 607.

MAKINO, J. & FUNATO, Y. 2004 *ApJ* **602**, 93.

MARCHANT, A. B. & SHAPIRO, S. L. 1980 *ApJ* **239**, 685.

MARCONI, A., RISALITI, G., GILLI, R., HUNT, L. K., MAIOLINO, R., & SALVATI, M. 2004 *MNRAS* **351**, 169.

MATSUBAYASHI, T., MAKINO, J., & EBISUZAKI, T. 2007 *ApJ* **656**, 879.

MAYER, L., KAZANTZIDIS, S., MADAU, P., COLPI, M., QUINN, T., & WADSLEY, J. 2007 *Science* **316**, 1874.

McLAUGHLIN, D. E., KING, A. R., & NAYAKSHIN, S. 2006 *ApJ* **650**, L37.

MERRITT, D. 1984 *ApJ* **276**, 26.

MERRITT, D. 2002 *ApJ* **568**, 998.

MERRITT, D. 2004 *Phys. Rev. Letts.* **92**, 201304.

MERRITT, D. 2006a *ApJ* **648**, 976.

MERRITT, D. 2006b *Rept. Prog. Phys.* **69**, 2513.

MERRITT, D., BERCZIK, P., & LAUN, F. 2007 *AJ* **133**, 553.

MERRITT, D. & CRUZ, F. 2001 *ApJ* **551**, L41.

MERRITT, D., GRAHAM, A. W., MOORE, B., DIEMAND, J., & TERZIĆ, B. 2006b *AJ* **132**, 2685.

MERRITT, D., MIKKOLA, S., & SZELL, A. 2007 *ApJ* **671**, 53.

MERRITT, D., MILOSAVLJEVIĆ, M., FAVATA, M., HUGHES, S. A., & HOLZ, D. E. 2004 *ApJ* **607**, L9.

MERRITT, D., NAVARRO, J. F., LUDLOW, A., & JENKINS, A. 2005 *ApJ* **624**, L85.

MERRITT, D. & POON, M. Y. 2004 *ApJ* **606**, 788.

MERRITT, D., STORCHI-BERGMANN, T., ROBINSON, A., BATCHELDOR, D., AXON, D., & CID FERNANDES, R. 2006a *MNRAS* **367**, 1746.

MERRITT, D. & SZELL, A. 2006 *ApJ* **648**, 890.

MERRITT, D. & WANG, J. 2005 *ApJ* **621**, L101.

MIHOS, J. C. & HERNQUIST, L. 1994 *ApJ* **437**, L47.

MIKKOLA, S. & VALTONEN, M. J. 1992 *MNRAS* **259**, 115.

MILOSAVLJEVIĆ, M. & MERRITT, D. 2001 *ApJ* **563**, 34.

MILOSAVLJEVIĆ, M. & MERRITT, D. 2003 *ApJ* **596**, 860.

MILOSAVLJEVIĆ, M., MERRITT, D., & HO, L. C. 2006 *ApJ* **652**, 120.

MILOSAVLJEVIĆ, M., MERRITT, D., REST, A., & VAN DEN BOSCH, F. C. 2002 *MNRAS* **331**, L51.

MURPHY, B. W., COHN, H. N., & DURISEN, R. H. 1991 *ApJ* **370**, 60.

NAVARRO, J. F., ET AL. 2004, *MNRAS* **349**, 1039.

PERETS, H. B. & ALEXANDER, T. 2008 *ApJ* **677**, 146.

PERETS, H. B., HOPMAN, C., & ALEXANDER, T. 2007 *ApJ* **656**, 709.

POON, M. Y. & MERRITT, D. 2001 *ApJ* **549**, 192.

POON, M. Y. & MERRITT, D. 2004 *ApJ* **606**, 774.

PORTEGIES ZWART, S. F. & MCMILLAN, S. L. W. 2002 *ApJ* **576**, 899.

PRETORIUS, F. 2005 *Phys. Rev. Letts.* **95**, 121101.

QUINLAN, G. D. 1996 *New Astron.* **1**, 35.

REID, M. J. & BRUNTHALER, A. 2004 *ApJ* **616**, 872.

RODRIGUEZ, C., TAYLOR, G. B., ZAVALA, R. T., PECK, A. B., POLLACK, L. K., & ROMANI, R. W. 2006 *ApJ* **646**, 623.

SAKAMOTO, K., OKUMURA, S. K., ISHIZUKI, S., & SCOVILLE, N. Z. 1999 *ApJS* **124**, 403.

SCHÖDEL, R., ET AL. 2007 *A&A* **469**, 125.

SÉRSIC, J. L. 1968 *Atlas de Galaxies Australes.* Cordoba: Observatorio Astronomico.

SESANA, A., HAARDT, F., & MADAU, P. 2006 *ApJ* **651**, 392.

SESANA, A., HAARDT, F., & MADAU, P. 2007 *ApJ* **660**, 546.

SESANA, A., HAARDT, F., MADAU, P., & VOLONTERI, M. 2004 *ApJ* **611**, 623.

SHAPIRO, S. L. 1977 *ApJ* **217**, 281

SHAPIRO, S. L. & MARCHANT, A. B. 1978 *ApJ* **225**, 603.

SHLOSMAN, I., BEGELMAN, M. C., & FRANK, J. 1990 *Nature* **345**, 679.

SPITZER, L. 1987 *Dynamical Evolution of Globular Clusters.* Princeton University Press.

TICHY, W. & MARRONETTI, P. 2007 *Phys. Rev. D* **76**, 061502.

TREMAINE, S. 1995 *AJ* **110**, 628.

TREMAINE, S. 1997. In *Unsolved Problems in Astrophysics* (eds. J. N. Bahcall & J. P. Ostriker), p. 137. Princeton University Press.

TRUJILLO, I., ERWIN, P., ASENSIO RAMOS, A., & GRAHAM, A. W. 2004 *AJ* **127**, 1917.

VALTONEN, M. J. 1996, *Comm. Astrophys.* **18**, 191.

VALTONEN, M. J., MIKKOLA, S., HEINAMAKI, P., & VALTONEN, H. 1994 *ApJS* **95**, 69.

VOLONTERI, M., HAARDT, F., & MADAU, P. 2003 *ApJ* **582**, 559.

WEHNER, E. H. & HARRIS, W. E. 2006 *ApJ* **644**, L17.

YOUNG, P. J. 1977 *ApJ* **215**, 36.

YU, Q. 2002 *MNRAS* **331**, 935.

YU, Q. & TREMAINE, S. 2003 *ApJ* **599**, 1129.

ZHAO, H., HAEHNELT, M. G., & REES, M. J. 2002 *New Astron.* **7**, 385.

ZIER, C. 2006 *MNRAS* **371**, L36.

ZIER, C. 2007 *MNRAS* **378**, 1309.

Black hole formation and growth: Simulations in general relativity

By STUART L. SHAPIRO[1,2]

[1]Department of Physics, University of Illinois at Urbana-Champaign, Urbana, IL 61801, USA

[2]Department of Astronomy and National Center for Supercomputing Applications, University of Illinois at Urbana-Champaign, Urbana, IL 61801, USA

Black holes are popping up all over the place: in compact binary x-ray sources and GRBs, in quasars, AGNs and the cores of all bulge galaxies, in binary black holes and binary black hole–neutron stars, and maybe even in the Large Hadron Collider! Black holes are strong-field objects governed by Einstein's equations of general relativity. Hence general relativistic, numerical simulations of dynamical phenomena involving black holes may help reveal ways in which black holes can form, grow, and be detected in the universe. To convey the state-of-the art, we summarize several representative simulations here, including the collapse of a hypermassive neutron star to a black hole following the merger of a binary neutron star, the magnetorotational collapse of a massive star to a black hole, and the formation and growth of supermassive black hole seeds by relativistic MHD accretion in the early universe.

1. Introduction

Black holes are 'sighted' everywhere in the universe these days. Originally located in compact binary x-ray sources in the 1970s, the cosmic presence of black holes has expanded considerably in recent decades. They now are believed to be the central engines that power quasars, active galactic nuclei (AGNs) and gamma-ray bursts (GRBs). They are identified in the cores of all bulge galaxies. They are presumed to form significant populations of compact binaries, including black hole–black hole binaries (BHBHs) and black hole–neutron star binaries (BHNSs). Black holes may even show up soon in the Large Hadron Collider!

Gravitationally, black holes are strong-field objects whose properties are governed by Einstein's theory of relativistic gravitation—general relativity. General relativistic simulations of gravitational collapse to black holes, BHBH mergers and recoil, black hole accretion, and other astrophysical phenomena involving black holes may help reveal how, when, and where black holes form, grow, and interact in the physical universe. As a consequence, such simulations can help identify the ways in which black holes can best be detected.

To illustrate how our understanding of black hole phenomena is sharpened by large-scale simulations in general relativity, we summarize in this paper the results of several recent computational investigations. The first few involve the formation of black holes from stellar collapse, while the last one concerns supermassive black hole growth via disk accretion. Most of these simulations utilize the tools of numerical relativity to solve Einstein's field equations of general relativity. So we shall begin with a brief overview of this important, rapidly maturing field.

2. Numerical relativity and the 3 + 1 formalism

Most current work in numerical relativity is performed within the framework of the 3 + 1 decomposition of Einstein's field equations using some adaptation of the standard ADM equations (Arnowitt, Deser, & Misner 1962). In this framework, spacetime is sliced

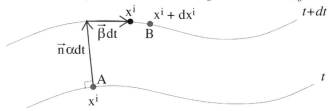

FIGURE 1. 3 + 1 decomposition of spacetime.

up into a sequence of space-like hypersurfaces of constant time t, appropriate for solving an initial-value problem. Consider two such time slices separated by an infinitesimal interval dt as shown in Figure 1. The spacetime metric measures the invariant interval between neighboring points A and B on the two slices according to

$$ds^2 = -\alpha^2 dt^2 + \gamma_{ij} \left(dx^i + \beta^i dt \right) \left(dx^j + \beta^j dt \right) \ , \tag{2.1}$$

where γ_{ij} is the spatial 3-metric on a time slice, α is the lapse function determining the proper time between the slices as measured by a time-like normal observer n^a at rest in the slice, and β^i is the shift vector, a spatial 3-vector that describes the relabeling of the spatial coordinates of points in the slice. The gravitational field satisfies the Hamiltonian and momentum *constraint equations* on each time slice, including the initial slice at $t = 0$:

$$R + K^2 - K_{ij} K^{ij} = 16\pi\rho \ \ (\text{Hamiltonian}) \ , \tag{2.2}$$
$$D_j \left(K^{ij} - \gamma^{ij} K \right) = 8\pi S^i \ \ (\text{momentum}) \ . \tag{2.3}$$

Here $R = R^i_{\ i}$ is the scalar curvature on the slice, R_{ij} is the 3-Ricci tensor, K_{ij} is the extrinsic curvature, K is its trace, and D_j is the covariant derivative operator on the slice. The quantities ρ and S^i are the mass and momentum densities of the matter, respectively; such matter source terms are formed by taking suitable projections of the matter stress-energy tensor T^{ab} with respect to the normal observer. Included in this stress-energy tensor are contributions from all the nongravitational sources of mass-energy, which we are simply calling the "matter" (e.g., baryons, electromagnetic fields, neutrinos, etc.).

A gravitational field satisfying the constraint equations on the initial slice can be determined at future times by integrating the *evolution equations*,

$$\partial_t \gamma_{ij} = -2\alpha K_{ij} + D_i \beta_j + D_j \beta_i \ , \tag{2.4}$$
$$\partial_t K_{ij} = \alpha \left(R_{ij} - 2K_{ik} K^k_{\ j} + K K_{ij} \right) - D_i D_j \alpha \tag{2.5}$$
$$+ \beta^k \partial_k K_{ij} + K_{ik} \partial_j \beta^k + K_{kj} \partial_i \beta^k - 8\pi\alpha \left(S_{ij} - \frac{1}{2} \gamma_{ij} \left(S - \rho \right) \right) \ ,$$

where S and S_{ij} are additional matter source terms. The evolution equations guarantee that the field equations will automatically satisfy the constraints on all future time slices identically, provided they satisfy them on the initial slice. Of course, this statement applies to the analytic set of equations and not necessarily to their numerical counterparts.

Note that the 3+1 formalism prescribes no equations for α and β^i. These four functions embody the four-fold gauge (coordinate) freedom inherent in general relativity. Choosing them judiciously, especially in the presence of black holes, is one of the main challenges of numerical relativity.

2.1. *The BSSN scheme*

During the past decade, significant improvement in our ability to numerically integrate Einstein's equations stably in full $3 + 1$ dimensions has been achieved by recasting the original ADM system of equations. One such reformulation is the so-called BSSN scheme (Shibata & Nakamura 1995; Baumgarte & Shapiro 1999). In this scheme, the physical metric and extrinsic curvature variables are replaced in favor of the conformal metric and extrinsic curvature, in the spirit of the "York-Lichnerowicz" split (Lichnerowicz 1944; York 1971):

$$\tilde{\gamma}_{ij} = e^{-4\phi}\gamma_{ij}, \quad \text{where} \quad e^{4\phi} = \gamma^{1/3} , \tag{2.6}$$

$$\tilde{A}_{ij} = \tilde{K}_{ij} - \frac{1}{3}\tilde{\gamma}_{ij}K . \tag{2.7}$$

Here a tilde ˜ denotes a conformal quantity and γ is the determinant of γ_{ij}. At the same time, a connection function $\tilde{\Gamma}^i$ is introduced according to

$$\tilde{\Gamma}^i \equiv \tilde{\gamma}^{jk}\tilde{\Gamma}^i{}_{jk} = -\partial_j\tilde{\gamma}^{ij} . \tag{2.8}$$

The quantities that are independently evolved in this scheme are now $\tilde{\gamma}_{ij}, \tilde{A}_{ij}, \phi, K$ and $\tilde{\Gamma}^i$. The advantage is that the Riemann operator appearing in the evolution equations [cf. eq. (2.5)] takes on the form,

$$\tilde{R}_{ij} = -\frac{1}{2}\underbrace{\tilde{\gamma}^{lm}\partial_m\partial_l\tilde{\gamma}_{ij}}_{\text{'Laplacian'}} + \underbrace{\tilde{\gamma}_{k(i}\partial_{j)}\tilde{\Gamma}^k}_{\text{remaining 2nd derivs}} + \cdots . \tag{2.9}$$

Thus the principal part of this operator, $\tilde{\gamma}^{lm}\partial_m\partial_l\tilde{\gamma}_{ij}$ is that of a Laplacian acting on the components of the metric $\tilde{\gamma}_{ij}$. All the other second derivatives of the metric have been absorbed in the derivatives of the connection functions. The coupled evolution equations for $\tilde{\gamma}_{ij}$ and $\tilde{A}_{ij}$ [cf. eqs. (2.4) and (2.5)] then reduce essentially to a wave equation,

$$\partial_t^2\tilde{\gamma}_{ij} \sim \partial_t\tilde{A}_{ij} \sim \tilde{R}_{ij} \sim \nabla^2\tilde{\gamma}_{ij} . \tag{2.10}$$

Wave equations not only reflect the hyperbolic nature of general relativity, but can be implemented numerically in a straight-forward and stable manner. By now, numerous simulations have demonstrated the dramatically improved stability achieved in the BSSN scheme over the standard ADM equations, and considerable effort has gone into explaining the improvement on theoretical grounds (see, e.g., Baumgarte & Shapiro 2003 for discussion and references). Many of the recent BHBH merger calculations have been performed using this scheme, beginning with Campanelli et al. (2006) and Baker et al. (2006; but see Pretorius 2005 for an alternative approach). The same is true for the simulations described below.

3. Binary neutron star mergers and hypermassive stars

The protagonist of several different astrophysical scenarios probed by recent numerical simulations is a *hypermassive* star, typically a hypermassive neutron star (HMNS). A hypermassive star is an equilibrium fluid configuration that supports itself against gravitational collapse by *differential* rotation. Uniform rotation can increase the maximum mass of a nonrotating, spherical equilibrium star by at most ∼20%, but differential rotation can achieve a much higher increase (Baumgarte, Shapiro, & Shibata 2000; Morrison, Baumgarte, & Shapiro 2004). Dynamical simulations using the BSSN scheme demonstrate (Baumgarte, Shapiro, & Shibata 2000) that hypermassive stars can be constructed that are *dynamically* stable, provided the ratio of rotational kinetic to gravitational potential energy, β, is not too large; for $\beta \gtrsim 0.24$ the configuration is subject to a nonaxisymmetric dynamical bar instability (Shibata, Baumgarte, & Shapiro 2000; Saijo et al. 2001). However, all hypermassive stars are *secularly* unstable to the redistribution of angular momentum by viscosity, magnetic braking, gravitational radiation, or any other agent that dissipates internal shear. Such a redistribution tends to drive a hypermassive star to uniform rotation, which cannot support the mass against collapse. Hence hypermassive stars are transient phenomena. Their formation following, for example, a NSNS merger, or core collapse in a massive, rotating star, may ultimately lead to a 'delayed' collapse to a black hole on secular (dissipative) timescales. Such a collapse will be accompanied inevitably by a delayed gravitational wave burst (Baumgarte, Shapiro, & Shibata 2000; Shapiro 2000).

The above scenario has become very relevant in light of the most recent and detailed simulations in full general relativity of NSNS mergers. State-of-the-art, fully relativistic simulations of NSNSs have been performed by Shibata and his collaborators (Shibata, Taniguchi, & Uryū 2003, 2005; Shibata 2005; Shibata & Taniguchi 2006). They consider mergers of $n = 1$ polytropes, as well as configurations obeying a more realistic nuclear equation of state (EOS). They treat mass ratios Q_M in the range $0.9 \leqslant Q_M \leqslant 1$, consistent with the range of Q_M in observed binary pulsars with accurately determined masses (Thorsett & Chakrabarty 1999; Stairs 2004). The key result is that there exists a critical mass $M_{crit} \sim 2.5$–2.7 $M_\odot$ of the binary system above which the merger leads to prompt collapse to a black hole, and below which the merger forms a hypermassive remnant. With the adopted EOS, the HMNS remnant undergoes delayed collapse in about $\sim$100 ms and emits a delayed gravitational wave burst. Most interesting, prior to collapse, the remnant forms a triaxial bar when a realistic EOS is adopted (see Figure 2) and the bar emits quasiperiodic gravitational waves at a frequency $f \sim 3$–4 kHz. Such a signal may be detectable by Advanced LIGO. It is interesting that for the adopted EOS, the mass M_{crit} is close to the value of the total mass found in each of the observed binary pulsars. Given that the masses of the individual stars in a binary can be determined by measuring the gravitational wave signal emitted during the adiabatic, in-spiral epoch prior to plunge and merger, the detection (or absence) of any quasiperiodic emission from the hypermassive remnant prior to delayed collapse may significantly constrain models of the nuclear EOS.

The possibility that a HMNS remnant forms following a NSNS merger had been fore-shadowed in earlier Newtonian simulations (Rasio & Shapiro 1994, 1999; Zhuge, Centrella, & McMillan 1996), in post-Newtonian simulations (Faber & Rasio 2000, 2002) and in conformally flat general relativistic simulations (Faber, Grandclément, & Rasio 2004). However, the recent fully relativistic simulations by Shibata, Taniguchi & Uryū (2003, 2005), Shibata (2005) and Shibata & Taniguchi (2006) provide the strongest theoretical evidence of this phenomenon to date, although the details undoubtedly depend on the adopted EOS. Triaxial equilibria can arise only in stars that can support sufficiently high values of β exceeding the classical bifurcation point at $\beta \approx 0.14$; reaching such high values requires EOSs with adiabatic indices exceeding $\Gamma \approx 2.25$ in Newtonian configurations, and comparable values in relativistic stars. It is not yet known whether the true nuclear EOS in neutron stars is this stiff, or what agent for redistributing angular momentum in a hypermassive star dominates (e.g., magnetic fields, turbulent viscosity or gravitational radiation). But it is already evident that hypermassive stars are likely to form from some mergers and that they will survive many dynamical timescales before undergoing delayed collapse. The recent measurement (Nice et al. 2005) of the mass of a neutron star in a neutron star-white dwarf binary of $M = 2.1$ $M_\odot$ establishes an

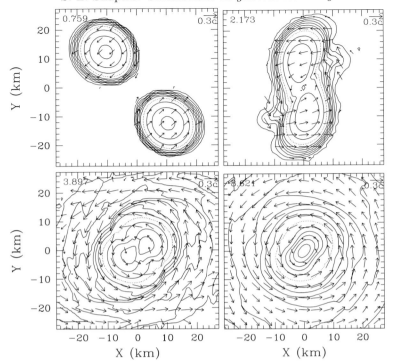

FIGURE 2. Formation of triaxial HMNS remnant following NSNS merger in 2.7 $M_\odot$ system. Snapshots of density contours are shown in the equatorial plane. The number in the upper left-hand corner denotes the elapsed time in ms; the initial orbital period is 2.11 ms. Vectors indicate the local velocity field (from Shibata, Taniguchi and Uryū 2005).

observational lower limit for the maximum mass of a neutron star; such a high value suggests that mergers in typical NSNSs may form hypermassive stars more often than undergoing prompt collapse.

3.1. *Collapse of a magnetized HMNS*

The HMNSs found above may survive for many rotation periods. However, as we stated, on longer timescales magnetic fields will transport angular momentum and this will trigger gravitational collapse. Two important magnetohydrodynamic (MHD) mechanisms which transport angular momentum are magnetic braking (Baumgarte, Shapiro, & Shibata 2000; Shapiro 2000; Cook, Shapiro, & Stephens 2003; Liu & Shapiro 2004) and the magnetorotational instability (MRI; Velikhov 1959; Chandrasekhar 1960; Balbus & Hawley 1991, 1998). Magnetic breaking transports angular momentum on the Alfvén time scale, $\tau_A \sim R/v_A \sim 1 (B/10^{14} \text{ G})^{-1}$ s, where R is the radius of the HMNS. MRI occurs wherever angular velocity Ω decreases with cylindrical radius ϖ. This instability grows exponentially with an e-folding time of $\tau_{\mathrm{MRI}} \approx 4 (\partial\Omega/\partial\ln\varpi)^{-1}$, independent of the field strength. For typical HMNSs considered here, $\tau_{\mathrm{MRI}} \sim 1$ ms. The length scale of the fastest growing unstable MRI modes, λ_{MRI}, does depend on the field strength: $\lambda_{\mathrm{MRI}} \sim 3$ m $(\Omega/4000 \text{ s}^{-1})^{-1} (B/10^{14}\text{G}) \ll R$. When the MRI saturates, turbulence consisting of small-scale eddies often develops, leading to angular momentum transport on a timescale much longer than τ_{MRI}. The computational challenge of evolving an HMNS is having sufficient spatial grid to resolve the MRI wavelength and sufficient integration time to follow the evolution on the long Alfvén timescale.

To determine the final fate of the HMNS, it is necessary to carry out MHD simulations in full general relativity. Such simulations have only recently become possible. Duez et al. (2005) and Shibata & Sekiguchi (2005) have developed new codes to evolve the coupled set of Einstein-Maxwell-MHD equations self-consistently. Our two codes have since been used to simulate the evolution of magnetized HMNSs (Duez et al. 2006a,b), and implications for short GRBs have been investigated (Shibata et al. 2006). Both codes give very similar results.

We assume axisymmetry and equatorial symmetry in all of these simulations. We use uniform computational grids with sizes up to 500×500 spatial zones in cylindrical coordinates. To model the remnant formed in binary merger simulations, we use as our initial data an equilibrium HMNS constructed from a $\Gamma = 2$ polytrope with mass M 1.7 times the spherical mass limit, or 1.5 times the limit for a uniformly rotating star built out of the same EOS. These are the typical values expected for a HMNS formed from a NSNS merger, for which $M \sim 2 \times 1.4\ M_\odot = 2.8\ M_\odot$. The differential rotation profile is chosen so that the ratio of equatorial to central Ω is $\sim 1/3$, comparable to what is found in simulations of NSNS mergers. (We find that an HMNS with a more realistic hybrid EOS rather than a polytrope evolves similarly; Duez et al. 2006b; Shibata et al. 2006) We add a seed poloidal magnetic field with strength proportional to the gas pressure. The initial magnetic pressure is set much smaller than the gas pressure, but not so small that λ_{MRI} cannot be resolved. Therefore, we set $\lambda_{\mathrm{MRI}} \approx R/10$, corresponding to $B \approx 10^{16}$ G and $\max(B^2/P) \sim 10^{-3}$. The resulting Alfvén timescale is about 16 central rotation periods, or $600M$.

In our evolutions, the effects of magnetic winding are reflected in the generation of a toroidal B field which grows linearly with time during the early phase of the evolution, and saturates on the Alfvén timescale. The effects of MRI are observed in an exponential growth of the poloidal field on the λ_{MRI} scale, a growth which saturates after a few rotation periods. The magnetic fields cause angular momentum to be transported outward, so that the core of the star contracts while the outer layers expand. After about 66 rotation periods, the core collapses to a black hole. Using the technique of black hole excision (see Duez, Shapiro, & Yo 2004 and references therein) to remove the interior of the black hole, with its nasty spacetime singularity, and replace it with suitable boundary conditions on all the variables just inside the horizon, we continue the evolution to a quasi-stationary state. The final state consists of a black hole of irreducible mass $\sim 0.9M$ surrounded by a hot accreting torus with rest mass $\sim 0.1M$ and a magnetic field collimated along the rotation axis; see Figure 3. At its final accretion rate, the torus should survive for ~ 10 ms. The torus is optically thick to neutrinos, and we estimate that it will emit $\sim 10^{50}$ ergs in neutrinos before being accreted. We also find that the cone outside the black hole centered along the rotation axis is very baryon poor. All these properties make this system a promising central engine for a short-hard GRB.

4. Black holes as central engines for GRBs

The combined observations of *BATSE*, *Swift*, *HETE-2*, *Chandra* and *HST* indicate that GRBs comprise at least two classes: long-soft and short-hard. Long-soft GRBs have characteristic timescales τ in the range $\tau \sim 2$–10^3 s. They are found in star-forming regions (spirals) and some are observed to be associated with supernovae. The favored model for the progenitor of a long-soft GRB is the collapse of a massive, rotating, magnetized star to a black hole ('collapsar'; MacFadyen & Woosley 1999). By contrast, short-hard GRBs have characteristic timescales in the range $\tau \sim 10$ ms $- 2$ s. They are identified in low star-forming regions (ellipticals) where associations with supernovae can be excluded.

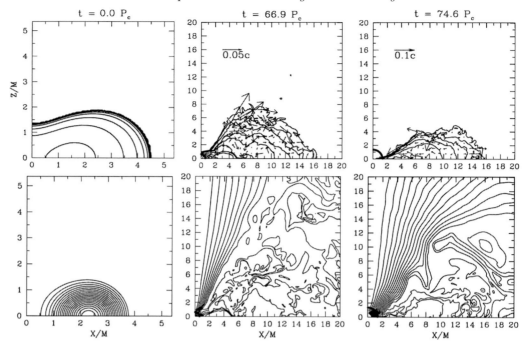

FIGURE 3. Collapse of a magnetized HMNS to a black hole. The upper panels show snapshots of the rest-mass density contours and velocity vectors on the meridional plane. The lower panels show the field lines (lines of constant vector potential A_ϕ) for the poloidal magnetic field at the same times as the upper panels. The density contours are drawn for $\rho/\rho_{max,0} = 10^{-0.3i-0.09}$ ($i = 0$–12). The field lines are drawn for $A_\phi = A_{\phi,min} + (A_{\phi,max} - A_{\phi,min})i/20$ ($i = 1$–19), where $A_{\phi,max}$ and $A_{\phi,min}$ are the maximum and minimum value of A_ϕ respectively at the given time. The thick solid curves in the lower left corner denote the apparent horizon (from Duez et al. 2006a).

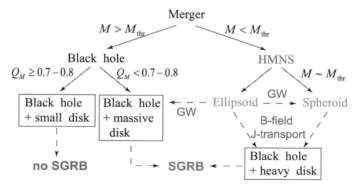

FIGURE 4. Plausible routes for the formation of a short-hard gamma-ray burst (SGRB) central engine following the merger of a binary neutron star. Here "GW" ("B-field J-transport") denotes angular momentum dissipation dominated by gravitational wave emission (magnetic fields; from Shibata & Taniguchi 2006).

The favored model for their progenitors are either NSNS or BHNS mergers. Alternative routes by which NSNS mergers can result in the generation of a short-hard GRB are traced in Figure 4. These alternatives have emerged from detailed simulations in general relativity. The HMNS route has already been summarized in Section 3.

The inspiral and merger of NSNSs and BHNSs have important implications for the detection of gravitational waves with Advanced LIGO. Recent estimates for the rates for detectable NSNS mergers are promising—in the neighborhood of 20–30 yr^{-1} (O'Shaughnessy et al. 2008). Simulations in general relativity that are now underway should be helpful in preparing for the exciting possibility of the simultaneous detection of a gravitational wave *and* GRB from the *same* source in the near future.

5. Magnetorotational collapse of massive stars to black holes

Recently, we performed simulations in axisymmetry of the magnetorotational collapse of very massive stars in full general relativity (Liu, Shapiro, & Stephens 2007). Our simulations are directly applicable to the collapse of supermassive stars with masses $M \gtrsim 10^3 \ M_\odot$ and to very massive Population III stars. They are also relevant for core collapse in massive Population I stars, since in all of these cases the governing EOS up to the appearance of a black hole can be approximated by an adiabatic $\Gamma = 4/3$ law (although its physical origin is different). These simulations may help explain the formation of the central engine in the collapsar model of long-soft GRBs (MacFadyen & Woosley 1999). Moreover, some long-soft GRBs observed at very high redshift might be related to the gravitational collapse of very massive Pop III stars (Schneider, Guetta, & Ferrara 2002; Bromm & Loeb 2006). Hence these simulations may also provide direct insights into the formation of GRB central engines arising from first-generation stars.

The simulations of Liu, Shapiro, & Stephens (2007) model the initial configurations by $n = 3$ polytropes, uniformly rotating near the mass-shedding limit and at the onset of radial instability to collapse. These simulations extend the earlier results of Shibata & Shapiro (2002) by incorporating the effects of a magnetic field and by tracking the evolution for a much longer time after the appearance of a central black hole. The ratio of magnetic to rotational kinetic energy in the initial stars is chosen to be small (1% and 10%). We find that such magnetic fields do not affect the *initial* collapse significantly. The core collapses to a black hole, after which black hole excision is employed to continue the evolution long enough for the hole reach a quasi-stationary state. We find that the black hole mass is $M_h = 0.95M$ and its spin parameter is $J_h/M_h^2 = 0.7$, with the remaining matter forming a torus around the black hole. The *subsequent* evolution of the torus does depend on the strength of the magnetic field. We freeze the spacetime metric ("Cowling approximation") and continue to follow the evolution of the torus after the black hole has relaxed to quasi-stationary equilibrium. In the absence of magnetic fields, the torus settles down following ejection of a small amount of matter due to shock heating. When magnetic fields are present, the field lines gradually collimate along the hole's rotation axis. MHD shocks and the MRI generate MHD turbulence in the the the torus and stochastic accretion onto the central black hole (see Figure 5). When the magnetic field is strong, a wind is generated in the torus, and the torus undergoes radial oscillations that drive episodic accretion onto the hole. These oscillations produce long-wavelength gravitational waves potentially detectable by LISA. The final state of magnetorotational collapse always consists of a central black hole surrounded by a collimated magnetic field and a hot, thick accretion torus. This system is a viable candidate for the central engine of a long-soft gamma-ray burst.

6. Cosmological growth of supermassive black holes

Growing evidence indicates that supermassive black holes (SMBHs) with masses in the range 10^6–$10^{10} \ M_\odot$ exist and are the engines that power AGNs and quasars. There

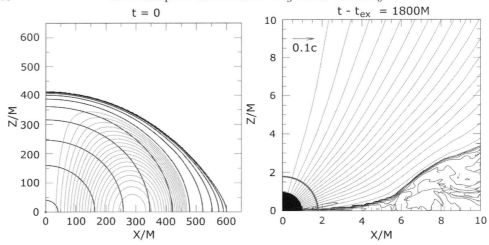

FIGURE 5. Magnetoroational collapse of a massive star to a black hole. Snapshots of meridional rest-mass density contours, velocity vectors and poloidal magnetic field lines for the initial and endpoint configurations for $n = 3$ collapse. Field lines coincide with contours of vector potential A_φ and are drawn for $A_\varphi = A_{\varphi,\max}(j/20)$ with $j = 1, 2, \cdots, 19$ where $A_{\varphi,\max}$ is the maximum value of A_φ. In the final, post-excision model ($t_{\mathrm{ex}} = 29150M$), the density levels are drawn for $\rho_0 = 100\rho_c(0)10^{-0.3j}$ ($j = 0$–10). The thick arc near the lower left corner of the right-hand frame denotes the apparent horizon and the shaded region the excision domain. (Adapted from Liu, Shapiro, & Stephens 2007.)

is also ample evidence that SMBHs reside at the centers of many—and perhaps most—galaxies, including the Milky Way. The highest redshift of a quasar discovered to date is $Z_{\mathrm{QSO}} = 6.43$, corresponding to QSO SDSS 1148+5251 (Fan et al. 2003). Accordingly, if they are the energy sources in quasars (QSOs), the first SMBHs must have formed prior to $Z_{\mathrm{QSO}} = 6.43$, or within $t = 0.87$ Gyr after the Big Bang in the concordance ΛCDM cosmological model. This requirement sets a significant constraint on black hole seed formation and growth mechanisms in the early universe. Once formed, black holes grow by a combination of mergers and gas accretion.

The more massive the initial seed, the less time is required for it to grow to SMBH scale, and the easier it is to have a SMBH in place by $Z \geqslant 6.43$. One possible progenitor that readily produces a SMBH is a supermassive star (SMS) with $M \gg 10^3\ M_\odot$ (Rees 1984; Shapiro 2004). SMSs can form when gaseous structures build up sufficient radiation pressure to inhibit fragmentation and prevent normal star formation; plausible cosmological scenarios have been proposed that can lead to this situation (Gnedin 2001; Bromm & Loeb 2003). Alternatively, the seed black holes that later grow to become SMBHs may originate from the collapse of Pop III stars $\lesssim 10^3\ M_\odot$ (Madau & Rees 2001). To achieve the required growth to $\sim 10^9\ M_\odot$ by $Z_{\mathrm{QSO}} \gtrsim 6.43$, it may be necessary for gas accretion, if restricted by the Eddington limiting luminosity, to occur at relatively *low* efficiency of rest-mass to radiation energy conversion ($\lesssim 0.2$; Shapiro 2005 and references therein), as we discuss below.

The efficiency of black hole accretion, and the resulting rate of black hole growth, is significantly affected by the spin of the black hole. The spin evolution of a black hole begins at birth. If the hole arises from the collapse of a massive or supermassive star, then it is likely to be born with a spin parameter in the range $0 \leqslant a/M \lesssim 0.8$. The lower limit applies if the progenitor star (or core) is nonrotating, the upper limit if it is spinning uniformly at the mass-shedding limit at the onset of collapse, as found in the simulations

a/M	ϵ_M	Spin Equilibrium?	Characterization
0.0	0.057	no	standard thin disk; nonspinning BH
0.95	0.19	yes	turbulent MHD disk
0.998	0.32	yes	standard thin disk; photon recapture
1.0	0.42	yes	standard thin disk; max spin BH

TABLE 1. Rest mass-to-radiation conversion efficiency vs. black hole spin

of Shibata & Shapiro (2002) and Liu, Shapiro, & Stephens (2007) and discussed in Section 5. Major mergers with other black holes of comparable mass will cause the black hole to spin up suddenly to $a/m \approx 0.8$–0.9, as the merged remnant acquires almost all of the mass and angular momentum that characterizes a circular orbit, quasi-stationary BHBH binary at the innermost stable circular orbit (ISCO). Once secular gravitational radiation loss drives the binary past the ISCO, the black holes undergo a rapid dynamical plunge and coalescence, with little additional loss of energy and angular momentum. This anticipated behavior has now been confirmed by numerical simulations of BHBH mergers (see Baker et al., this volume, and references therein). These simulations also show that gravitational radiation reaction can induce a large kick velocity ($\gtrsim 1000$ km s^{-1}) in the remnants following mergers. While in principle these large kick velocities pose a great hazard for the growth of black hole seeds to SMBHs by $Z \sim 6$, large kicks are possible only if the spins of the black hole binary companions are appreciable and their masses are comparable. In the end, gravitational recoil does not pose a significant threat to the formation of the SMBH population observed locally, although high mass seeds are favored (Volonteri 2007).

Minor mergers with many smaller black holes, isotropically distributed, cause the black hole to spin down: $a/m \sim M^{-7/3}$ (Hughes & Blandford 2003; Gammie, Shapiro, & McKinney 2004). The reason is that the ISCO and specific angular momentum of black holes orbiting counter-clockwise is larger than for holes orbiting clockwise, hence the net effect of isotropic capture is spindown.

However, it is likely that most of the mass of a supermassive black hole has been acquired by gas accretion, not mergers. Such a conclusion can be inferred from the observation that the luminosity density of quasars is roughly 0.1–0.2 of the local SMBH mass density (Soltan 1982; Yu & Tremaine 2002), an equality that arises naturally from growth via gaseous disk accretion. Steady gas accretion will quickly drive the black hole to spin equilibrium, with a spin parameter that depends on the nature of the flow. If accretion occurs via a relativistic "standard thin disk," then the hole will be driven to the Kerr limiting value, $a/M = 1$ (Bardeen 1970). Correcting for the recapture of some of the emitted photons from such a disk reduces the equilibrium spin value to 0.998 (Thorne 1974). If, however, the accretion is driven by MRI turbulence in a relativistic MHD disk, then recent simulations indicate (McKinney & Gammie 2004; Gammie, Shapiro, & McKinney 2004; De Villiers et al. 2005) that the equilibrium spin will fall to ~ 0.95. The small differences between these equilibrium spin parameters is deceptive, for they correspond to very different rest mass-to-radiation conversion efficiencies, as shown in Table 1.

The significance of these different values is that the growth of a black hole with time via steady accretion depends *exponentially*, on the rest-mass-to-radiation conversion efficiency, ϵ_M, as we will now recall. Define the rest-mass-to-radiation conversion efficiency ϵ_M and the luminosity efficiency ϵ_L according to

$$\epsilon_M \equiv L/\dot{M}_0 c^2 = \epsilon_M(a/M) \quad \epsilon_L \equiv L/L_E , \quad (6.1)$$

where M is the black hole mass, M_0 is the accreted rest-mass, and L_E is the Eddington luminosity given by

$$L_E = \frac{4\pi M \mu_e m_p c}{\sigma_T} \approx 1.3 \times 10^{46} \mu_e M_8 \text{ erg s}^{-1} \ . \tag{6.2}$$

Here τ is the growth timescale,

$$\tau = \frac{Mc^2}{L_E} \approx 0.45 \mu_e^{-1} \text{ Gyr} \ , \tag{6.3}$$

and μ_e is the mean molecular weight per electron. With the above definitions, the growth rate of a black hole due to accretion is

$$\frac{dM}{dt} = (1 - \epsilon_M)\frac{dM_0}{dt} = \left[\frac{\epsilon_L(1 - \epsilon_M)}{\epsilon_M}\right]\frac{M}{\tau} \ . \tag{6.4}$$

Integrating equation (6.4) for steady accretion with constant efficiencies trivially yields the mass amplification of a black hole with time,

$$M(t)/M(t_i) = \exp\left[\frac{\epsilon_L(1 - \epsilon_M)}{\epsilon_M}\frac{(t - t_i)}{\tau}\right] \ , \tag{6.5}$$

showing the exponential dependence on the efficiency factors.

Now a possible clue to the upper limit of ϵ_L is provided by the broad-line quasars in a Sloan Digital Sky Survey sample of 12,698 nearby quasars in the redshift interval $0.1 \leqslant z \leqslant 2.1$. This survey supports the value $\epsilon_L \approx 1$ as a physical upper limit (McLure & Dunlop 2004). Hence to maximize mass amplification via steady accretion, one must accrete with a small value of ϵ_M at the Eddington limit, $\epsilon_L \approx 1$.

Figure 6 evaluates equation (6.5) for black hole seeds that form at redshift Z_i from the collapse of Pop III stars in the range $100 \lesssim M/M_\odot \lesssim 600$, and subsequently grow to $10^9 \, M_\odot$ by $Z_{\rm QSO} = 6.43$. The ΛCDM concordance cosmological model is assumed. Of particular relevance is the case of "merger-assisted" mass amplification, whereby mergers account for a typical growth of $\sim 10^4$ in black hole mass, the remainder being by gas accretion (see, e.g., Yoo & Miralda-Éscude 2004). The figure shows that for $Z_i \gtrsim 40$, the required growth of the seed to SMBH status is easily achieved for a relativistic MHD accretion disk, but is only marginally possible for a standard thin disk that accounts for photon recapture, and not at all possible for a standard thin disk that drives the black hole to maximal spin. However, if the initial black hole seed is less than $600 \, M_\odot$, accretion via a standard thin disk appears to be ruled out altogether. The fact that a black hole driven to spin equilibrium by a turbulent MHD disk accretes with low enough efficiency to grow to supermassive size by $Z_{\rm QSO} \approx 6.43$ is potentially significant. It points to the need for further relativistic MHD simulations of black hole accretion with ever greater physical sophistication, including the full effects of radiation transfer.

These same conclusions also hold if the black hole seed forms much earlier than $Z_i \approx 40$, and they may be tightened if the seed forms later. In fact, it may be likely that the seed forms later, at $Z_i \lesssim 40$, given that even $4 - \sigma$ peaks in the density perturbation spectrum for the progenitor halo of SDSS 1148+5251 do not collapse until $Z \sim 30$ in the ΛCDM concordance cosmology (see, e.g., Figure 5 in Barkana & Loeb 2001.) Moreover, the potential wells of the earliest halos are quite shallow (~ 1 km s^{-1}) and may not be able to retain enough gas to form stars. Nevertheless, the effect of altering the date of birth of the black hole seed is not very great unless $Z_i \lesssim 20$–25, as is evident from Figure 3.

Should a quasar be discovered at $Z_{\rm QSO}$ substantially above 6.43, it would not be understood easily in the context of supermassive black hole growth by gas accretion from a seed arising from the collapse of a stellar-mass, Pop III progenitor.

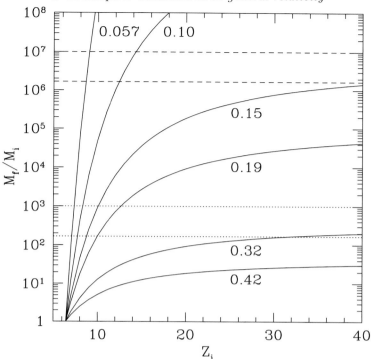

FIGURE 6. Black hole accretion mass amplification M_f/M_i versus redshift Z_i of the initial seed. Here we plot the amplification achieved by redshift $Z_f = 6.43$, the highest known quasar redshift, corresponding to 1148+5251. Each solid curve is labeled by the adopted constant radiation efficiency, ϵ_M; the luminosity is assumed to be the Eddington value ($\epsilon_L = 1$). The horizontal dashed lines bracket the range of amplification required for accretion alone to grow a seed black hole of mass $100 \leqslant M/M_\odot \leqslant 600$ formed from the collapse of a Pop III star to $10^9\, M_\odot$. The horizontal dotted lines bracket the required accretion amplification range assuming that mergers account for a growth of 10^4 in black hole mass, the remainder being by gas accretion (from Shapiro 2005).

The analysis presented here is illustrative only—the results may change as the treatment is refined. The main point of this example is to emphasize that our understanding of structure formation in the early universe as it pertains to the formation and growth of supermassive black holes (global physics) depends in part on resolving some of the important details of relativistic BHBH recoil and relativistic black hole accretion (local physics). To understand these details, in turn, requires large-scale simulations in full general relativity, which are now possible and underway.

7. Conclusions

We have discussed a number of black hole scenarios that require numerical simulations in full general relativity for true understanding. Some of these processes involve the formation of black holes, others concern their subsequent growth and interactions. The important point is that numerical relativity is at last mature enough to probe these issues reliably. Numerical relativity can now serve as an important tool to simulate stellar collapse to black holes, the in-spiral, merger and recoil of BHBHs, NSNSs and BHNSs, the generation of gravitational waves from stellar collapse and binary mergers, accretion onto black holes, and countless other phenomena involving black holes and their strong gravitational fields. With such a tool, these processes now can be investigated at a fundamental level without many of the *ad hoc* assumptions and approximations required in previous treatments. This should lead to an improved qualitative picture, as well as more reliable quantitative results.

We are grateful to T. Baumgarte, C. Gammie, Y-T. Liu, and M. Shibata for useful discussions. S.L.S. is supported by NSF grants PHY-0205155, PHY-0345151, and PHY-0650377; and NASA Grants NNG04GK54G and NNX07AG96G to the University of Illinois.

REFERENCES

ARNOWITT, R., DESER, S., & MISNER, C. W. 1962 In *Gravitation: An Introduction to Current Research* (ed. L. Witten). p. 127. Wiley.

BAKER, J. G., CENTRELLA, J., CHOI, D. I., KOPPITZ, M., & VAN METER, J. 2006 *Phys. Rev. Lett.* **96**, 11102.

BALBUS, S. A. & HAWLEY, J. P. 1991 *ApJ* **376**, 214.

BALBUS, S. A. & HAWLEY, J. P. 1998 *Rev. Mod. Phys.* **70**, 1.

BARDEEN, J. M. 1970 *Nature* **226**, 64.

BARKANA, R. & LOEB, A. 2001 *Phys. Repts.* **349**, 125.

BAUMGARTE, T. W. & SHAPIRO, S. L. 1999 *Phys. Rev. D* **59**, 024007.

BAUMGARTE, T. W. & SHAPIRO, S. L. 2003 *Phys. Repts.* **376**, 41.

BAUMGARTE, T. W., SHAPIRO, S. L., & SHIBATA, M. 2000 *ApJ* **528**, L29.

BROMM, V. & LOEB, A. 2003 *ApJ* **596**, 34.

BROMM, V. & LOEB, A. 2006 *ApJ* **642**, 382.

CAMPANELLI, M., LOUSTO, C. O., MARRONETTI, P., & ZLOCHOWER, Y. 2006 *Phys. Rev. Lett* **96**, 111101.

CHANDRASEKHAR, S. 1960 *Proc. Nat. Acad. Sci.* **46**, 253.

COOK, J. N., SHAPIRO, S. L., & STEPHENS, B. C. 2003 *ApJ* **599**, 1272.

DE VILLIERS, J.-P., HAWLEY, J. F., KROLIK, J. H. & HIROSE, S. 2005 *ApJ* **620**, 878.

DUEZ, M. D., LIU, Y. T., SHAPIRO, S. L., SHIBATA, M., & STEPHENS, B. C. 2006a *Phys. Rev. Lett.* **96**, 031101.

DUEZ, M. D., LIU, Y. T., SHAPIRO, S. L., SHIBATA, M., & STEPHENS, B. C. 2006b *Phys. Rev. D* **73**, 104015.

DUEZ, M. D., LIU, Y. T., SHAPIRO, S. L., & STEPHENS, B. C. 2005 *Phys. Rev. D* **72**, 024028.

DUEZ, M. D., SHAPIRO, S. L., & YO, H.-J. 2004 *Phys. Rev. D* **69**, 104016.

FABER, J. A., GRANDCLÉMENT, P., & RASIO, F. A. 2004 *Phys. Rev. D* **69**, 124038.

FABER, J. A. & RASIO, F. A. 2000 *Phys. Rev. D* **62**, 064012.

FABER, J. A. & RASIO, F. A. 2002 *Phys. Rev. D* **65**, 084042.

FAN, X., ET AL. 2003 *AJ* **125**, 1649.

GAMMIE, C. F., SHAPIRO, S. L., & MCKINNEY 2004 *ApJ* **602**, 312.

GNEDIN, O. Y. 2001 *Class. Quan. Grav.* **18**, 3983.

HUGHES, S. A. & BLANDFORD, R. D. 2003 *ApJ* **585**, L101.

LICHNEROWICZ, A. 1944 *J. Math. Pure Appl.* **23**, 37.

LIU, Y. T. & SHAPIRO, S. L. 2004 *Phys. Rev. D* **69**, 044009.

LIU, Y. T., SHAPIRO, S. L., & STEPHENS, B. 2007 *Phys. Rev. D* **76**, 084087.

MACFADYEN, A. I. & WOOSLEY, S. E. 1999 *ApJ* **524**, 262.

MADAU, P. & REES, M. 2001 *ApJ* **551**, L27.

MCKINNEY, J. C. & GAMMIE, C. F. 2004 *ApJ* **611**, 977.

MCLURE, R. J. & DUNLOP, J. S. 2004 *MNRAS* **352**, 1390.

MORRISON, I. A., BAUMGARTE, T. W., & SHAPIRO, S. L. 2004 *ApJ* **610**, 941.

NICE, D. J., SPLAVER, E. M., STAIRS, I. H., LÖHMER, O., JESSNER, A., KRAMER, M., & CORDES, J. M. 2005 *ApJ* **634**, 1242.

O'SHAUGHNESSY, R., KIM, C., KALOGERA, V., & BELCZYNSKI, K. 2008 *ApJ* **672**, 479.

PRETORIUS, F. 2005 *Phys. Rev. Lett.* **95**, 121101.

RASIO, F. A. & SHAPIRO, S. L. 1994 *ApJ* **432**, 242.

RASIO, F. A. & SHAPIRO, S. L. 1999 *Class. Quant. Grav.* **16**, R1.

REES, M. J. 1984 *Ann. Rev. Astro. ApJ.* **22**, 471.

SAIJO, M., SHIBATA, M., BAUMGARTE, T. W., & SHAPIRO, S. L. 2001 *ApJ* **548**, 991.

SCHNEIDER, R., GUETTA, D., & FERRARA, A. 2002 *MNRAS* **334**, 173.

SHAPIRO, S. L. 2000 *ApJ* **544**, 397.

SHAPIRO, S. L. 2004. In *Coevolution of Black Holes and Galaxies* (ed. L. C. Ho). Carnegie Observatories Astrophysics Series, Vol. I, p. 103. Cambridge University Press.

SHAPIRO S. L. 2005 *ApJ* **620**, 59.

SHIBATA, M. 2005 *Phys. Rev. Lett.* **94**, 201101.

SHIBATA, M., BAUMGARTE, T. W., & SHAPIRO, S. L. 2000 *ApJ* **542**, 453.

SHIBATA, M., DUEZ, M. D., LIU, Y. T., SHAPIRO, S. L., & STEPHENS, B. C. *Phys. Rev. Lett.* **96**, 031102.

SHIBATA, M. & NAKAMURA, T. 1995 *Phys. Rev. D* **52**, 5428.

SHIBATA, M. & SEKIGUCHI, Y. I. 2005 *Phys. Rev. D* **72**, 044014.

SHIBATA, M. & SHAPIRO, S. L. 2002 *ApJ* **572**, L39.

SHIBATA, M. & TANIGUCHI, K. 2006 *Phys. Rev. D* **73**, 064027.

SHIBATA, M., TANIGUCHI, K., & URYŪ, K. 2003 *Phys. Rev. D* **68**, 084020.

SHIBATA, M., TANIGUCHI, K., & URYŪ, K. 2005 *Phys. Rev. D* **71**, 084021.

SOLTAN, A. 1982 *MNRAS* **200**, 115.

STAIRS, I. H. 2004 *Science* **304**, 547.

THORNE, K. S. 1974 *ApJ* **191**, 507.

THORSETT, S. E. & CHAKRABARTY, D. 1999 *ApJ* **512**, 288.

VELIKHOV, E. P. 1959 *Sov. Phys.-JETP* **232**, 995.

VOLONTERI, M. 2007 *ApJ* **663**, L5.

YOO, J. & MIRALDA-ÉSCUDE, J. 2004 *ApJ* **614**, L25.

YORK, J. W., JR. 1971 *Phys. Rev. Lett.* **26**, 1656.

YU, Q. & TREMAINE, S. 2002 *MNRAS* **335**, 965.

ZHUGE, X., CENTRELLA, J. M. & MCMILLAN, S. L. W. 1996 *Phys. Rev. D* **54**, 7261.

Estimating the spins of stellar-mass black holes

By **JEFFREY E. McCLINTOCK, RAMESH NARAYAN,**
AND **REBECCA SHAFEE**

Harvard-Smithsonian Center for Astrophysics, 60 Garden Street, Cambridge, MA 02138, USA

We describe a program that we have embarked on to estimate the spins of stellar-mass black holes in x-ray binaries. We fit the continuum x-ray spectrum of the radiation from the accretion disk using the standard thin disk model, and extract the dimensionless spin parameter $a_* = a/M$ of the black hole as a parameter of the fit. We have obtained results on three systems, 4U 1543$-$47 ($a_* = 0.7$–0.85), GRO J1655$-$40 (0.65–0.8), and GRS 1915$+$105 (0.98–1), and have nearly completed the analysis of two additional systems. We anticipate expanding the sample of spin estimates to about a dozen over the next several years.

1. Introduction

The first black hole (BH), Cygnus X–1, was identified and its mass estimated in 1972. We now know of about 40 stellar-mass black holes in x-ray binaries in the Milky Way and neighboring galaxies. The masses of 21 of these, which range from $\sim$5–15 $M_\odot$, have been measured by observing the dynamics of their binary companion stars (Remillard & McClintock 2006; Orosz et al. 2007). In addition, it has become clear that virtually every galaxy has a supermassive black hole with $M \sim 10^6$–10^{10} $M_\odot$ in its nucleus. A few dozen of these supermassive BHs have reliable mass estimates, which have been obtained via dynamical observations of stars and gas in their vicinity (Begelman 2003).

With many mass measurements now in hand, the next logical step is to measure spin. This would mark a major milestone since, once we have both a BH's mass and spin, we will have achieved a complete description of the object. Furthermore, spin is arguably the more important parameter. Mass simply supplies a scale, whereas spin changes the geometry and fundamentally conditions the ways in which a BH interacts with its environment.

Unfortunately, spin is much harder to measure than mass. The effects of spin are revealed only in the regime of strong gravity close to the hole, where the sole probe available to us is the accreting gas. Thus, we must make accurate observations of the radiation emitted by the inner regions of the accretion disk, and we must have a reliable model of the emission. Until recently, there was no credible measurement of BH spin.

The situation has changed within the last couple of years. Following up on the pioneering work of Zhang, Cui, & Chen (1997), the first breakthrough came with estimates of the spin parameter $a_* \equiv a/M$ reported by our group (see Table 1) for three stellar-mass BHs (Shafee et al. 2006; McClintock et al. 2006): GRO J1655–40, 4U 1543–47, and GRS 1915+105. These spin estimates were obtained by modeling the continuum x-ray spectrum from the accretion disk surrounding the BHs. Following our work, the spin of a supermassive BH was estimated by an independent method, modeling the profile of the Fe K line (Brenneman & Reynolds 2006).

This paper is organized as follows. In Section 2 we describe the continuum-fitting method and comment on our efforts to establish our methodology. In Section 3 we review the extensive evidence for the existence of a stable inner accretion-disk radius, which provides a strong empirical foundation for the continuum-fitting method of determining

BH Binary System	$M/M_\odot$	a_*	Reference
4U 1543–47	9.4 ± 1.0	0.7–0.85	Shafee et al. (2006)
GRO J1655–40	6.30 ± 0.27	0.65–0.8	Shafee et al. (2006)
GRS 1915+105	14 ± 4.4	0.98–1	McClintock et al. (2006)

TABLE 1. Spin estimates of stellar-mass black holes

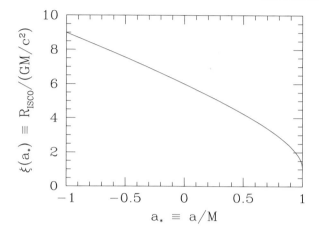

FIGURE 1. Shows the dependence of the quantity, $\xi = R_{\mathrm{ISCO}}/(GM/c^2)$, on the BH spin parameter, $a_* \equiv a/M = cJ/GM^2$, where M and J are the mass and angular momentum of the BH (Shapiro & Teukolsky 1983). The spin parameter is restricted to the range $-1 \leqslant a_* \leqslant 1$; negative values correspond to the BH counter-rotating with respect to the orbit.

spin. The importance of measuring spin is briefly described in Section 4. In Section 5 we discuss work in progress and future prospects, and we offer our conclusions.

2. The method: Fitting the x-ray continuum spectrum

A definite prediction of relativity theory is the existence of an innermost stable circular orbit (ISCO) for a test particle orbiting a BH. Once a particle is inside this radius, it suddenly plunges into the hole. Gas in a geometrically thin accretion disk has negligible pressure support in the radial direction and behaves for many purposes like a test particle. Thus, the gas spirals in (through the action of viscosity) via a series of nearly circular orbits until it reaches the ISCO, at which point it plunges into the BH. In other words, the disk is effectively truncated at an inner edge located at the ISCO.

In our method, we estimate the radius of the inner edge of the disk by fitting the x-ray continuum spectrum, and identify this radius with R_{ISCO}, the radius of the ISCO. Since the dimensionless ratio $\xi \equiv R_{\mathrm{ISCO}}/(GM/c^2)$ is solely a monotonic function of the BH spin parameter a_* (Figure 1), knowing its value allows one immediately to infer the BH spin parameter a_*. The variations in R_{ISCO} are large: e.g., for a BH of 10 $M_\odot$, R_{ISCO} ranges from 90 km down to 15 km as a_* increases from zero to unity, which implies that we should, in principle, be able to estimate a_* with good precision.

The idealized thin disk model (Novikov & Thorne 1973) describes an axisymmetric, radiatively efficient accretion flow in which, for a given BH mass M, mass accretion rate $\dot{M}$ and BH spin parameter a_*, we can calculate precisely the total luminosity of the disk, $L_{\mathrm{disk}} = \eta \dot{M} c^2$, where the radiative efficiency factor η is a function only of a_*, as well as the profile of the radiative flux $F_{\mathrm{disk}}(R)$ emitted as a function of radius R. Moreover, the accreting gas is optically thick, and the emission is thermal and blackbody-like, making it straightforward to compute the spectrum of the emission. Most importantly, as discussed above, the inner edge of the disk is located at the ISCO of the BH space-time. By analyzing the spectrum of the disk radiation and combining it with knowledge of the distance D to the source and the mass M of the BH, we can obtain a_*. This is the principle behind our method of estimating BH spin, which was first described by Zhang et al. (1997; see also Gierliński, Maciolek-Niedzẃiecki & Ebisawa 2001).

In practice, as we describe below, the method involves fitting x-ray spectral data to a fully relativistic model of the disk emission and obtaining a_* as a fit parameter. However, one can qualitatively understand the method by noting that it effectively seeks to measure the radius of the ISCO. Before discussing how this is done, we remind the reader how one measures the radius R_* of a star. Given the distance D to the star, the radiation flux F_{obs} received from the star, and the temperature T of the continuum radiation, the luminosity of the star is given by

$$L_* = 4\pi D^2 F_{\mathrm{obs}} = 4\pi R_*^2 \sigma T^4 \ . \tag{2.1}$$

Thus, from F_{obs} and T, we can obtain the solid angle $\pi(R_*/D)^2$ subtended by the star, and if the distance is known, we immediately obtain the stellar radius R_*. Of course, for accurate results we must allow for limb darkening and other non-blackbody effects in the stellar emission by computing a stellar atmosphere model.

The same principle applies to an accretion disk, but with some differences. First, since $F_{\mathrm{disk}}(R)$ varies with radius, the radiation temperature T also varies with R. But the precise variation is known for the idealized thin disk, so it is easily incorporated into the model. Second, since the bulk of the emission is from the inner regions of the disk, the effective area of the radiating surface is directly proportional to the square of the disk inner radius, $A_{\mathrm{eff}} = C R_{\mathrm{ISCO}}^2$, where the constant C is known. Third, the observed flux F_{obs} depends not only on the luminosity and the distance, but also on the inclination i of the disk to the line-of-sight.† Allowing for these differences, one can write a relation for the disk problem similar in spirit to Eq. (2.1), but with additional geometric factors that are readily calculated from the disk model. Therefore, in analogy with the stellar case, given F_{obs} and a characteristic T (from x-ray observations), one obtains the solid angle subtended by the ISCO: $\pi \cos i \, (R_{\mathrm{ISCO}}/D)^2$. If we know i and D, we obtain R_{ISCO}, and if we also know M, we obtain a_*. This is the basic idea of the method.

We note in passing that for the method to succeed it is essential to have accurate measurements of the BH mass M, inclination of the accretion disk i, and distance D as inputs to the continuum-fitting process (Shafee et al. 2006; McClintock et al. 2006). This dynamical work is not discussed here, although roughly half of our total effort is directed toward securing these dynamical data (e.g., Orosz et al. 2007).

Given accurate information on M, i and D, there are three main issues that must be dealt with before applying the method:

(1) We must carefully trace rays from the surface of the orbiting disk to the observer in the Kerr metric of the rotating BH in order to accurately compute the observed flux and spectrum. To this end, our group has developed a model called KERRBB (Li et al. 2005), which has been incorporated into XSPEC (Arnaud 1996) and is now publicly available for fitting x-ray data.

---

† We assume that the spin of the BH is approximately aligned with the orbital angular momentum vector of the binary; there is no strong contrary evidence despite the often-cited examples of GRO J1655–40 and SAX J1819.3–2525 (see Section 2.2 in Narayan & McClintock 2005).

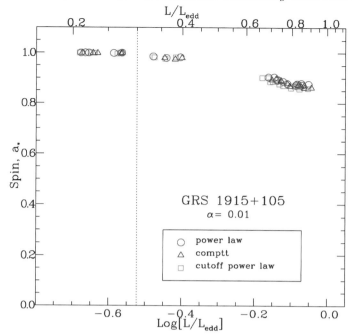

FIGURE 2. Shows the estimated spin parameter a_* of the BH in GRS 1915+105, as a function of the Eddington-scaled luminosity $L/L_{\rm Edd}$. The spectral data were analyzed using KERRBB2 combined with three different models of the high energy Comptonized radiation (shown by different symbols). For $L/L_{\rm Edd} < 0.3$ (to the left of the vertical dotted line), all the estimates of a_* are consistent with a value nearly equal to unity. The result is insensitive to the precise Comptonization model used in the analysis. (Taken from McClintock et al. 2006).

(2) We need an accurate model of the disk atmosphere for computing the spectral hardening factor f (see Section 4). We use the advanced models of our collaborator Shane Davis (Davis et al. 2005) and this element is thus well in hand. Specifically, we have computed tables of f versus $L/L_{\rm Edd}$ for a wide range of models. Further, we have incorporated these into a new version of KERRBB dubbed KERRBB2 (McClintock et al. 2006), which allows us to fit directly for the spin parameter a_* and the mass accretion rate $\dot{M}$.

(3) Most importantly, the accretion disk around the BH must be well described by the standard geometrically thin and optically thick disk model, whose validity is assumed by KERRBB and KERRBB2. To ensure this, we restrict our attention strictly to observations in the thermal state (optically thick emission) and limit ourselves to luminosities below 30% of the Eddington limit (McClintock et al. 2006; Shafee, Narayan & McClintock 2008).

Beyond these three issues, we must ultimately push theory to its limits in order to understand accretion processes near the ISCO and to obtain the most accurate model of $F_{\rm disk}(R)$ that can be achieved (see Section 3).

For a full description of the mechanics of our current continuum-fitting methodology, we refer the reader to Section 4 in McClintock et al. (2006). In brief, we first select rigorously defined thermal-state x-ray data (Section 4; Remillard & McClintock 2006). We then fit the broadband x-ray continuum spectrum, using our fully relativistic model of a thin accretion disk (KERRBB2) in Kerr space-time, which includes all relativistic effects (Li et al. 2005) and an advanced treatment of spectral hardening (Section 4; Davis et al.

2005). The model also includes self-irradiation of the disk ("returning radiation"), the effects of limb darkening, and the effect of a torque of any magnitude at the inner edge of the disk, although our published results are based on zero torque, which is justified in Shafee et al. (2008). As noted above, our new hybrid code KERRBB2 allows us to fit directly for the two parameters of interest: the spin a_* and the mass accretion rate $\dot{M}$. Using the known radiative efficiency factor η of the disk for a given a_*, and the fitted value of $\dot{M}$, we compute for each observation the Eddington-scaled luminosity, $L/L_{\rm Edd}$, and consider only those observations for which $L/L_{\rm Edd} < 0.3$ (Section 3; Shafee et al. 2008). Finally, we present our results in the form of plots of a_* versus $\log(L/L_{\rm Edd})$.

As an example, Figure 2 shows our results on GRS 1915+105 (McClintock et al. 2006). Over the luminosity range $L/L_{\rm Edd} < 0.3$, the data are consistent with a single value of a_* close to unity. Allowing for statistical errors and uncertainties in the input values of M, i and D, we estimate a_* to lie in the range 0.98–1 (Table 1). For luminosities closer to Eddington, the a_* estimates obtained using our method are lower, as also found by Middleton et al. (2006), who analyzed three observations with luminosities between $0.4L_{\rm Edd}$ and $1.4L_{\rm Edd}$ (cf. McClintock et al. 2006, Figure 12). Neither the cause for the decrease nor its magnitude are presently understood. However, it is not surprising that our model, which assumes a geometrically thin accretion disk, should fail at luminosities close to Eddington when the disk is likely to be very thick.

The results we published on 4U 1543–47 and GRO J1655–40 in Shafee et al. (2006) were obtained with KERRBB. We have re-analyzed the same data using KERRBB2, which gives a slightly larger range of uncertainty for the derived values of a_*. The spin values listed in Table 1 correspond to the more recent analysis.

3. Establishing the continuum-fitting method

Given our straightforward methodology and our in-depth experience in determining the spins of three BHs, we are confident that we can achieve our goal of amassing a total of a dozen or so measurements of BH spin during the next three to four years. Equally important, however, are our efforts to demonstrate that our methodology is sound. The largest systematic error in the BH spin estimates reported so far arise from uncertainties in the validity of the disk model we employ. Thus, it is obviously crucial to pursue detailed theoretical studies of the physics of BH accretion flows near the ISCO.

Recently, we obtained encouraging preliminary results (Shafee et al. 2008) based on a hydrodynamic study showing that the errors in our spin estimates due to viscous torque and dissipation near the ISCO are quite modest for disk luminosities $\lesssim 30\%$ of the Eddington limit. This is the luminosity limit that we had already adopted in our earlier work (McClintock et al. 2006). We are presently working to extend these hydro models to full GR MHD, where magnetic stresses may possibly cause important deviations from the standard thin disk model (e.g., Krolik 1999; Gammie 1999; Krolik & Hawley 2002).

In addition to this fundamental theoretical work, we are engaged in a broader effort to assess all scenarios that can ultimately impact upon our estimates of BH spin. Two examples: (i) With J. C. Lee, we are examining the possible effects of warm absorbers (i.e., photoionized gas) on our spin estimates via an analysis of HETG grating spectra; and (ii) we are in the process of making a stringent test of our spin model by obtaining a VLBA parallax distance and improved radial velocities for the microquasar GRS 1915+105 (see Section 6.4 in McClintock et al. 2006).

4. A basis for optimism

Among the several spectral states of accreting BHs, the *thermal state* (see Table 2 in Remillard & McClintock 2006), formerly known as the high soft state, is central to the work proposed here. A feature of this state is that the x-ray spectrum is dominated by a soft blackbody-like component which is emitted by (relatively) cool optically thick gas in the accretion disk. In addition, there is a minor nonthermal tail component of emission, which probably originates from a hot optically thin corona. In practice, this poorly understood Comptonized component of emission contributes $\lesssim 10\%$ of the flux in a 2–20 keV band (e.g., *RXTE*) and an even much smaller fraction in an 0.5–10 keV band (e.g., *ASCA* and *Chandra*), which captures nearly all of the ~ 1 keV thermal spectrum. Thus, the only spectra we consider—thermal-state spectra—are largely free of the uncertain effects of Comptonization (e.g., Figure 2). These observed spectra are believed to very closely match the classic thin accretion disk models of the early 1970s (Shakura & Sunyaev 1973; Novikov & Thorne 1973).

There is a long history of evidence suggesting that fitting the x-ray continuum is a promising approach to measuring BH spin. This history begins in the mid-1980s with the simple non-relativistic multicolor disk model (Mitsuda et al. 1984; Makishima et al. 1986), which returns the color temperature $T_{\rm in}$ at the inner-disk radius $R_{\rm in}$. In their review paper on BH binaries, Tanaka & Lewin (1995) summarize examples of the steady decay (by factors of 10–100) of the thermal flux of transient sources during which $R_{\rm in}$ remains quite constant (see their Figure 3.14). They remark that the constancy of $R_{\rm in}$ suggests that this fit parameter is related to the radius of the ISCO. More recently, this evidence for a constant inner radius in the thermal state has been presented for a number of sources via plots showing that the bolometric luminosity of the thermal component is approximately proportional to $T_{\rm in}^4$ (Kubota, Makishima, & Ebisawa 2001; Kubota & Makishima 2004; Gierliński & Done 2004; Abe et al. 2005; McClintock et al. 2009).

We now demonstrate that the case for the constancy of the inner disk radius is further strengthened if one considers the effects of spectral hardening, which we determine via the state-of-the-art disk atmosphere models of Davis et al. (2005). At the high disk temperatures typically found in BH disks ($T_{\rm in} \sim 10^7$ K), non-blackbody effects are important and one replaces $T_{\rm in}$ by the effective temperature $T_{\rm eff} = T_{\rm in}/f$, where f is a "spectral hardening factor" (Shimura & Takahara 1995; Merloni, Fabian, & Ross 2000; Davis et al. 2005). In Figure 3, we illustrate the effects of spectral hardening on the relationship between luminosity and temperature for two BH transients (see also Davis, Done, & Blaes 2006). The figure extends results that are presented in Figure 8 in McClintock et al. (2009). The top two panels show the Eddington-scaled luminosities of the two BH transients during their entire outburst cycles. The bold plotting symbols denote the rigorously defined thermal-state data (see Table 2 in Remillard & McClintock 2006). In the lower panels, we consider only these thermal-state spectral data, and we ignore the remaining data that are strongly Comptonized and for which the models are very uncertain.

Panels *b* show plots of Eddington-scaled luminosity versus the color temperature $T_{\rm in}$; the dashed lines show an $L/L_{\rm Edd} \propto T_{\rm in}^4$ relation (McClintock et al. 2006). Note that the observed luminosity rises more slowly than $T_{\rm in}^4$, which appears to suggest that $R_{\rm in}$ is not constant. Panel *c* shows an appropriate model of the spectral hardening factor f as a function of luminosity. Using this relationship, we replotted the luminosity data shown in panels *b* versus $T_{\rm eff}$, thereby obtaining the results shown in panels *d*. Here one finds that the luminosity is closely proportional to $T_{\rm eff}^4$, which provides strong evidence for the presence of a *stable inner disk radius*. Obviously, this non-relativistic analysis

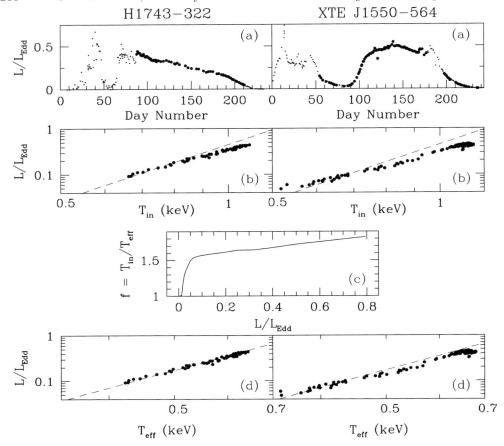

FIGURE 3. Evidence for the constancy of the inner disk radius and an illustration of the effects of spectral hardening. Shown are thermal-state data collected for H1743–322 in 2003 and XTE J1550–564 in 1998–1999 in hundreds of pointed observations using the *RXTE* PCA detector (McClintock et al. 2009). (*a*) The evolution of the luminosities of the two transients throughout their complete eight-month outburst cycles. The luminosities are scaled to the Eddington limit; for mass and distance estimates, see McClintock et al. (2009). (*b*) Luminosity versus the color temperature; the log-log slope of the dashed line is 4. (*c*) The spectral hardening factor $f \equiv T_{in}/T_{eff}$ versus luminosity computed from the disk atmosphere model of Davis et al. (2005) using BHSPEC in XSPEC (Arnaud 1996). This model was computed for a PCA response matrix in the 2–20 keV band, $M = 10\,M_{\odot}$ and $i = 70°$ (McClintock et al. 2009), and $a_* = 0.5$. The model depends only weakly on the assumed value of the spin parameter. (*d*) Luminosity versus the effective temperature $T_{eff} = T_{in}/f$, derived from the model results shown in panel *c*. Note how the data here hug the dashed T^4 line much more closely than in panels *b*.

cannot provide a secure value for the radius of the ISCO, nor even establish that this stable radius is the ISCO. Nevertheless, the presence of a fixed radius indicates that the continuum-fitting method is a well-founded approach to measuring BH spin.

5. Importance of measuring spin

In order to model the ways that an accreting BH can interact with its environment, one must know its spin. For example, the many proposals relating relativistic jets to BH spin (Blandford & Znajek 1977; Meier 2003; McKinney & Gammie 2004; Hawley & Krolik

2006) will remain mere speculation until sufficient data on BH spins have been amassed and models are tested and confirmed. Likewise, measurements of spin are comparably important for testing stellar-collapse models of gamma-ray burst sources (Woosley 1993; MacFadyen & Woosley 1999; Woosley & Heger 2006). Knowledge of spin is also crucial for the development of gravitational-wave astronomy, and our Shafee et al. (2006) paper has already motivated the first computation of waveforms for coalescing BHs that includes the effects of spin (Campanelli, Lousto, & Zlochower 2006). There are several other obvious applications of spin data, such as crucial input to models of BH formation and BH-binary evolution (Lee, Brown, & Wijers 2002; Moreno-Mendez et al. 2008) and to models of the powerful low-frequency QPOs (1–30 Hz) and complex, non-thermal BH states and their evolution (Remillard & McClintock 2006). Finally, we note that the high spins we have measured to date were very likely imparted to these BHs during the process of their formation (see Section 6.2 in McClintock et al. 2006).

6. Conclusions and future prospects

We have completed a thorough and precise dynamical study of the only known eclipsing BH, M33 X–7 (Pietsch et al. 2006), which is the most massive stellar BH known, $M = 15.65 \pm 1.45\ M_\odot$ (Orosz et al. 2007). Furthermore, the mass of the secondary star is $M_2 = 70.0 \pm 6.9\ M_\odot$, which puts it among the most massive stars whose masses are well determined. We have recently completed a paper on the spin of this BH based on ∼2 Msec of *Chandra* ACIS data (Liu et al. 2008). We are also in the process of estimating the spin of XTE J1550–564 using *RXTE* PCA data, and we anticipate estimating the spins of more than half a dozen other stellar-mass BHs during the next 3–4 years.

An especially exciting prospect is the possibility of obtaining independent estimates of spin via either the Fe K line profile (Reynolds & Nowak 2003; Brenneman & Reynolds 2006; Miller 2007) or high-frequency (50–450 Hz) QPOs (Török et al. 2005; Remillard & McClintock 2006), which are observed for some of these sources. Because spin is such a critical parameter, we and many others are planning to vigorously pursue these additional avenues, as this will provide arguably the best possible check on our results. Future x-ray polarimetry missions may provide yet an additional channel for measuring spin (e.g., Connors, Stark, & Piran 1980).

We conclude with a list of questions that motivate us. What range of spins will we find? Will GRS 1915+105 stand alone, or will we find other examples of extreme spin? As we continue to refine our models and our measurements of M, i and D, will we consistently find values of $a_* < 1$, or will we be challenged by apparent and unphysical values of the spin parameter that exceed unity? Will there be large differences in spin between the class of young, persistent systems with their massive secondaries (M33 X–7, LMC X–1 and LMC X–3) and the ancient transient systems with their low-mass secondaries? What constraints will these spin results place on BH formation, evolutionary models of BH binaries, models of relativistic jets and gamma-ray bursts, etc.? What will be the implications of these spin measurements for the emerging field of gravitational-wave astronomy in the Advanced LIGO era? How will this new knowledge help shape the observing programs of *GLAST*, *Black Hole Finder Probe*, *Constellation-X*, and *XEUS*?

REFERENCES

ABE, Y., FUKAZAWA, Y., KUBOTA, A., KASAMA, & D., MAKISHIMA, K. 2005 *PASJ* **57**, 629.
ARNAUD, K. A. 1996. In *Astronomical Data Analysis Software and Systems V.* (eds. G. H. Jacoby & J. Barnes). ASP Conf. Ser. 101, p. 17. ASP.
BEGELMAN, M. C. 2003 *Science* **300**, 1898.

BLANDFORD, R. D. & ZNAJEK, R. L. 1977 *MNRAS* **179**, 433.

BRENNEMAN, L. W. & REYNOLDS, C. S. 2006 *ApJ* **652**, 1028.

CAMPANELLI, M., LOUSTO, C. O., & ZLOCHOWER, Y. 2006 *Phys. Rev. D* **74**, 041501(1–5).

CONNORS, P. A., STARK, R. F., & PIRAN, T. 1980 *ApJ* **235**, 224.

DAVIS, S. W., BLAES, O. M., HUBENY, I., & TURNER, N. J. 2005 *ApJ* **621**, 372.

DAVIS, S. W., DONE, C., & BLAES, O. M. 2006 *ApJ* **647**, 525.

GAMMIE, C. F. 1999 *ApJ* **522**, L57.

GIERLIŃSKI, M. & DONE, C. 2004 *MNRAS* **347**, 885.

GIERLIŃSKI, M., MACIOLEK-NIEDZŹWIECKI, A., & EBISAWA, K. 2001 *MNRAS* **325**, 1253.

HAWLEY, J. F. & KROLIK, J. H. 2006 *ApJ* **641**, 103.

KROLIK, J. H. 1999 *ApJ* **515**, L73.

KROLIK, J. H. & HAWLEY, J. F. 2002 *ApJ* **573**, 754.

KUBOTA, A. & MAKISHIMA, K. 2004 *ApJ* **601**, 428.

KUBOTA, A., MAKISHIMA, K., & EBISAWA, K. 2001 *ApJ* **560**, L147.

LEE, C.-H., BROWN, G. E., & WIJERS, R. A. M. J. 2002 *ApJ* **575**, 996.

LI, L.-X., ZIMMERMAN, E. R., NARAYAN, R., & MCCLINTOCK, J. E. 2005 *ApJS* **157**, 335.

LIU, J., MCCLINTOCK, J. E., NARAYAN, R., DAVIS, S. W., & OROSZ, J. A. 2008 *ApJ* **679**, L37.

MACFADYEN, A. I. & WOOSLEY, S. E. 1999 *ApJ* **524**, 262.

MAKISHIMA, K., MAEJIMA, Y., MITSUDA, K., BRADT, H. V., REMILLARD, R. A., TUOHY, I. R., HOSHI, R., & NAKAGAWA, M. 1986 *ApJ* **308**, 635.

MCCLINTOCK, J. E., REMILLARD, R. A., RUPEN, M. P., TORRES, M. A. P., STEEGHS, D., LEVINE, A. M., & OROSZ, J. A. 2009 *ApJ* **698**, 1398.

MCCLINTOCK, J. E., SHAFEE, R., NARAYAN, R., REMILLARD, R. A., DAVIS, S. W., & LI, L.-X. 2006 *ApJ* **652**, 518.

MCKINNEY, J. C. & GAMMIE, C. F. 2004 *ApJ* **611**, 977.

MEIER, D. L. 2003 *New Astron. Rev.* **47**, 667.

MERLONI, A., FABIAN, A. C., & ROSS 2000 *MNRAS* **313**, 193.

MIDDLETON, M., DONE, C., GIERLIŃSKI, M., & DAVIS, S. W. 2006 *ApJ* **373**, 1004.

MILLER, J. M. 2007 *ARA&A* **45**, 441.

MITSUDA, K., INOUE, H., KOYAMA, K., ET AL. 1984 *PASJ* **36**, 741.

MORENO-MENDEZ, E., BROWN, G. E., LEE, C.-H., & WALTER, F. M. 2008; arXiv:astro-ph/0612461v4.

NARAYAN, R. & MCCLINTOCK, J. E. 2005 *ApJ* **623**, 1017.

NOVIKOV, I. D. & THORNE, K. S. 1973. In *Blackholes* (eds. C. DeWitt & B. DeWitt). p. 343. Gordon & Breach.

OROSZ, J. A., MCCLINTOCK, J. E., NARAYAN, R., ET AL. 2007 *Nature* **449**, 872.

PIETSCH, W., HABERL, F., SASAKI, M., GAETZ, T. J., PLUCINSKY, P. P., GHAVAMIAN, P., LONG, K. S., & PANNUTI, T. G. 2006 *ApJ* **646**, 420.

REMILLARD, R. A. & MCCLINTOCK, J. E. 2006 *ARA&A* **44**, 49.

REYNOLDS, C. S. & NOWAK, M. A. 2003 *Phys. Rep.* **377**, 389.

SHAFEE, R., MCCLINTOCK, J. E., NARAYAN, R., DAVIS, S. W., LI, L.-X., & REMILLARD, R. A. 2006 *ApJ* **636**, L113.

SHAFEE, R., NARAYAN, R., & MCCLINTOCK, J. E. 2008 *ApJ* **676**, 549.

SHAKURA, N. I. & SUNYAEV, R. A. 1973 *A&A* **24**, 337.

SHAPIRO, S. L. & TEUKOLSKY, S. A. 1983 *Black Holes, White Dwarfs, and Neutron Stars.* Wiley.

SHIMURA, T. & TAKAHARA, F. 1995 *ApJ* **445**, 780.

TANAKA, Y. & LEWIN, W. H. G. 1995. In *X-ray Binaries* (eds. W. H. G. Lewin, J. van Paradijs, & E. P. J. van den Heuvel). p. 126. Cambridge Univ. Press.

TÖRÖK, G., ABRAMOWICZ, M. A., KLUZŹNIAK, W., STUCHLÍK, Z. 2005 *A&A* **436**, 1.

WOOSLEY, S. E. 1993 *ApJ* **405**, 273.

WOOSLEY, S. E. & HEGER, A. 2006 *ApJ* **637**, 914.

ZHANG, S. N., CUI, W., & CHEN, W. 1997 *ApJ* **482**, L155.

Stellar relaxation processes near the Galactic massive black hole

By TAL ALEXANDER

Faculty of Physics, Weizmann Institute of Science, P.O. Box 26, Rehovot 76100, Israel;
William Z. and Eda Bess Novick Career Development Chair

The massive black hole (MBH) in the Galactic Center (GC) and the stars around it form a unique stellar dynamics laboratory for studying how relaxation processes affect the distribution of stars and compact remnants and lead to close interactions between them and the MBH. Recent theoretical studies suggest that processes beyond "minimal" 2-body relaxation may operate and even dominate relaxation and its consequences in the GC. I describe loss-cone refilling by massive perturbers, strong mass segregation and resonant relaxation; review observational evidence that these processes play a role in the GC; and discuss some cosmic implications for the rates of gravitational wave emission events from compact remnants inspiraling into MBHs, and the coalescence timescales of binary MBHs.

1. Introduction

The $M_{\bullet} \sim 4 \times 10^6 \, M_{\odot}$ MBH in the GC and the stars around it are the closest and observationally most accessible of such systems (Eisenhauer et al. 2005; Ghez et al. 2005). Observations of the GC thus offer a unique opportunity to study in great detail the effects of the MBH and its extreme environment on star formation, stellar evolution and stellar dynamics, and the interactions of stars and compact remnants with the MBH.

Here the focus is stellar relaxation processes. Relaxation plays an important role in a wide range of phenomena that involve close interactions with an MBH (the "loss-cone problem," Section 1.1). Such processes include gravitational wave (GW) emission by compact remnants inspiraling into an MBH—"extreme-mass ratio inspiral events" (EMRIs; see review by Amaro-Seoane et al. 2007), tidal flares from tidal disruption events (Frank & Rees 1976), tidal capture and tidal scattering of stars (Alexander & Morris 2003; Alexander & Livio 2001), 3-body exchanges with binaries leading to the capture of stars on tight orbits around the MBH and the ejection of hyper velocity stars (HVSs) out of the galaxy (Hills 1988), the orbital decay and coalescence of a binary MBHs (and the "last parsec stalling problem," see review by Merritt & Milosavljević 2005) and perhaps also the formation of ultra-luminous x-ray sources in star clusters following stellar capture around an intermediate-mass black hole (IMBH; Hopman et al. 2004). Relaxation processes are possibly linked to the presence and properties of unusual stellar populations that are observed near MBHs, such as the central "S-star" cluster, the stellar disks in the GC (Eisenhauer et al. 2005; Paumard et al. 2006) and the stellar disk in M31 (Bender et al. 2005).

Dynamical relaxation by star-star interactions is inherent to the discreteness of stellar systems. In the absence of additional mechanisms to randomize stars in phase-space, standard 2-body stellar relaxation assures a minimal degree of randomization, albeit one that could be too slow to be of practical interest. This review will discuss relaxation processes beyond standard stellar relaxation, which operate on much shorter timescales, or else operate in a qualitatively different way: massive perturbers (Section 2), strong mass segregation (Section 3) and resonant relaxation (Section 4).

The dynamical properties of the GC, specifically its short 2-body relaxation time and high stellar density, are probably not typical of galaxies in general (Section 1.2). However, dynamical processes that can be probed by GC observations have implications beyond the GC. In particular, the Milky Way is the archetype of the subset of galaxies with low-mass MBHs that are key targets for planned space-borne gravitational wave detectors, such as the *Laser Interferometer Space Antenna* (*LISA*). GC studies may help us understand the effect of such relaxation processes on the open questions of the cosmic EMRI event rate and the EMRI orbital characteristics.

Before turning to a discussion of the non-standard relaxation processes that are expected to operate in the GC, it is useful to briefly review the dynamics leading to close interactions with an MBH (loss-cone theory) and the dynamical conditions in the GC.

1.1. *Infall and inspiral into an MBH*

Stars can fall into the MBH either by losing orbital energy, so that the orbit shrinks down to the size of the last stable circular orbit ($r_{\mathrm{LSCO}} = 3r_s$ for a non-rotating MBH, where the event horizon is at the Schwarzschild radius $r_s = 2GM_\bullet/c^2$), or by losing orbital angular momentum so that the orbit becomes nearly radial and unstable (periapse $r_p < 2r_s$ for a star with zero orbital energy falling into a non-rotating MBH).† The timescale to lose energy by 2-body scattering, $T_E \equiv |E/\dot{E}|$ is of the order of the relaxation time,

$$T_E \sim T_R \sim (M_\bullet/M_\star)^2 \tau_{\mathrm{dyn}}(r)/N_\star(<r) \log N_\star(<r)\,, \qquad (1.1)$$

where $N_\star(<r)$ is the number of stars inside r, $\tau_{\mathrm{dyn}}(r) \sim \sqrt{r^3/GM_\bullet}$ is the dynamical time and spherical symmetry and a Keplerian velocity dispersion are assumed, $\sigma^2 \sim GM_\bullet/r$. The maximal angular momentum available for an orbit with energy E is that of a circular orbit, $J_c(E) = GM_\bullet/\sqrt{2E}$ (using here the stellar dynamical sign convention $E \equiv -v^2/2 - \phi(r) > 0$). The timescale for losing angular momentum, $T_J \equiv |J/\dot{J}|$, can be much shorter than T_E when $J < J_c$, since

$$T_J = [J/J_c(E)]^2 T_E\,. \qquad (1.2)$$

As a consequence, almost all stars that reach the MBH and are ultimately destroyed by a close interaction with it, do so by being scattered to low-J "loss-cone" orbits near radial orbits with $J < J_{lc} \simeq \sqrt{2GM_\bullet q}$, where q is the maximal periapse required for the close interaction of interest to occur (Frank & Rees 1976; Lightman & Shapiro 1977). The rate of close interaction events, Γ_{lc}, is set by the replenishment rate of stars into the loss cone. When the replenishment mechanism is diffusion in phase space by 2-body scattering, $\Gamma_{\mathrm{lc}} \propto T_R^{-1}$, which is typically a very low rate. Close to the MBH, at high-E, where the relative size of the loss cone in phase-space is large ($J_{lc}/J_c \propto \sqrt{qE}$), relaxation is too slow to replenish the lost stars, and the loss cone is, on average, empty. Farther out, at low E, where the loss cone is small, relaxation can replenish the lost stars, the loss cone is full (isotropic distribution of stars) and the local replenishment rate is maximal. Nevertheless, the contribution to the total replenishment rate from the low-E, full loss-cone regions of phase-space, where the timescales are longer and the stellar densities lower, remains small compared to that from the empty loss-cone regions at high-E (Lightman & Shapiro 1977).

The observational and theoretical interest in such close interactions motivated numerous investigations of alternative efficient loss-cone replenishment mechanisms, such as 2-body relaxation in non-spherically symmetric potentials (Magorrian & Tremaine 1999;

---

† If the stars are tidally disrupted before falling in the MBH, the relevant distance scale is the tidal disruption radius $r_t \sim R_\star(M_\bullet/M_\star)^{1/3} > r_s$, rather than the event horizon r_s.

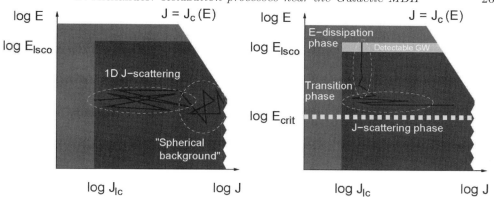

FIGURE 1. A schematic representation of the phase-space $(\log E, \log J)$ trajectories leading a star to the MBH. Each segment of the random-walk trajectory represents the change in the phase coordinates over some fixed time step Δt. The shaded areas on top $(E > E_{\rm LSCO})$ and on the left $(J < J_{\rm lc})$ are regions of phase space where no stable orbits exist. The diagonal boundary on the right is the maximal angular momentum $J_c(E)$. Left: Infall without dissipation. A star with initially high J is scattered with roughly equal relative magnitude in E and J. Eventually a random kick will send it to a low-J orbit, where J-scattering is much faster than E-scattering, making it plunge directly into the MBH. Right: Inspiral with dissipation. Energy dissipation by the emission of GW can lead to very rapid orbital decay on low-J orbits, faster than the mean time between scattering events, thus enabling the star to reach a short-period orbit with detectable GW emission (narrow horizontal shaded strip on top). Statistically, nearly all stars with initial energy $E > E_{\rm crit}$ will ultimately inspiral into the MBH, while nearly all stars with $E < E_{\rm crit}$ will ultimately plunge into the MBH, following a trajectory similar to the one depicted in the left panel.

Berczik et al. 2006), chaotic orbits in triaxial potentials (Norman & Silk 1983; Merritt & Poon 2004; Gerhard & Binney 1985; Holley-Bockelmann & Sigurdsson 2006), relaxation by massive perturbers (Zhao et al. 2002; Perets et al. 2007), resonant relaxation (Rauch & Tremaine 1996; Rauch & Ingalls 1998; Hopman & Alexander 2006a; Levin 2007), or perturbations by a massive accretion disk or a secondary IMBH (Polnarev & Rees 1994; Levin et al. 2005).

Close interactions with an MBH fall in two dynamical categories (Alexander & Hopman 2003): infall processes, such as tidal disruption, where the star is destroyed promptly on its first close encounter with the MBH, and inspiral processes, such as GW EMRI events, where multiple consecutive close encounters are required for the orbit to gradually decay. The infall takes about an orbital period, the time to fall from the point of deflection to the center, whereas the inspiral process takes much longer, depending on the energy extraction efficiency of the dissipational mechanism involved (for example, GW emission, tidal heating or drag against a massive accretion disk). In most cases, the dissipated energy is a steeply decreasing function of the periapse† and so the inspiral time scales with the number of periapse passages, and hence with the initial orbital period.

An infall or inspiral event can occur only if the star, once deflected into the loss cone, avoids being re-scattered out of it—and in the case of inspiral, also avoids being scattered directly into the MBH. Because inspiral processes are slow, stars can avoid re-scattering, complete the inspiral and decay to an interesting, very short-period orbit with high emitted dissipative power, only if they are deflected into the loss cone from an initially short-period orbit, with $E > E_{\rm crit}$. Figure 1 shows a schematic description

† E.g., the GW energy emitted per orbit scales as $\Delta E \propto (M_\star c^2/M_\bullet)(r_p/r_s)^{-7/2}$ (Peters 1964).

of the phase-space evolution of infalling and inspiraling stars, and the emergence of a critical energy scale. For inspiral by GW emission into an $M_\bullet \sim O(10^6\,M_\odot)$ MBH, E_{crit} corresponds to an initial distance scale of $r_{crit} \sim 0.01$ pc (the *ansatz* $r \leftrightarrow E = GM_\bullet/2a$, is assumed here, where a is the Keplerian semi-major axis). The EMRI event rate is then approximately (Hopman & Alexander 2005)

$$\Gamma_{lc} \sim N_{GW}(<r_{crit})/T_R(r_{crit}) \propto N_{GW}(<r_{crit})N_\star(<r_{crit})/\tau_{dyn}(r_{crit}) , \qquad (1.3)$$

where $N_{GW}(<r)$ is the number of potential GW sources (compact remnants) within distance r of the MBH. A critical energy can be similarly defined for infall processes. Because infall is much faster than inspiral, E_{crit} is much lower and r_{crit} much larger. For example, the critical radius for tidal disruption in the GC is $r_{crit} \sim$ few pc (Lightman & Shapiro 1977; Syer & Ulmer 1999; Magorrian & Tremaine 1999). Most of the stars that infall or inspiral originate near r_{crit}.

Equation (1.3) shows that the degree of central concentration of compact remnants strongly affects the EMRI event rate. Mass segregation therefore substantially increases the predicted EMRI event rate from inspiraling $\mathcal{O}(10\,M_\odot)$ stellar black holes (SBHs), which are the most massive, long-lived objects in the population (Hopman & Alexander 2006b; Section 3). Similarly, the capture of compact remnants very near the MBH by 3-body exchanges between the MBH and binaries (Section 2) can also strongly affect the EMRI rate (Perets et al. 2007). The dependence of Γ_{lc} on T_R is not trivial, since r_{crit} itself depends on T_R: the shorter the relaxation time, the faster stars are scattered into the loss cone, but also out of it. Detailed analysis shows that the two effects cancel out for $n_\star \propto r^{-3/2}$ stellar cusps. Since in most galactic nuclei the logarithmic slope of the density profile is not much different from $-3/2$, the EMRI rate is expected to be roughly independent of the relaxation time (Hopman & Alexander 2005). It should be emphasized that this result applies only to 2-body relaxation, and needs to be re-examined if other loss-cone replenishment mechanisms dominate the dynamics.

1.2. *The dynamical state of the stellar system around the Galactic MBH*

The stellar system around the Galactic MBH is expected to be in a state of dynamical relaxation in a high density cusp. This is a direct consequence of the low mass of the Galactic MBH and of the $M_\bullet/\sigma$ relation, the tight observed correlation between the mass of central MBHs and the typical velocity dispersion in the bulges of their host galaxies, $M_\bullet \propto \sigma^\beta$, where $4 \lesssim \beta \lesssim 5$ (Ferrarese & Merritt 2000; Gebhardt et al. 2000). $\beta = 4$ is assumed here for simplicity; the conclusions below are reinforced if $\beta > 4$.

The MBH radius of dynamical influence is conventionally defined as $r_h \sim GM_\bullet/\sigma^2 \propto M_\bullet^{1/2}$. The mass in stars within the radius of influence is of the order of the mass of the MBH, so their number is $N_h \sim M_\bullet/M_\star$, where $M_\star$ is the mean stellar mass, and the average stellar density within r_h is $\bar{n}_h \sim N_h/r_h^3$. The two-body relaxation time at r_h is $T_R \sim (M_\bullet/M_\star)^2 \tau_h/N_h$. It then follows that $T_R \propto M_\bullet^{5/4}$ and $\bar{n}_h \propto M_\bullet^{-1/2}$. Evaluated for the Galactic MBH, $T_R \sim \mathcal{O}(1\,\text{Gyr}) < t_H$ (the Hubble time) and $\bar{n}_h \sim \mathcal{O}(10^5\,\text{pc}^{-3})$. The short relaxation time implies that the system will return to its relaxed steady state following a major perturbation, such as a merger with a second MBH (Merritt & Szell 2006; Merritt et al. 2007). Note that $T_R > t_h$ is for an MBH only a few times more massive than the Galactic MBH. The GC is thus a member of a relatively small subset of galaxies with high-density relaxed stellar cusps.

A relaxed stellar system is expected to settle into a power-law cusp distribution, $n_\star \propto r^{-\alpha}$, (Section 3). The high stellar density in a steeply rising cusp allows star-star and star-MBH interactions to become frequent enough to be dynamically relevant and observationally interesting [Eq. (1.3)]. For example, the rates of both tidal disruption events (Wang & Merritt 2004) and EMRI inspiral events (Hopman & Alexander 2005) scale inversely with the MBH mass, $\Gamma \propto N_h/T_R \propto M_\bullet^{-1/4}$.

2. Massive perturbers

2.1. *Massive perturbers and the loss cone*

The relaxation time [Eq. (1.1)] is proportional to $(M_\star^2 n_\star)^{-1}$. This can be readily understood by considering the "$\Gamma \sim nv\Sigma$" collision rate between stars of mass $M_\star$ and mean space density in volume V, $n_\star = N_\star/V$, where the cross-section $\Sigma \sim \pi r_c^2$ is evaluated for collisions at the capture radius $r_c = 2GM_\star/v^2$, the minimal radius for a soft encounter with a typical velocity v. The rate of scattering by stars is then $\Gamma_\star \sim n_\star M_\star^2/v^3 \sim T_R^{-1}$; integration over all collision radii increases the rate only by a logarithmic Coulomb factor. When the system also contains a few very massive objects such as giant molecular clouds (GMCs), stellar clusters, or IMBHs (if such exist), these massive perturbers (MPs) of mass $M_p \gg M_\star$ and space density $n_p = N_p/V \ll n_\star$ will scatter stars at the capture radius $r_c = G(M_\star + M_p)/v^2$ at a rate of $\Gamma_p \sim n_p(M_\star + M_p)^2/v^3 \sim n_p M_p^2/v^3$. MPs could well dominate the relaxation even if they are very rare, as long as

$$\mu_2 \equiv M_p^2 N_p/M_\star^2 N_\star > 1\,. \tag{2.1}$$

Efficient relaxation by MPs was first suggested by Spitzer & Schwarzschild (1951, 1953) to explain stellar velocities in the Galactic disk. Its relevance for replenishing the loss cone was subsequently investigated in the context of solar system dynamics for the scattering of Oort cloud comets to the Sun (Hills 1981; Bailey 1983), and more recently as a mechanism for establishing the $M_\bullet/\sigma$ correlation by fast accretion of stars and dark matter (Zhao et al. 2002). Here the focus is on the consequences of MPs for the replenishment of the loss cone, and the implications for stellar populations in the Galaxy (Perets et al. 2007), the coalescence of binary MBHs (Perets & Alexander 2008) and for the cosmic rates of EMRIs (Perets et al. 2007).

Loss-cone replenishment by MPs can be described by the standard loss-cone formalism (e.g., Young 1977) with only few modifications (Perets et al. 2007). The large size of the MPs is taken into account by accordingly decreasing the Coulomb logarithm; the orbital averaging of phase-space diffusion due to scattering by stars is done incoherently (sum of squares), while for rare MPs, where there may be on average less than one scattering event per orbital period, the averaging is done coherently (square of sums).

The relative contributions of relaxation by stars and relaxation by MPs to the total loss-cone replenishment rate depend on the size of $r_{\rm crit}$ relative to the spatial distribution of the MPs—$r_{\rm crit}$ increases with the loss-cone size, and in the case of inspiral, also with the efficiency of the dissipative process. MPs are extended objects which cannot survive in the strong tidal field of the MBH—though IMBHs could be the one exception. Generally, MPs in galactic centers could also be affected by an intense central radiation field, whether the AGN's or the stars', or by outflows associated with accretion on an MBH. These processes introduce an inner cutoff $r_{\rm MP}$ to the MP distribution. A plausible estimate is $r_{\rm MP} \gtrsim \mathcal{O}(r_h)$. This is the case in the GC, where the clumpy circumnuclear gas ring lies outside the central 1.5 pc, on a scale comparable to r_h. The event rates of processes such as tidal disruption of single stars ($r_{\rm crit} \sim r_h$) or GW EMRI ($r_{\rm crit} \ll r_h$), where stellar relaxation by itself efficiently fills the loss cone at $r_{\rm crit} < r < r_{\rm MP}$, will not be much enhanced by additional relaxation due to MPs; the stellar distribution function (DF) cannot be more random than isotropic. In contrast, the event rates of processes whose

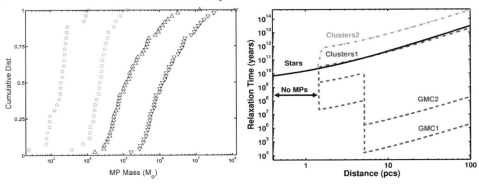

FIGURE 2. The observed MPs in the GC and their effect on the relaxation time. Left: The observed mass function of molecular cloud massive perturbers in the GC (adapted from Perets et al. 2007 with permission from the *Astrophysical Journal*). Lower (○) and upper (virial; □) mass estimates for the molecular clumps in the inner ∼5 pc, based on the molecular line observations of Christopher et al. (2005), and lower (△) and upper (virial; ▽) mass estimates for the GMCs in the inner ∼100 pc of the Galaxy, based on the molecular line observations of Oka et al. (2001). Right: The relaxation time as function of distance from the Galactic MBH due to stars alone, the upper (GMC1) and lower (GMC2) mass estimates of the molecular clumps and GMCs and due to upper (Clusters1) and lower (Clusters2) estimates on the number and masses of stellar clusters. The sharp transitions at $r = 1.5$ and 5 pc are artifacts of the non-continuous MP distribution assumed here. GMCs dominate the relaxation in the GC.

loss cone is large, and which would have remained empty beyond r_{MP} in the absence of MPs, can be increased by orders of magnitude by the presence of MPs. Most of the enhancement is due to MPs near r_{MP} (Perets et al. 2007).

Here we consider two processes with large loss cones, where MPs play an important role: the tidal disruption of stellar binaries of total mass M_{12} and semi-major axis a_{12} that interact with the MBH at a distance $r_p < r_t \sim a_{12}(M_\bullet/M_{12})^{1/3}$, leading to the capture of one star around the MBH and the ejection of the other as a HVS (Hills 1988), and the orbital decay of a binary MBH of total mass M_{12}, mass ratio $M_2/M_1 = Q < 1$ and semi-major axis a_{12} by interactions with stars at a distance $r_p \lesssim \mathcal{O}(a_{12})$ (the "slingshot effect"; Begelman et al. 1980).

2.2. *Massive perturbers in the Galactic Center*

MPs in the GC include GMCs, stellar clusters and possibly IMBHs, if these exist. Direct observational evidence (Figure 2) indicates that the dominant MPs on the $r \sim 5$–100 pc scale are $\mathcal{O}(100)$ GMCs in the mass range 10^4–$10^8\,M_\odot$, with rms mass of $\sim 10^7\,M_\odot$ and a typical size of $R_p \sim 5$ pc (the quoted range includes an order of magnitude uncertainty in the mass determination; Oka et al. 2001; Güsten & Philipp 2004), and on the $r \sim 1.5$–5 pc scale, $\mathcal{O}(10)$ molecular clumps† with masses in the range 10^3–$10^5\,M_\odot$, with rms mass of $\sim 10^4\,M_\odot$ and a typical size of $R_p \sim 0.25$ pc (Christopher et al. 2005). The ~ 10 observed stellar clusters (Figer et al. 1999; Borissova et al. 2005) may compete with stellar relaxation (Perets & Alexander 2008). Compared to the $\sim 2 \times 10^8 \sim 1\,M_\odot$ stars in the

---

† The division of a quasi-continuous medium into individual clouds is somewhat arbitrary, since several sub-clumps can be identified as a single cloud, depending on the spatial resolution of the observations and the adopted definition of a cloud. For a fixed total MP mass, $M = M_p N_p$ within a region of size r, the relaxation time scales with N_p as $T_R \propto (M_p^2 N_p)^{-1} = M^{-2} N_p$; the more massive and less numerous the clouds, the shorter T_R. The value of T_R thus depends on the way clouds are counted. Obviously, the statistical treatment of relaxation is valid only for $N_p \gg 1$ and $R_p \ll r$.

central 100 pc (Figer et al. 2004), the GMCs are expected to decrease the relaxation time by a factor $\mu_2 \sim 50$–5×10^7 [Eq. (2.1)]. Figure 2 shows a more detailed estimate of the local relaxation time for the various molecular cloud models, taking into account, among other considerations, the Coulomb factors. The relaxation time is indeed substantially decreased, by factors of 10–10^7 relative to that by stars alone, depending on distance from the center, and on the GMC mass estimates. If IMBHs do exist, then the effects of accelerated relaxation will be even stronger than predicted here, and probably extend all the way to the center.

2.3. *Galactic and cosmic implications*

With stellar relaxation alone, the empty loss-cone region of MBH-binary interactions is large ($r_t \propto a_{12}$) and extends out to >100 pc. However, the MPs that exist in the Galaxy on that scale accelerate relaxation, efficiently fill the loss cone, and thus increase the binary disruption rate by several orders of magnitude, making binary disruptions dynamically and observationally relevant (Perets et al. 2007). Such events, which result in the energetic ejection of one star, and the capture of the other on a close orbit around the MBH, have various possible implications. Disruptions of binaries by the Galactic MBH (Hills 1988; Yu & Tremaine 2003; Gualandris et al. 2005; Bromley et al. 2006) were suggested to be the origin of the hyper-velocity B stars† ($v \gtrsim 500\,\mathrm{km\,s^{-1}}$), observed tens of kpc away from the GC (Hirsch et al. 2005; Brown et al. 2005; Edelmann et al. 2005; Brown et al. 2006a), and the origin of the puzzling "S stars" (Gould & Quillen 2003; Ginsburg & Loeb 2006), a cluster of ~ 10–30 main-sequence B stars ($4\,M_\odot \lesssim M_\star \lesssim 15\,M_\odot$, main-sequence lifespan $t_\star \sim$ few $\times\, 10^7$–few $\times\, 10^8$ yr) on random tight orbits around the MBH in the central few $\times\, 0.01$ pc (Eisenhauer et al. 2005; Ghez et al. 2005). Compact objects captured this way could eventually become zero-eccentricity GW sources (Miller et al. 2005), in contrast to high-eccentricity sources typical of single-star inspiral (Hopman & Alexander 2005). These two classes of sources are expected to emit very different gravitational wave-forms.

Dynamical arguments and simulations show that on average, ~ 0.75 of MBH-binary encounters lead to capture, and that the mean semi-major axis of the captured star is related to that of the original binary by (Hills 1988, 1991)

$$\langle a \rangle \sim (M_\bullet/M_{12})^{2/3}\, a_{12}\,, \tag{2.2}$$

which implies a very high initial eccentricity, $1 - e = r_t/\langle a \rangle = (M_{12}/M_\bullet)^{1/3} \sim \mathcal{O}(0.01)$. The tidal capture process can be viewed as a mapping between the properties of field binaries far from the MBH, and the orbital properties of the captured stars: wide binaries result in wide captured orbits, and vice versa (Figure 3). The mean velocity of the ejected star at infinity (neglecting the galactic potential) is

$$\langle v_\infty^2 \rangle \sim \sqrt{2} G M_{12}^{2/3} M_\bullet^{1/3}/a_{12}\,. \tag{2.3}$$

This translates, for example, to $v_\infty \sim 2000\,\mathrm{km\,s^{-1}}$ for a $2 \times 4\,M_\odot$ B-star binary with $a_{12} = 0.2$ AU, well above the escape velocity from the Galaxy.

Of particular interest is the connection between the HVSs and the S stars that is implied by the binary tidal disruption scenario. The stellar binary mass ratio distribution is peaked around ~ 1 (Duquennoy & Mayor 1991; Kobulnicky et al. 2006), and so the observed similarity in the spectral type of the S stars and HVSs is consistent with this scenario. Figure 4 shows an estimate of the number of tidally captured S stars for

---

† HVS candidates are chosen for spectroscopy by color, to maximize the contrast against the halo population, and so are pre-selected to have B-type spectra (e.g., Brown et al. 2006b).

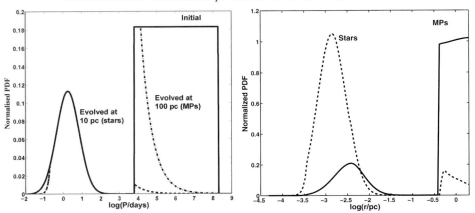

FIGURE 3. A schematic representation of the mapping of the initial binary period distribution to the semi-major axis of the tidally captured star. Left: The bimodal initial period distribution for old white dwarf/main sequence binaries (adapted from Willems & Kolb 2004), and its subsequent evolution due to GW coalescence (for the shortest periods) and to slow evaporation by field stars at 100 pc and faster evaporation at 10 pc. Right: the resulting semi-major axis distribution of the captured stars, due to scattering by stars, which occurs on the $\mathcal{O}(10\,\mathrm{pc})$ scale, and due to scattering by MPs, which occurs on the $\mathcal{O}(100\,\mathrm{pc})$ scale.

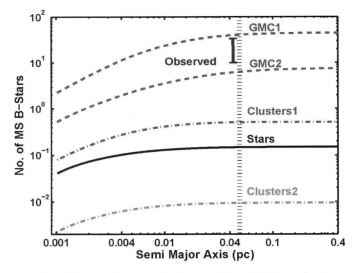

FIGURE 4. A comparison between the cumulative number of S stars (main-sequence B stars) observed orbiting the Galactic MBH on randomly oriented orbits (vertical bar), and the predicted number captured by 3-body tidal interactions of the MBH with binaries deflected to the center by massive perturbers, for different massive perturbers models (Perets et al. 2007). The observed extent of the S-star cluster is indicated by the vertical hashed line.

different MP populations (Perets et al. 2007), based on the observed orbital properties of young massive binaries ($a \sim 0.20^{+0.60}_{-0.15}$ AU) and their fraction among young massive stars in the field ($f_2 \sim 0.75$; Kobulnicky et al. 2006), and on a model for the stellar density distribution in the inner ~ 100 pc of the Galaxy (isothermal, normalized by the observations of Genzel et al. 2003) and a mass function model (continuous star formation with a universal initial mass function, IMF, Figer et al. 2004, see also Figure 6). The typical binary was modeled as a $2 \times 7.5\,M_\odot$ binary (main-sequence B stars with a lifespan of $t_\star \sim 5 \times 10^7$ yr). Dynamical evaporation is negligible for such short-lived binaries. The steady state number of captured S stars is then $\langle N_\star \rangle = \Gamma_{lc} t_\star$. Figure 4 shows that with stellar relaxation alone, tidal capture cannot explain the S-star population. However, relaxation by GMCs is consistent with the observed number of S stars, as well as with the spatial extent of the cluster of ~ 0.04 pc, which reflects the hardness of young massive binaries in the field [Eq. (2.2)]. It is also consistent with the fact that the S cluster does not include any star earlier then O8V/B0V. Such short-lived binaries are very rare in the field, and their mean number in the S cluster is predicted to be $\langle N_\star \rangle < 1$.

The MP-induced binary tidal disruption scenario also predicts that there should be 10–50 hyper-velocity $\sim 4\,M_\odot$ B stars at distances between 20 and 120 kpc from the GC. This is consistent with the total number of 43 ± 31 extrapolated by Brown et al. (2006a), based on the HVSs detected at these distances in their field of search. The tidal disruption scenario predicts an isotropic distribution of HVSs around the GC, and a random ejection history, in contrast to models where the ejection is related to a discrete binary MBH merger event (Yu & Tremaine 2003; Haardt et al. 2006; Baumgardt et al. 2006; Levin 2006). The HVSs observed to date are consistent with an isotropic HVS distribution uniformly distributed in ejection time (Brown et al. 2006a) and thus support the tidal disruption scenario.

The tidal disruption scenario can naturally explain many of the properties of the S stars and HVSs, but it has two potential flaws. (i) The predicted high eccentricities of the captured stars are larger than those observed for a few of the S stars ($e \sim 0.4$; Eisenhauer et al. 2005). However, the low observed eccentricities are expected to evolve after the capture by rapid resonant relaxation (Section 4). (ii) The lifespan of the most massive and shortest lived S stars ($t_\star \sim 2 \times 10^7$ yr) is shorter by a factor $\lesssim 10$ than the MP-accelerated relaxation time in the inner ~ 5 pc (Figure 2), where a substantial fraction of the binaries are scattered from. Thus if a binary in those regions starts on a near-circular orbit, MP-induced relaxation is not fast enough to scatter it to a $J < J_{lc}$ orbit [Eq. (1.2)] within its lifetime. However, as the timescale discrepancy is not large, and as it affects only the most massive binaries in the central few pc, where the determination of T_R is ambiguous (see footnote on page 266), this does not appear to be a fatal flaw of this scenario. It does, however, highlight the importance of observationally quantifying the relaxation time in the GC and the distribution and properties of the field binaries.

Low-mass binaries are also deflected to the MBH by MPs and tidally disrupted at rates as high as $\sim 10^{-4}\,\mathrm{yr}^{-1}$ (Perets et al. 2007). Neither the faint captured low-mass stars nor the late-type HVSs are detectable at this time. However, the captured stars affect the inner cusp dynamics in a way that may have implications for cosmic GW EMRI events. Binary disruption is effectively a local "source term" that modifies the flow of stars in phase space [cf. Eq. (3.3)], setting a diverging flow into the MBH and away from it, which modifies the steady-state spatial distribution. Detailed calculations, which take into account the period distribution of low-mass binaries and the effects of binary evaporation, indicate that MP-induced tidal captures of white dwarfs close to the MBH efficiently competes against mass-segregation, which tends to lower the density of the low-mass white dwarfs there and raise the density of massive stellar BHs (Section 3, Figure 8). As a result, the cosmic rate of GW EMRI events involving white dwarfs is predicted to be at least comparable to that involving stellar BHs (Perets et al. 2007).

The proximity of the GC allows GW bursts from the fly-by of stars near the MBH to be detected (Rubbo et al. 2006). MP-induced tidal binary disruptions increase the stellar density close to the MBH and therefore the rate of GW bursts increases significantly. In particular, the rate of GW bursts from white dwarfs increases from $\sim 0.1\,\mathrm{yr}^{-1}$ (Hopman et al. 2007) to a detectable rate of $\sim 2\,\mathrm{yr}^{-1}$ (Perets et al. 2007).

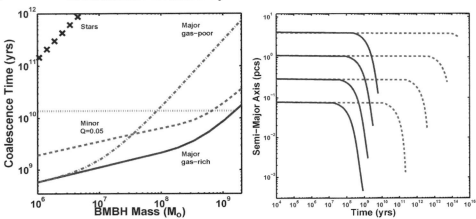

FIGURE 5. Accelerated binary MBH mergers in the presence of MPs (Perets & Alexander 2008).
Left: The time to coalescence as a function of binary MBH mass, for different merger scenarios
distinguished by the mass ratio Q between the two MBHs and the MP contents of host galaxies.
The age of the universe is indicated by the dotted horizontal line. Stellar relaxation alone cannot
supply a high enough rate of stars for the slingshot mechanism to complete the merger within a
Hubble time. However, in minor mergers ($Q = 0.05$) and major gas-rich mergers ($Q = 1$) with
MPs, merger is possible within a Hubble time for all but the most massive MBHs. Right: The
evolution of the binary MBH semi-major axis as a function of time for major mergers ($Q = 1$)
in the presence of MPs (solid line) and stellar relaxation alone (dashed line), for binary MBH
masses of 10^6, 10^7, 10^8 and 10^9 $M_\odot$ (from the bottom up).

Binary MBHs form in the aftermath of galactic mergers, when the two MBHs sink
by dynamical friction to the center of the merged galaxy. Once the binary hardens, the
orbital decay continues by 3-body interactions with stars that are deflected to the center
and extract energy from the binary, until the orbit becomes tight enough for efficient GW
emission, which rapidly leads to coalescence. Simulations show that when the loss cone
is replenished by stellar relaxation alone, the interaction rate is too slow for the binary
MBH to coalesce within a Hubble time (e.g., Berczik et al. 2005; see review by Merritt &
Milosavljević 2005; Figure 5). This "last parsec stalling problem" appears to contradict
the circumstantial evidence that most galactic nuclei contain only a single MBH (Berczik
et al. 2006; Merritt & Milosavljević 2005), and furthermore implies few such very strong
GW sources for *LISA*. One route† for resolving the stalling problem is by accelerated
MP-induced loss-cone replenishment (Perets & Alexander 2008).

Figure 5 shows the time to coalescence, as function of the binary MBH mass, for different merger and MP scenarios, based on a combination of extrapolation of the Galactic
MP population to early type galaxies, on extra-galactic observations of molecular gas
in galactic centers, and on results from galactic merger simulations. The results show
that MPs allow binary MBHs in gas-rich galaxies to coalesce within a Hubble time over
nearly the entire range of M_{12}. The situation with respect to gas-poor galaxies is less
clear, since it is harder to reliably model the MPs there (probably clusters rather than
GMCs). However, even for such galaxies, MPs allow coalescence within a Hubble time
up to masses of $M_{12} \lesssim 10^8$ $M_\odot$.

† Other possible routes are by interactions with gas in "wet mergers" (Ivanov et al. 1999;
Escala et al. 2005; Dotti et al. 2007), by interactions with a third MBH (Makino & Ebisuzaki
1994; Blaes et al. 2002; Iwasawa et al. 2006), or by accelerated loss-cone replenishment in a
non-axisymmetric potential, (Yu 2002; Berczik et al. 2006), or in a steep cusp (Zier 2006).

Efficient binary MBH coalescence by MPs has various implications. It increases the cosmic rate of GW events from MBH–MBH mergers, it increases the "mass deficit" in the galactic core (the stellar mass ejected from the core by the slingshot effect; Milosavljević et al. 2002; Ravindranath et al. 2002; Graham 2004; Ferrarese et al. 2006), it leads to the ejection of hyper-velocity stars to the inter-galactic space, but it suppresses the formation of triple MBH systems and the ejection of MBHs into intergalactic space (Saslaw et al. 1974; Blaes et al. 2002; Hoffman & Loeb 2007; Iwasawa et al. 2006).

3. Strong mass segregation

3.1. *The Bahcall-Wolf solution of moderate mass-segregation*

The 2-body relaxation timescale around the Galactic MBH, $T_R \sim \mathcal{O}(1\,\mathrm{Gyr})$, is short enough for the old stellar population there to relax to a universal steady-state configuration, independently of the initial conditions. This configuration was investigated by Bahcall & Wolf (1976, 1977). The Bahcall-Wolf solution predicts that in the Keplerian potential near an MBH, stars of mass $M_\star$ in a multi-mass population, $M_1 < M_\star < M_2$, have a DF that is approximately a power-law of the specific orbital energy ϵ, $f_M(\epsilon) \propto \epsilon^{p_M}$, where $p_M \propto M_\star$ with a proportionality constant $p_M/M_\star \simeq 1/(4M_2)$. In a Keplerian potential, this DF corresponds to a density cusp $n_M(r) \propto r^{-\alpha_M}$, where $\alpha_M = 3/2 + p_M$. Elementary considerations show that $\alpha = 7/4$ ($p = 1/4$) for a single mass population (e.g., Binney & Tremaine 1987, Section 8.4–7). This follows from the conservation of the orbital energy that is extracted from stars that are scattered into the MBH, and transferred outward by the ambient scattering stars in a steady-state, distance-independent current, $\mathrm{d}E(r)/\mathrm{d}t \sim N_\star(<r)E_\star(r)/T_R(r) \sim r^{7/2-2\alpha} = \mathrm{const}$ (using the relations $N_\star(<r) \propto r^{3-\alpha}$, $E_\star \sim r^{-1}$ and $T_R \propto r^{\alpha-3/2}$, Section 1.2). The Bahcall-Wolf solution reproduces this result for a single mass population, and predicts that it should apply also to the heaviest stars in a multi-mass population. The Bahcall-Wolf solution thus implies that at most $\Delta\alpha = 1/4$ between the lightest and heaviest stars in the population. The predicted degree of segregation is moderate.

Theoretical considerations, results from dynamical simulations, and GC observations hint that the moderate segregation solution should not and does not always hold, even in relaxed systems. As formulated, the solution depends only on the stellar masses, but not on the mass function. However, this cannot apply generally, since in the limit where the massive objects are very rare, they are expected to sink efficiently to the center by dynamical friction, and create a cusp much steeper than $\alpha = 7/4$. As shown below (Section 3.2), models of the present-day mass function in the central few pc of the GC suggest that the massive objects are relatively rare. Dynamical simulations of mass segregation in the GC based on such a mass function (Figure 8) indeed show steep cusps ($\alpha > 2$) for the heaviest masses. Finally, the observed surface density distribution of GC stars in the magnitude bin $14.75 < K < 15.75$, which corresponds to the low-mass ($0.5 \lesssim M_\star \lesssim 2\,M_\odot$) Red Clump/horizontal branch giants (Figure 6), is substantially flatter than that of the higher-mass giants ($M_\star \sim 3\,M_\odot$) that populate the adjacent bins of brighter and fainter magnitudes (Figure 6; Schödel et al. 2007). The sign of this trend is as expected for mass segregation, but the size of the effect is much larger than predicted by the Bahcall-Wolf moderate segregation solution. However, it can be explained in terms of mass-segregation if $\Delta\alpha \gtrsim 1$ (Levi 2006). While none of these hints for strong mass-segregation is decisive in itself, and other explanations are possible, taken together they motivate a re-examination of the mass-segregation solution in a relaxed system.

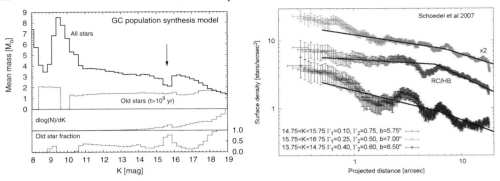

FIGURE 6. Left: A theoretical population model for the central few pc of the GC (Alexander & Sternberg 1999; Alexander 2005), assuming continuous star formation over the past 10 Gyr (Figer et al. 2004) with a "universal" IMF (Miller & Scalo 1979). Bottom panel: The fraction of old stars (defined here as stars with main-sequence lifespan of > 1 Gyr) in the population, as a function of the K-band magnitude in the GC (for $DM + A_K = 17.2$ mag, Eisenhauer et al. 2005). Middle panel: The K-band luminosity function. Top panel: The mean mass of all stars and of the old stars only, as function of the K-band magnitude. The concentration of old Red Clump/horizontal branch giants around $K \sim 15.5$ is clearly seen as an excess in the luminosity function, as an increase the fraction of old stars and as a decrease in the mean stellar mass relative to stars both immediately brighter and fainter (Schödel et al. 2007). Right: The observed azimuthally-averaged stellar surface number density around the Galactic MBH as function of projected angular distance, $\Sigma(R)$ ($R = 1''$ corresponds to ~ 0.04 pc in the GC) in three adjacent K-magnitude bins, centered around the bin associated with the Red Clump giants ($14.75 < K < 15.75$), with broken power-law fits $\Sigma \propto R^{-\Gamma}$ (Schödel et al. 2007, adapted with permission from *Astronomy and Astrophysics*). Top: $15.75 < K < 16.75$ (density multiplied by 2 for display purposes). Middle: $14.75 < K < 15.75$ (the Red Clump/horizontal branch range). Bottom: $13.75 < K < 14.75$.

3.2. *The relaxational self-coupling parameter*

Assume for simplicity a stellar system with a two-mass population of light stars of mass M_L, total initial number N_L and local density $n_L(r)$ and heavy stars of mass M_H, total initial number N_H and local density $n_H(r)$. The self-interaction rate is then $\Gamma_{LL} \propto n_L M_L^2 / v^3$ for the light stars and $\Gamma_{HH} \propto n_H M_H^2 / v^3$ for the heavy stars (Section 2). In the limit where the heavy stars are test particles ($n_H/n_L \ll M_L^2/M_H^2$, or equivalently $\Gamma_{HH}/\Gamma_{LL} \ll 1$), the heavy stars interact mostly with the light ones, lose energy and sink to the center by dynamical friction. Conversely, in the limit $\Gamma_{HH}/\Gamma_{LL} \gg 1$, the heavy stars interact mostly with each other, effectively decouple from the light stars and establish an $\alpha = 7/4$ cusp typical of a single mass population. This suggests that the *global* relaxational self-coupling parameter [cf. Eq. (2.1)], defined as

$$\mu_2 \equiv N_H M_H^2 / N_L M_L^2 \,, \tag{3.1}$$

can be used to determine whether the system settles into the moderate (Bahcall-Wolf) mass-segregation solution ($\mu_2 > 1$) or the strong mass-segregation solution ($\mu_2 < 1$). This hypothesis is borne out by the numerical results presented below† (Alexander & Hopman 2009; Section 3.3). For a continuous mass distribution, μ_2 can be generalized to

$$\mu_2 \equiv \int_{M_0}^{M_2} M_\star^2 (\mathrm{d}N/\mathrm{d}M_\star)\mathrm{d}M_\star \left/ \int_{M_1}^{M_0} M_\star^2 (\mathrm{d}N/\mathrm{d}M_\star)\mathrm{d}M_\star \,, \right. \tag{3.2}$$

† In the limit $M_H/M_L \gg 1$, it may be necessary to take explicitly into account the dynamical friction timescale in order to obtain a more accurate segregation criterion. Here μ_2 is adopted for its simplicity.

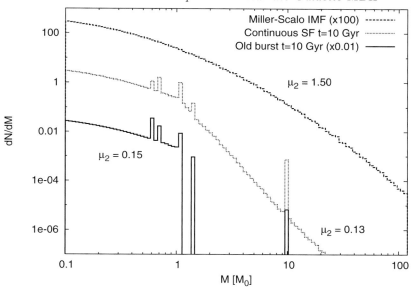

FIGURE 7. The predicted values of the global relaxational self-coupling parameter μ_2 for a "universal" Miller & Scalo (1979) IMF (top line, shifted by $\times 100$ for display purposes), an evolved mass function assuming continuous stars formation over 10 Gyr (middle line), and an evolved star-formation burst 10 Gyr old (bottom line, shifted by $\times 0.01$ for display purposes; Alexander & Hopman 2009). The mass functions of the old populations develop excesses in the ~ 0.6–$1.4\,M_\odot$ range due to the accumulation of white dwarfs and neutron stars, and in the $\sim 10\,M_\odot$ range due to the accumulation of stellar black holes (here represented by a simplified discrete mass spectrum, see Alexander 2005, Table 2.1).

the ratio between the second moments of the mass distribution of the heavy $(M_\star > M_0)$ and light $(M_\star < M_0)$ stars, for some suitable choice of the light/heavy boundary mass M_0.

The value of μ_2 depends on the population's present-day mass function. So-called universal IMFs, which extend all the way from the brown dwarf boundary $M_1 \sim 0.1\,M_\odot$ to $M_2 \gtrsim 100\,M_\odot$ (e.g., the Salpeter 1955 IMF, and its subsequent refinements, the Miller & Scalo 1979 and Kroupa 2001 IMFs), result in evolved populations, old star-bursts or continuously star forming populations, that naturally separate into two mass scales, the $\mathcal{O}(1\,M_\odot)$ scale of low-mass main-sequence dwarfs, white dwarfs and neutrons stars, and the $\mathcal{O}(10\,M_\odot)$ scale of stellar black holes, and typically have $\mu_2 < 1$ (Figure 7). Such evolved populations are thus well approximated by the simple 2-mass population model. In particular, the volume-averaged stellar population in the central few pc of the GC is reasonably well approximated by a 10-Gyr-old, continuously star-forming population with a universal IMF† (Alexander & Sternberg 1999; Figure 6). Generally, 10-Gyr-old, continuously star-forming populations with a power-law IMF, $dN/dM_\star \propto M_\star^{-\gamma}$, have $\mu_2 < 1$ for $\gamma \gtrsim 2$, and $\mu_2 > 1$ for $\gamma \lesssim 2$. Since the critical value $\gamma = 2$ is close to the generic Salpeter index $\gamma = 2.35$, it is quite possible that both the moderate and strong segregation solutions are realized around galactic MBHs, depending on the system-to-

† Note that recent analysis of late-type giants in the GC suggests that the IMF in the inner ~ 1 pc of the GC could typically be a flat $\gamma \sim 0.85$ power-law (Maness et al. 2007). This would imply $\mu_2 \gg 1$ in the inner ~ 1 pc, possibly a volume-averaged $\mu_2 > 1$ in the inner few pc (the "collection basin" for stellar BHs; Miralda-Escudé & Gould 2000), and hence moderate segregation.

system scatter in the IMF (and perhaps also realized in clusters around IMBHs, if such exist).

3.3. *Solutions of the Fokker-Planck energy equation*

The steady-state configuration of stars around an MBH can be described in terms of the diffusion of stars in phase space, from an infinite reservoir of unbound stars with a given mass function (the host galaxy, far from the MBH), to an absorbing boundary at high energy where stars are destroyed (the MBH event horizon, tidal disruption radius, or collisional destruction radius).

Bahcall & Wolf (1976, 1977) simplified the full Fokker-Planck treatment in (E, J) phase space by integrating over J so as to reduce it to E only, by assuming a Keplerian potential, and by recasting it in the form of a particle-conservation equation. In dimensionless form these can be written as (Hopman & Alexander 2006b)

$$\frac{\partial}{\partial \tau} g_M(x, \tau) = -x^{5/2} \frac{\partial}{\partial x} Q_M(x, \tau) - R_M(x, \tau) \,, \tag{3.3}$$

where M, x and τ are the dimensionless mass, energy, and time, respectively, g_M is the dimensionless DF, Q_M is the flow integral, which expresses the diffusion rate of stars by 2-body scattering to energies above x, and $R_M \propto g_M/T_R$ is the J-averaged effective loss-cone term. Q_M and R_M are non-linear functions of the set of DFs $\{g_M\}$. The equations are solved for $\{g_M\}$ by finite difference methods starting from an arbitrary initial DF and integrating forward in time until steady state is reached, subject to the boundary conditions that no stars exist at energies above some destruction energy x_D, $g_M(x > x_D) = 0$, and that the unbound stars are drawn from an isothermal distribution with a given mass function, $g_M(x < 0) = N_M \exp(Mx)$. Bahcall & Wolf (1977) showed that the stellar space density distribution,

$$n_M(r) \propto \int_{-\infty}^{r/r_h} g_M(x) \sqrt{r/r_h - x} \, \mathrm{d}x \,, \tag{3.4}$$

does not depend strongly on the exact form of the loss-cone term, and proceeded to use in their mass-segregation calculations a simplified version of Eq. (3.3) by setting $R_M = 0$. This approximation can be justified by noting that while the existence of a loss cone drastically increases the flow rate of stars into the MBH, it typically affects only a small volume in phase space near $J \sim 0$. This translates to small changes only in the J-integrated DF $g_M(x)$, mostly for $x \to x_D$, and even smaller changes in $n_M(r)$ due to the smoothing effect of the $g_M(x) \to n_M(r)$ transformation [Eq. (3.4)]. Here we adopt this approximation to allow direct comparison with the Bahcall & Wolf (1977) results, after verifying that the inclusion of the loss-cone term indeed does not significantly change the derived stellar cusps (cf. Figures 8 and 9).

We calculated a suite of such Fokker-Planck mass-segregation models for 2-mass populations with different mass ratios M_H/M_L and mass functions N_H/N_L, spanning a very wide range of the global relaxational self-coupling parameter values,[†] $10^{-3} < \mu_2 < 10^3$ (Alexander & Hopman 2009). The DFs are not exact power laws, and the logarithmic slopes $p_M(x) = \mathrm{d} \log g_M/\mathrm{d} \log x$ depend somewhat on energy, especially near the boundaries. However, analysis of the results is considerably simplified by the fact that the values of $p_M(x)$ vary monotonically with μ_2 at all x, and so the order ranking of p_M for different models does not depend on the choice of x. Figure 9 shows p_L and p_H at

---

† It is unlikely that real stellar system will have relaxational self-coupling parameters $\mu_2 \ll 0.1$. However, the study of such models is useful for understanding the mathematical properties of the solution.

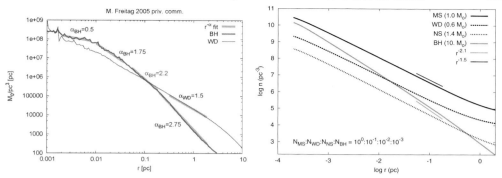

FIGURE 8. Numerical models of the mass distribution in the GC showing strong segregation. Right: The density distribution of $10\,M_\odot$ stellar BHs and $0.7\,M_\odot$ white dwarfs at 10 Gyr in an approximate model of the GC, with the evolved universal IMF of Figure 5, derived by M. Freitag (priv. comm., reproduced here with permission. See also Freitag et al. 2006), using an implementation of the Hénon method (Freitag & Benz 2002). The low-mass white dwarfs settle into a $\alpha_L \simeq 1.5$ power-law cusp. The distribution of massive stellar BHs can be approximated by a piece-wise broken power law, with $\alpha_H \sim 2.2$ at $r \sim 0.1$ pc. Left: The space density of a simplified 4-component population model for the GC, as given by the solution of the Fokker-Planck equations with a loss-cone term (adapted from Hopman & Alexander 2006b, with permission from the *Astrophysical Journal*). The logarithmic slope of the density cusp of the stellar BHs at $r = 0.1$ pc is $\alpha \simeq 2.1$, as compared to $\alpha \simeq 1.5$ for the lighter species.

$x = 10$, which corresponds to $r \sim 0.1$ pc in the GC. This choice samples $g_M(x)$ in a representative region, far from either boundaries at $x = 0$ and $x_D = 10^4$, and translates to an observationally relevant region in the GC, which is close enough to the MBH to be nearly Keplerian, but still contains a large number of observed stars to allow meaningful statistics (cf. Figure 6).

Figure 9 shows that for $\mu_2 > 1$, the Fokker-Planck calculations recover the Bahcall-Wolf solution: $p_H \simeq 1/4$ irrespective of the mass ratio, and $p_L \simeq (1/4)(M_L/M_H)$. However, for $\mu_2 < 1$ there is a marked qualitative change in the nature of the solutions, as anticipated by the analysis in Section 3.2. The more the light stars dominate the population (the smaller μ_2), the more they approach the single population solution $p_L = 1/4$ (Section 3.1). The heavy stars settle to a much steeper cusp with $p_H > 1/4$. Figure 9 also shows the grid of models explored by Bahcall & Wolf (1977), which, while large, covers only the $\mu_2 > 1$ range. This explains why the strong segregation branch of the solutions escaped their notice; their one model with $\mu_2 \simeq 0.6$ has a low mass ratio $M_H/M_L = 1.5$, where the two solution branches are not very different.

The $\mu_2 < 1$ models explored here follow the $p_M \propto M_\star$ relation noted by Bahcall & Wolf (1977) for the approximate mass-segregation solutions without a loss-cone term. Therefore, in those models where the limit $p_L \to 1/4$ is reached (for $M_H/M_L = 1.5$, 3), the heavy stars reach the asymptotic value $p_H \to (1/4)(M_H/M_L)$. It remains to be seen whether this result also holds for higher mass ratios, and for the full Fokker-Planck equation [Eq. (3.3)] with the loss-cone term.

A realistic evolved stellar system, such as the GC, is expected to have a maximal mass ratio of at least $M_H/M_L = 10$ and $\mu_2 \simeq 0.15$ (Figure 7). The mass-segregation calculations indicate that for these parameters the stellar BHs are expected to form an $\alpha_H \simeq 2.1$–2.2 cusp (Figures 8, 9), significantly steeper than the $\alpha_H = 1.75$ predicted by the Bahcall-Wolf solution of moderate segregation. It is encouraging that this logarithmic slope is very close to that found in numerical simulations (Figure 8) and that it is broadly consistent with what is needed to explain the observed trend in the stellar

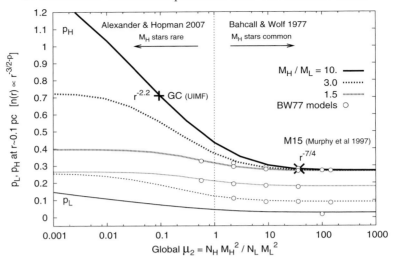

FIGURE 9. Fokker-Planck mass-segregation results. The logarithmic slopes p_H and p_L of the DFs of the heavy stars (thick lines) and light stars (narrow lines), evaluated at ($r \sim 0.1$ pc in the GC), as function of the global relaxational self-coupling parameter μ_2, for mass ratios of $M_H/M_L = 1.5, 3, 10$ (Alexander & Hopman 2009). The logarithmic slope of the stellar density of massive stars in the GC, assuming a universal IMF and continuous star formation history ($\mu_2 \sim 0.13$, Figure 7, $\alpha_H = 3/2 + p_H \simeq 2.2$) is indicated by a cross on the left, and for globular cluster M15 (assuming it harbors an IMBH), on the right (estimated at $\mu_2 \sim 37$, based on the mass function model of Murphy et al. 1997, $\alpha_H \simeq 1.75$). The results for the models calculated by Bahcall & Wolf (1977) are indicated by circles.

surface density distributions in the GC in terms of mass segregation (Figure 6). In other systems the moderate segregation solution may apply. For example, if the globular cluster M15 contains an IMBH (e.g., Gerssen et al. 2002), then a tentative determination of its present-day mass function (Murphy et al. 1997) suggests a high relaxational self-coupling parameter, $\mu_2 \sim 40$ and a relatively shallow $\alpha = 7/4$ cusp of stellar BHs. Full-scale numeric simulations that are free of the restrictive assumptions of the analytic approach adopted here—Keplerian potential, fixed boundary conditions, approximate treatment of the loss cone and a fixed 2-mass stellar population—are needed to verify and test these predictions in more detail.

Strong segregation will affect the cosmic rates of EMRI events. A detailed analysis of the anticipated change relative to the various discrepant published rate estimates depends on their specific assumptions (e.g., the assumed mass function, slope of the cusp, normalization of the stellar number density), and is outside the scope of this review.

4. Resonant relaxation

4.1. *Resonant relaxation dynamics*

The effect of 2-body relaxation on a test star is incoherent: the star experiences randomly oriented, uncorrelated perturbations from the ambient stars, and as a result its orbit deviates in a random-walk fashion from its original phase-space coordinates (in a stationary spherical smoothed potential where E and J would have been conserved in the continuum limit, $\Delta E \propto \sqrt{t}$ and $\Delta J \propto \sqrt{t}$ due to 2-body interactions0. Resonant relaxation (RR; Rauch & Tremaine 1996; Rauch & Ingalls 1998) is a form of accelerated relaxation of the orbital angular momentum, which occurs when approximate symmetries in the potential

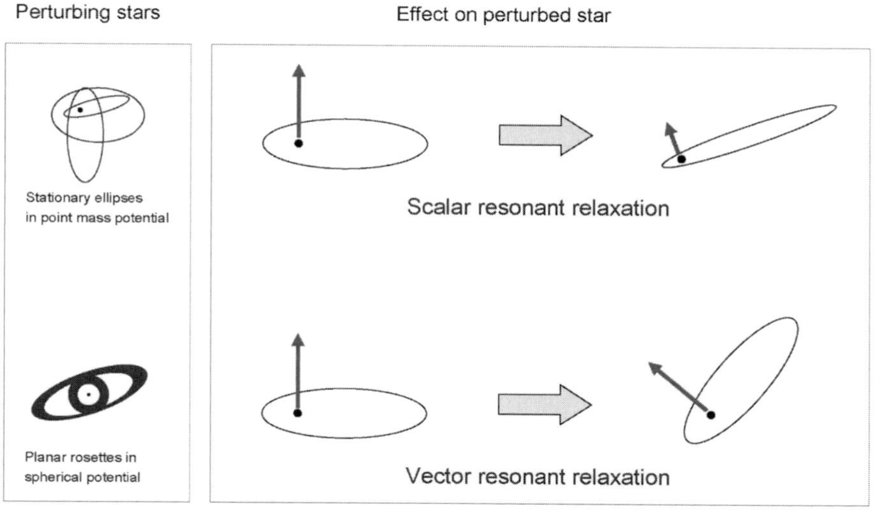

Perturbing stars   Effect on perturbed star

FIGURE 10. A sketch comparing the symmetries leading to scalar and vector RR. Top: The torques by fixed elliptical "mass wires" in a Keplerian potential lead to rapid changes in both the direction and magnitude of the orbital angular momentum of a test star. Bottom: The torques by fixed "mass annuli" in a non-Keplerian spherical potential lead to rapid changes in the direction, but not in the magnitude of the orbital angular momentum of a test star.

restrict the orbital evolution of the perturbing stars. This happens in the almost Keplerian potential near an MBH, where the orbits are approximately fixed ellipses—the potential of the enclosed stellar mass far from the MBH, or General Relativistic (GR) precession near the MBH, eventually leads to deviations from pure elliptical orbits—or in a non-Keplerian, but nearly spherically symmetric potential around an MBH, where the orbits approximately conserve their angular momentum and move on rosette-like planar orbits. The fluctuations in the potential due to stellar motions eventually lead to deviations from strictly planar orbits. As long as the symmetry is approximately conserved, on times shorter than the coherence timescale t_ω, the orbit of a test star with semi-major axis a experiences correlated (coherent) perturbations,† which can be described as a constant residual torque exerted by the superposed potentials of the $N_\star(<a)$ randomly oriented elliptical "mass wires" in a Keplerian potential, or "mass annuli" in a non-Keplerian spherical potential, that represent the orbitally-averaged mass distribution of individual perturbing stars. The magnitude of the residual torque is then $\dot{J} \sim N_\star^{1/2}(<a)GM_\star/a$ and the change in the angular momentum of the test star increases linearly with time, $\Delta J \sim \dot{J}t$ (for $t < t_w$). The orbital energy, on the other hand, remains unchanged, since the potential is constant.

RR in a Keplerian potential is called *scalar* RR since it changes both the magnitude and direction of **J**. Scalar RR can therefore change a circular orbit into an almost radial, MBH-approaching one. In contrast, RR in a non-Keplerian spherical potential is called *vector* RR since, for reasons of symmetry, it changes only the direction of **J**, but not its magnitude (Figure 10). Vector RR can randomize the orbital orientations, but does not play a role in supplying stars to the loss cone.

---

† RR is better described as "coherent relaxation." The term "resonant" refers to the equality of the radial and azimuthal orbital periods in a Keplerian potential, which results in closed ellipse orbits.

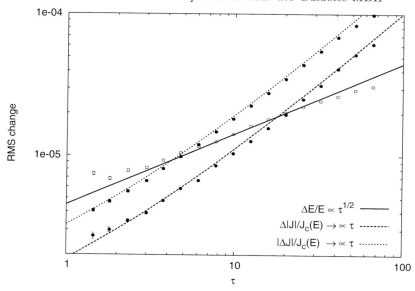

FIGURE 11. A correlation analysis of N-body simulations ($N = 200$) showing the relaxation of energy and of scalar and vector angular momentum around an MBH in the Keplerian limit ($M_\star/M_\bullet = 3 \times 10^{-7}$) for a thermal population of stars, as a function of the elapsed time-lag (Eilon, Kupi & Alexander 2009). The change in $\Delta E/E$, $\Delta J/J_c$ and $|\Delta \mathbf{J}|/J_c$ is plotted as function of the normalized time-lag $\tau = \Delta t/P$ in the range $1 \le \tau \le 100$. The mass precession coherence time of the system is $\tau_M = t_M/P \sim 1.7 \times 10^4$, and the potential fluctuation coherence time is $\tau_\phi = t_\phi/P \sim 1.2 \times 10^5$, so both scalar and vector RR are expected to grow linearly over the plotted range. More detailed analysis shows that the $\Delta J/J_c(E)$ is a function of both energy and angular momentum, which for $\tau \to 0$ scales as $\sqrt{\tau}$, and for $1 \ll \tau \ll \tau_w$ scales as τ, and that $|\Delta \mathbf{J}|/J_c$ is simply proportional to $\Delta J/J_c$. The correlation analysis is an efficient method for quantifying relaxation in N-body results (cf. Rauch & Tremaine 1996, Figure 1). The theoretical predictions for $\Delta J/J_c$ and $|\Delta \mathbf{J}|/J_c$ fit the numeric results very well. As expected, $\Delta E/E \propto \sqrt{\tau}$ at all time-lags.

On timescales longer than the coherence time, the orbital orientations of the perturbing stars drift, and coherence is lost. the maximal change in angular momentum during the linear coherence time, $\Delta J_\omega \sim \dot{J} t_\omega$ then becomes the "mean free path" for a random walk in J-space, whose time-step is t_ω. On timescales longer than the coherence time, the angular momentum changes incoherently $\propto \sqrt{t}$, but much faster than it would have in the absence of RR. The energy is unaffected by RR and always evolves incoherently $\propto \sqrt{t}$ on the long non-resonant relaxation timescale (Figure 11). The RR timescale T_{RR} is defined, like the incoherent 2-body relaxation timescale, as the time to change J by order J_c [Eqs. (1.1), (1.2)], $T_{RR} \sim (J_c/\Delta J_\omega)^2 t_\omega$, which can be expressed as (Hopman & Alexander 2006a)

$$T_{RR} = A^\omega_{RR} \frac{N_\star(<a)}{\mu^2(<a)} \frac{P^2(a)}{t_\omega} \simeq \frac{A^\omega_{RR}}{N_\star(<a)} \left(\frac{M_\bullet}{M_\star}\right)^2 \frac{P^2(a)}{t_\omega}, \qquad (4.1)$$

where μ is the relative enclosed stellar mass, $\mu = N_\star M_\star/(M_\bullet + N_\star M_\star)$, P is the radial orbital period, and the last approximate equality holds in the Keplerian regime. Here and below, the constants A^ω_{RR} are numerical factors of order unity that depend on the specifics of the coherence-limiting process, on the orbital characteristics of the test star, and probably also on the parameters of the stellar distribution. These constants are not well determined at this time.

The coherence time depends on the symmetry assumed and on the process that breaks it. For a non-relativistic near-Keplerian potential, the limiting process is precession due to the potential of the distributed stellar mass,†

$$t_\omega = t_M \sim \frac{M_\bullet}{N_\star(<a)M_\star} P(a) \,. \tag{4.2}$$

Remarkably, the resulting RR timescale does not depend on $N_\star$. Close to the MBH is much shorter than the non-coherent 2-body relaxation timescale (here denoted for emphasis as T_{NR}),

$$T_{RR}^M = A_{RR}^M \frac{M_\bullet}{M_\star} P(a) \sim \frac{N_\star(<a)M_\star}{M_\bullet} T_{NR} \,. \tag{4.3}$$

Yet closer to the MBH, it is GR precession that limits the coherence,

$$t_\omega = t_{GR} = \frac{8}{3} \left(\frac{J}{J_{\text{LSO}}} \right)^2 P(a) \,, \tag{4.4}$$

where $J_{\text{LSO}} = 4GM_\bullet/c$ is the last stable orbit for $\epsilon \ll c^2$. The GR precession is prograde, while that due to the distributed mass is retrograde, and so they may partially cancel each other. Their combined effect on the scalar RR timescale is

$$T_{RR}^s \simeq \frac{A_{RR}^s}{N_\star(<a)} \left(\frac{M_\bullet}{M_\star} \right)^2 P^2(a) \left| \frac{1}{t_M} - \frac{1}{t_{GR}} \right| \,. \tag{4.5}$$

Since t_M increases with r, while t_{GR} decreases with r, scalar RR is fastest at some finite distance from the MBH, which typically coincides with $\sim r_{\text{crit}}/2$ for *LISA* EMRI targets (Figure 13).

Precession does not affect vector RR. The coherence in a non-Keplerian spherical potential is limited by the change in the total gravitational potential $\phi = \phi_\bullet + \phi_\star$ caused by the fluctuations in the stellar potential, $\phi_\star$, due to the realignment of the stars as they rotate by π on their orbits,

$$t_\omega = t_\phi = \frac{\phi}{\dot\phi_\star} \sim \frac{N_\star^{1/2}(<a)}{2\mu} P(a) \simeq \frac{M_\bullet}{2N_\star^{1/2}(<a)M_\star} P(a) \,, \tag{4.6}$$

where the last approximate equality holds in the Keplerian regime. The vector RR timescale is obtained by substituting t_ϕ in Eq. (4.1),

$$T_{RR}^v = 2A_{RR}^v \frac{N_\star^{1/2}(<a)}{\mu(<a)} P(a) \simeq 2A_{RR}^v \left(\frac{M_\bullet}{M_\star} \right) \frac{P(a)}{N_\star^{1/2}(<a)} \,. \tag{4.7}$$

Vector RR is much faster than scalar RR (Figure 13).

4.2. Resonant relaxation and EMRI rates

The efficiency of scalar RR quickly decreases with distance from the MBH, since the coherence time falls as $M_\star(<r)/M_\bullet$ grows. At r_h, where $M_\star(<r_h)/M_\bullet \sim \mathcal{O}(1)$, scalar RR is almost completely quenched. Because $r_{\text{crit}} \sim r_h$ for tidal disruption (Lightman & Shapiro 1977; Section 1.1), RR does not significantly enhance the tidal disruption rate (Rauch & Tremaine 1996). In contrast, $r_{\text{crit}} \sim 0.01$ pc for EMRI events, where $M_\star(<r_{\text{crit}})/M_\bullet \ll 1$ and T_{RR}^s is near its minimum. Scalar RR therefore dominates the

---

† The enclosed stellar mass $N_\star M_\star$ changes the Keplerian period $P \propto M_\bullet^{-1/2}$ by $\Delta P/P = N_\star M_\star/2M_\bullet = \Delta\varphi/2\pi$. Identifying de-coherence with a phase drift $\Delta\varphi = \pi$ then implies $t_M \sim (\pi/\Delta\varphi)P$.

dynamics of the loss cone for GW EMRI events (Hopman & Alexander 2006a). Scalar RR accelerates the flow of stars in phase-space from large-J orbits to low-J orbits that approach the MBH and can lose orbital energy and angular momentum by the emission of GWs. However, if unchecked, RR would continue to rapidly drive the stars to plunging orbits that fall directly into the MBH. This is where GR precession is predicted to play an important role (Hopman & Alexander 2006a). Orbits with very small periapse, $r_p \sim$ few $\times r_s$, where GW emission becomes appreciable, are also orbits where the GR precession rate becomes large enough (t_{GR} becomes short enough) to quench RR, and allow the EMRI inspiral to proceed undisturbed. This subtle cancelation, which is critical for the observability of EMRI events, still has to be verified by direct simulations.

The effect of scalar RR can be included in an approximate way in the Fokker-Planck equation [Eq. (3.3)] as an additional loss-cone term $R_{RR} \propto \chi g/T^s_{RR}$, where the efficiency factor χ parameterizes the uncertainties that enter through the various order-unity factors A^ω_{RR} [Eq. 4.1)]. Such calculations show that the poorly determined value of the efficiency can strongly affect the predicted EMRI rates (Figure 12). As the efficiency rises, the EMRI rate first increases because stars are supplied faster to the loss cone, but when the efficiency continues to rise the stars are drained so rapidly into the MBH, that the EMRI rates are strongly suppressed.

Rauch & Tremaine (1996) explored the efficiency of RR by a few small-scale N-body simulations, and noted a large variance around the derived mean efficiency. Here we use their mean efficiency as the reference point ($\chi = 1$), but consider also values smaller and larger. Figure 12 shows the GW inspiral rate and direct plunge rate as function of the unknown efficiency χ, relative to the no-RR case ($\chi = 0$), derived from Fokker-Planck calculations of a single mass population. The EMRI rate rises to ~ 8 times more than is expected without RR, peaking at $1 \lesssim \chi \lesssim 2$, but then falls rapidly to zero at $\chi \gtrsim 10$. The strong χ-dependence of the EMRI rates provides strong motivation to determine the RR efficiency and its dependence on the parameters of the system both numerically (Gürkan & Hopman 2007; Eilon, Kupi, & Alexander 2009; Figure 11), and by direct observations of the only accessible system at present where RR effects may play a role—the stars around the Galactic MBH.

4.3. *Resonant relaxation and stellar populations in the GC*

The stellar population in the GC includes both young and old stars, and is composed of distinct sub-populations, each with its own kinematical properties (see Alexander 2005 for a review). As shown below, RR can naturally explain some of the systematic differences between the various dynamical components in the GC. Conversely, GC observations of these populations can then test the various assumptions and approximations that enter into analytic treatment of RR, and in particular constrain the poorly determined RR efficiency.

Figure 13 summarizes the typical distance scales and ages associated with these populations, and compares them with the various relaxation timescales. The calculation of the relaxation timescales are approximate since they assume a single mass population. The non-resonant 2-body relaxation timescale [Eq. (1.1)] is roughly independent of the radius in the GC, $T_{NR} \sim$ few $\times 10^9$ yr (assuming a mean stellar mass of $M_\star = 1\,M_\odot$; in a multi-mass system it is expected to decrease to $T_{NR} \sim 10^8$ yr in the inner 0.001 pc due to mass segregation, Hopman & Alexander 2006b). Because neither the RR efficiency, nor the mass function is known with confidence, the scalar RR timescale, T^s_{RR}, is shown for two different assumptions; $\chi M_\star = 1\,M_\odot$ and $\chi M_\star = 10\,M_\odot$. As discussed in Section 4.1, beyond $r \sim 0.1$ pc, $T^s_{RR} > T_{NR}$ due to mass precession, and the loss-cone replenishment is dominated by non-coherent relaxation. T^s_{RR} decreases toward the MBH, until it

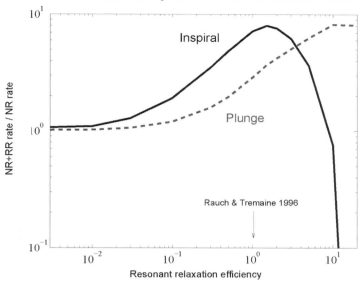

FIGURE 12. The relative rates of GW EMRI events and direct infall (plunge) events, as function of the unknown efficiency of RR, χ, normalized to $\chi = 1$ for the values derived by Rauch & Tremaine (1996).

reaches a minimum, where it starts increasing again due to GR precession. The distance scale where T^s_{RR} is shortest happens to coincide with the volume $r \lesssim r_{\rm crit}$, where most GW EMRI sources are expected to lie and where $T^s_{RR} \ll T_{NR}$, so RR dominates EMRI loss-cone dynamics (Section 4.2). In contrast to scalar RR, the vector RR timescale T^v_{RR} (shown here for an assumed $\chi M_\star = 1\,M_\odot$) decreases unquenched toward the MBH.

Dynamical populations and structures whose estimated age exceed these relaxation timescales must be relaxed. Those whose age cannot be determined, but whose lifespan exceeds the relaxation timescales may be affected, unless we are observing them at an atypical time soon after they were created. The youngest dynamical structure observed in the GC is the stellar disk (or possibly two non-aligned disks; Levin & Beloborodov 2003; Genzel et al. 2003; Paumard et al. 2006), which is composed of ~ 50 young massive OB stars with an age of $t_\star \sim 6 \pm 2$ Myr, on co-planar, co-rotating orbits that extend between ~ 0.04–0.5 pc. The inner edge of the disk is sharply defined and it coincides with the outer boundary of the S-stars cluster (Section 2.2). Figure 13 shows that the vector RR timescale equals the age of the stellar disk at its inner edge, and so is consistent with the spatial extent of the disk. Even if the S stars were initially the inner part of the disk (this does not appear likely given that they are systematically lighter than the disk stars), vector RR would have efficiently randomized their orbital inclination. However, their measured high eccentricities (Eisenhauer et al. 2005; Ghez et al. 2005) would then be hard to explain. If instead, the S stars are not-related to the disks, but were tidally captured around the MBH by 3-body exchange interactions (Section 2.3), then only their lifespan can be determined. Tidal capture leads to an extremely eccentric captured orbit [Eq. (2.2)]. Scalar RR could then randomize and decrease the eccentricities of at least a few of the older S stars closer to the MBH. Vector and scalar RR could also explain why the old evolved giants (with progenitor masses of $M_\star \sim 2$–$8\,M_\odot$; Genzel et al. 1994) at $r \gtrsim 0.1$ pc appear dynamically relaxed (Genzel et al. 2000), in spite of the fact that their lifespans are shorter than the non-coherent relaxation time.

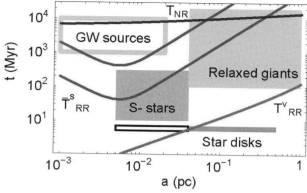

FIGURE 13. Evidence for resonant relaxation in the GC in the age vs. distance from the MBH plane. The spatial extent and estimated age of the various dynamical sub-populations in the GC (shaded areas) is compared with the non-resonant 2-body relaxation timescale (top line, for assumed mean mass of $M_\star = 1\,M_\odot$) and with the scalar RR timescale (two curved lines, top one for $\chi M_\star = 1\,M_\odot$, bottom one for $\chi M_\star = 10\,M_\odot$) and vector RR timescale (bottom line, for $\chi M_\star = 1\,M_\odot$). The populations include the young stellar rings in the GC (filled rectangle in the bottom right); the S stars, if they were born with the disks (open rectangle in the bottom left); the maximal lifespan of the S stars (filled rectangle in the middle left); the dynamically relaxed old red giants (filled rectangle in the top right); and the reservoir of GW inspiral sources, where the age is roughly estimated by the progenitor's age or the time to sink to the center (open rectangle in the top left). Stellar components that are older than the various relaxation times must be randomized. (Hopman & Alexander 2006a, reproduced with permission from the *Astrophysical Journal*).

It should be noted that the effect of RR on the stellar density distribution is not expected to be large even quite close to the MBH ($r \lesssim 0.1$ pc), unless the efficiency χ is very high, because the RR-induced depletion of the DF at high energies is smoothed by the transformation from the DF to $n_\star(r)$ [Eq. (3.4)] and by the contribution of unbound stars to the central density.

5. Summary

Relaxation processes play an important role in the GC, where the 2-body relaxation time is shorter than the age of the system and the stellar density is high. The scaling laws that follow from the $M_\bullet/\sigma$ relation imply that the same must hold for all galaxies with $M_\bullet \lesssim$ few $\times\, 10^7\,M_\odot$. Relaxation processes affect the distribution of stars and compact remnants, lead to close interactions between them and the MBH, and may be related to the unusual stellar populations that are observed in the GC. These are of relevance because of the very high quality stellar data coming from the GC, and because galactic nuclei with low-mass MBHs like the GC are expected to be important GW EMRI targets for the next generation of space-borne GW detectors. In addition, efficient relaxation mechanisms that operate and can be studied in the GC may play a role even in galactic nuclei with high-mass MBHs, where 2-body relaxation is unimportant.

Three processes beyond minimal two-body relaxation were discussed here: accelerated loss-cone replenishment by MPs, strong mass-segregation in evolved populations, and rapid RR. Evidence was presented that these processes operate and may even dominate relaxation and its consequences in the GC: The S stars and HVSs are consistent with relaxation by GMCs; there are hints for strong mass segregation in the central density suppression of the low-mass Red Clump giants and in numeric simulations of the GC, and RR appears to play a role in the truncation of the stellar disks and the orbital randomization of the S stars and the late-type giants. There are also cosmic implications: MPs enable the efficient merger of binary MBHs, and boost the rates of white dwarf EMRIs captured near the MBH by tidal disruptions of stellar binaries. Strong segregation, and in particular RR can strongly affect the EMRI rates from stellar BHs.

The stellar dynamics laboratory in the GC holds great promise for future progress in understanding these mechanisms and their implications.

REFERENCES

ALEXANDER, T. 2005 *Phys. Rep.* **419**, 65.

ALEXANDER, T. & HOPMAN, C. 2003 *ApJ* **590**, L29.

ALEXANDER, T. & HOPMAN, C. 2009 *ApJ* **697**, 1861.

ALEXANDER, T. & LIVIO, M. 2001 *ApJ* **560**, L143.

ALEXANDER, T. & MORRIS, M. 2003 *ApJ* **590**, L25.

ALEXANDER, T. & STERNBERG, A. 1999 *ApJ* **520**, 137.

AMARO-SEOANE, P., GAIR, J. R., FREITAG, M., MILLER, M. C., MANDEL, I., CUTLER, C. J., & BABAK, S. 2007 *Class. Quant. Grav.* **24**, R113.

BAHCALL, J. N. & WOLF, R. A. 1976 *ApJ* **209**, 214.

BAHCALL, J. N. & WOLF, R. A. 1977 *ApJ* **216**, 883.

BAILEY, M. E. 1983 *MNRAS* **204**, 603.

BAUMGARDT, H., GUALANDRIS, A., & PORTEGIES ZWART, S. 2006 *MNRAS* **372**, 174.

BEGELMAN, M. C., BLANDFORD, R. D., & REES, M. J. 1980 *Nature* **287**, 307.

BENDER, R. ET AL. 2005 *ApJ* **631**, 280.

BERCZIK, P., MERRITT, D., & SPURZEM, R. 2005 *ApJ* **633**, 680.

BERCZIK, P., MERRITT, D., SPURZEM, R., & BISCHOF, H.-P. 2006 *ApJ* **642**, L21.

BINNEY, J. & TREMAINE, S. 1987 *Galactic Dynamics*. Princeton University Press.

BLAES, O., LEE, M. H., & SOCRATES, A. 2002 *ApJ* **578**, 775.

BORISSOVA, J., IVANOV, V. D., MINNITI, D., GEISLER, D., & STEPHENS, A. W. 2005 *A&A* **435**, 95.

BROMLEY, B. C., KENYON, S. J., GELLER, M. J., BARCIKOWSKI, E., BROWN, W. R., & KURTZ, M. J. 2006 *ApJ* **653**, 1194.

BROWN, W. R., GELLER, M. J., KENYON, S. J., & KURTZ, M. J. 2005 *ApJ* **622**, L33.

BROWN, W. R., GELLER, M. J., KENYON, S. J., & KURTZ, M. J. 2006a *ApJ* **640**, L35.

BROWN, W. R., GELLER, M. J., KENYON, S. J., & KURTZ, M. J. 2006b *ApJ* **647**, 303.

CHRISTOPHER, M. H., SCOVILLE, N. Z., STOLOVY, S. R., & YUN, M. S. 2005 *ApJ* **622**, 346.

DOTTI, M., COLPI, M., HAARDT, F., & MAYER, L. 2007 *MNRAS* **379**, 956.

DUQUENNOY, A. & MAYOR, M. 1991 *A&A* **248**, 485.

EDELMANN, H., NAPIWOTZKI, R., HEBER, U., CHRISTLIEB, N., & REIMERS, D. 2005 *ApJ* **634**, L181.

EILON, E., KUPI, G., & ALEXANDER, T. 2009 *ApJ* **698**, 641.

EISENHAUER, F., ET AL. 2005 *ApJ* **628**, 246.

ESCALA, A., LARSON, R. B., COPPI, P. S., & MARDONES, D. 2005 *ApJ* **630**, 152.

FERRARESE, L., CÔTÉ, P., JORDÁN, A., PENG, E. W., BLAKESLEE, J. P., PIATEK, S., MEI, S., MERRITT, D., MILOSAVLJEVIĆ, M., TONRY, J. L., & WEST, M. J. 2006 *ApJS* **164**, 334.

FERRARESE, L. & MERRITT, D. 2000 *ApJ* **539**, L9.

FIGER, D. F., KIM, S. S., MORRIS, M., SERABYN, E., RICH, R. M., & MCLEAN, I. S. 1999 *ApJ* **525**, 750.

FIGER, D. F., RICH, R. M., KIM, S. S., MORRIS, M., & SERABYN, E. 2004 *ApJ* **601**, 319.

FRANK, J. & REES, M. J. 1976 *MNRAS* **176**, 633.

FREITAG, M., AMARO-SEOANE, P., & KALOGERA, V. 2006 *ApJ* **649**, 91.

FREITAG, M. & BENZ, W. 2002 *A&A* **394**, 345.

GEBHARDT, K., ET AL. 2000 *ApJ* **539**, L13.

GENZEL, R., HOLLENBACH, D., & TOWNES, C. H. 1994 *Reports of Progress in Physics* **57**, 417.

GENZEL, R., PICHON, C., ECKART, A., GERHARD, O. E., & OTT, T. 2000 *MNRAS* **317**, 348.

GENZEL, R., ET AL. 2003 *ApJ* **594**, 812.

GERHARD, O. E. & BINNEY, J. 1985 *MNRAS* **216**, 467.

GERSSEN, J., VAN DER MAREL, R. P., GEBHARDT, K., GUHATHAKURTA, P., PETERSON, R. C., & PRYOR, C. 2002 *AJ* **124**, 3270.

GHEZ, A. M., SALIM, S., HORNSTEIN, S. D., TANNER, A., LU, J. R., MORRIS, M., BECKLIN, E. E., & DUCHÊNE, G. 2005 *ApJ* **620**, 744.

GINSBURG, I. & LOEB, A. 2006 *MNRAS* **368**, 221.

GOULD, A. & QUILLEN, A. C. 2003 *ApJ* **592**, 935.

GRAHAM, A. W. 2004 *ApJ* **613**, L33.

GUALANDRIS, A., PORTEGIES ZWART, S., & SIPIOR, M. S. 2005 *MNRAS* **363**, 223.

GÜRKAN, M. A. & HOPMAN, C. 2007 *MNRAS* **379**, 1083.

GÜSTEN, R. & PHILIPP, S. D. 2004. In *The Dense Interstellar Medium in Galaxies* (eds. S. Pfalzner, C. Kramer, C. Staubmeier, & A. Heithausen). p. 253. Springer.

HAARDT, F., SESANA, A., & MADAU, P. 2006 *Memorie della Societa Astronomica Italiana* **77**, 653.

HILLS, J. G. 1981 *AJ* **86**, 1730.

HILLS, J. G. 1988 *Nature* **331**, 687.

HILLS, J. G. 1991 *AJ* **102**, 704.

HIRSCH, H. A., HEBER, U., O'TOOLE, S. J. & BRESOLIN, F. 2005 *A&A* **444**, L61.

HOFFMAN, L. & LOEB, A. 2007 *MNRAS* **377**, 957.

HOLLEY-BOCKELMANN, K. & SIGURDSSON, S. 2006 *MNRAS*, submitted; ArXiv:astro-ph/0601520.

HOPMAN, C. & ALEXANDER, T. 2005 *ApJ* **629**, 362.

HOPMAN, C. & ALEXANDER, T. 2006a *ApJ* **645**, 1152.

HOPMAN, C. & ALEXANDER, T. 2006b *ApJ* **645**, L133.

HOPMAN, C., FREITAG, M., & LARSON, S. L. 2007 *MNRAS* **378**, 129.

HOPMAN, C., PORTEGIES ZWART, S. F., & ALEXANDER, T. 2004 *ApJ* **604**, L101.

IVANOV, P. B., PAPALOIZOU, J. C. B., & POLNAREV, A. G. 1999 *MNRAS* **307**, 79.

IWASAWA, M., FUNATO, Y., & MAKINO, J. 2006 *ApJ* **651**, 1059.

KOBULNICKY, H. A., FRYER, C. L., & KIMINKI, D. C. 2006 *ApJ*, submitted; ArXiv:astro-ph/0605069.

KROUPA, P. 2001 *MNRAS* **322**, 231.

LEVI, M. 2006 Master's thesis, Weizmann Institute of Science.

LEVIN, Y. 2006 *ApJ* **653**, 1203.

LEVIN, Y. 2007 *MNRAS* **374**, 515.

LEVIN, Y. & BELOBORODOV, A. M. 2003 *ApJ* **590**, L33.

LEVIN, Y., WU, A., & THOMMES, E. 2005 *ApJ* **635**, 341.

LIGHTMAN, A. P. & SHAPIRO, S. L. 1977 *ApJ* **211**, 244.

MAGORRIAN, J. & TREMAINE, S. 1999 *MNRAS* **309**, 447.

MAKINO, J. & EBISUZAKI, T. 1994 *ApJ* **436**, 607.

MANESS, H., ET AL. 2007 *ApJ* **669**, 1024.

MERRITT, D., MIKKOLA, S., & SZELL, A. 2007 *ApJ* **671**, 53.

MERRITT, D. & MILOSAVLJEVIĆ, M. 2005 *Living Reviews in Relativity* **8**, 8.

MERRITT, D. & POON, M. Y. 2004 *ApJ* **606**, 788.

MERRITT, D. & SZELL, A. 2006 *ApJ* **648**, 890.

MILLER, G. E. & SCALO, J. M. 1979 *ApJS* **41**, 513.

MILLER, M. C., FREITAG, M., HAMILTON, D. P., & LAUBURG, V. M. 2005 *ApJ* **631**, L117.

MILOSAVLJEVIĆ, M., MERRITT, D., REST, A., & VAN DEN BOSCH, F. C. 2002 *MNRAS* **331**, L51.

MIRALDA-ESCUDÉ, J. & GOULD, A. 2000 *ApJ* **545**, 847.

MURPHY, B. W., COHN, H. N., LUGGER, P. M., & DRUKIER, G. A. 1997 *BAAS*, **29**, 1338.

NORMAN, C. & SILK, J. 1983 *ApJ* **266**, 502.

OKA, T., HASEGAWA, T., SATO, F., TSUBOI, M., MIYAZAKI, A., & SUGIMOTO, M. 2001 *ApJ* **562**, 348.

PAUMARD, T., ET AL. 2006 *ApJ* **643**, 1011.

PERETS, H. B. & ALEXANDER, T. 2008 *ApJ* **677**, 146.

PERETS, H. B., HOPMAN, C., & ALEXANDER, T. 2007 *ApJ* **656**, 709. Erratum **669**, 661.

PETERS, P. C. 1964 *Physical Review* **136**, 1224.

POLNAREV, A. G. & REES, M. J. 1994 *A&A* **283**, 301.

RAUCH, K. P. & INGALLS, B. 1998 *MNRAS* **299**, 1231.

RAUCH, K. P. & TREMAINE, S. 1996 *New Astronomy* **1**, 149.

RAVINDRANATH, S., HO, L. C. & FILIPPENKO, A. V. 2002 *ApJ* **566**, 801.

RUBBO, L. J., HOLLEY-BOCKELMANN, K. & FINN, L. S. 2006 *ApJ* **649**, L25.

SALPETER, E. E. 1955 *ApJ* **121**, 161.

SASLAW, W. C., VALTONEN, M. J., & AARSETH, S. J. 1974 *ApJ* **190**, 253.

SCHÖDEL, R., ET AL. 2007 *A&A* **469**, 125.

SPITZER, L. J. & SCHWARZSCHILD, M. 1951 *ApJ* **114**, 385.

SPITZER, L. J. & SCHWARZSCHILD, M. 1953 *ApJ* **118**, 106.

SYER, D. & ULMER, A. 1999 *MNRAS* **306**, 35.

WANG, J. & MERRITT, D. 2004 *ApJ* **600**, 149.

WILLEMS, B. & KOLB, U. 2004 *A&A* **419**, 1057.

YOUNG, P. J. 1977 *ApJ* **215**, 36.

YU, Q. 2002 *MNRAS* **331**, 935.

YU, Q. & TREMAINE, S. 2003 *ApJ* **599**, 1129.

ZHAO, H., HAEHNELT, M. G., & REES, M. J. 2002 *New Astronomy* **7**, 385.

ZIER, C. 2006 *MNRAS* **371**, L36.

Tidal disruptions of stars by supermassive black holes

By SUVI GEZARI

California Institute of Technology, MC 405-47, 1200 E. California Boulevard,
Pasadena, CA 91125, USA

A dormant supermassive black hole lurking in the nucleus of a galaxy will be revealed when a star approaches close enough to be torn apart by tidal forces, and a flare of radiation is emitted as the stream of stellar debris falls back onto the black hole. The luminosity, temperature, and decay of a tidal disruption flare are dependent on the mass and spin of the central black hole, and can be used to directly probe dormant black holes in distant galaxies for which the sphere of influence of the black hole is unresolved, and a dynamical measurement on the black hole mass is not possible. Here we present the discovery of tidal disruption flares in the ultraviolet with the *GALEX* Deep Imaging Survey, and compare the properties of the flares to the theoretical predictions.

1. Introduction

The tidal disruption of a star by a supermassive black hole (SMBH) has been proposed to be a unique probe for dormant black holes lurking in the nuclei of normal galaxies (Frank & Rees 1976; Lidskii & Ozernoi 1979). When a star passes close enough to a central black hole to be torn apart by tidal forces, the stellar debris falls back onto the black hole and produces a luminous accretion flare (Rees 1988). The detection of a tidal disruption flare is the only direct signpost for dormant black holes in the nuclei of galaxies for which the dynamical signature of the SMBH cannot be measured, i.e., when the sphere of influence of the black hole is unresolved ($R_{\text{inf}} = GM_{\text{BH}}/\sigma_\star^2$). The rate at which stars pass within the tidal disruption radius of a central supermassive black hole, $R_T \approx R_\star(M_{\text{BH}}/M_\star)^{1/3}$, depends on the flux of stars into loss cone orbits with $R_p < R_T$, where R_p is the distance of closest approach of the star's orbit. Stellar dynamical models of the nuclei of galaxies (Magorrian & Tremaine 1999; Wang & Merritt 2004) predict that this will occur once every 10^3–10^5 yr in a galaxy, depending on the galaxy's nuclear stellar density profile and the mass of the central black hole. There is a critical black hole mass above which R_T is smaller than the Schwarzschild radius (R_s), and the star is swallowed whole without disruption (Hills 1975), $M_{\text{crit}} = 1.15 \times 10^8 (r^3/m)^{1/2} \, M_\odot$, where $r_\star = R_\star/R_\odot$, and $m_\star = M_\star/M_\odot$.

After disruption, half of the stellar debris is ejected from the system, and less than half of the mass of the star remains bound to the black hole. Numerical simulations show that the fraction of debris that is eventually accreted can be as little as $0.1M_\star$ due to the strong compression of the gas stream during the second passage through pericenter which can impart a fraction of the gas with an escape velocity (Ayal et al. 2000). The start of the flare, t_0, occurs when the most tightly bound gas returns to pericenter (R_p) after the time of disruption, t_D. This time delay, $t_0 - t_D$ is dependent on the black hole mass as

$$t_0 - t_D = 0.11k^{-3/2}(R_p/R_T)^3 r_\star^{3/2} m_\star^{-1} M_6^{1/2} \text{ yr} \ , \tag{1.1}$$

where $M_6 = M_{\text{BH}}/10^6 \, M_\odot$, and $k = 1$ if the star has no spin on disruption, and $k = 3$ if it is spun-up to near break-up (Li et al. 2002). The accretion rate during the start of the "fallback" phase is predicted to result in a flare that peaks close to the Eddington luminosity, with a blackbody spectrum with $T_{\text{eff}} \approx (L_{\text{Edd}}/4\pi\sigma R_T^2)^{1/4} = 2.5 \times 10^5 M_6^{1/12} r_\star^{-1/2}$ K,

286

that peaks in the extreme-UV (Ulmer 1999). After t_0, the rate at which the bound debris falls back to pericenter declines as a function of time as $dM/dt \propto [(t - t_D)/(t_0 - t_D)]^{-5/3}$ (Phinney 1989; Evans & Kochanek 1989), which determines the characteristic power-law decay of the luminosity of the flare over the following months and years.

Although the theoretical work described above makes predictions for the rate, luminosity, temperature, and decay of tidal disruption events, there is limited observational evidence to test these predictions. The most convincing cases for a stellar disruption event occur from host galaxies with no evidence of an active galactic nucleus (AGN) for which an upward fluctuation in the accretion rate could also explain a luminous UV/x-ray flare. A UV flare from the nucleus of the elliptical galaxy NGC 4552 was proposed to be the result of the tidal stripping of a stellar atmosphere (Renzini et al. 1995), however the possible presence of a persistent, low-luminosity AGN detected in hard x-rays (Xu et al. 2005) makes this interpretation uncertain. The *ROSAT* All-Sky survey conducted in 1990–1991 sampled hundreds of thousands of galaxies in the soft x-ray band, and detected luminous (10^{42}–10^{44} ergs s^{-1}), soft [$T_{\mathrm{bb}} = (6–12) \times 10^5$ K] x-ray flares from several galaxies with no previous evidence for AGN activity, and with a flare rate of $\approx 1 \times 10^{-5}$ yr^{-1} per galaxy (Donley et al. 2002), that is in agreement with the theoretical stellar disruption rate. A decade later, follow-up *Chandra* and *XMM-Newton* observations of three of the galaxies demonstrated that they had faded by factors of 240–6000. Although this dramatic fading is consistent with the $(t - t_D)^{-5/3}$ decay of a tidal disruption flare (Komossa et al. 2004; Halpern et al. 2004; Li et al. 2002), the follow-up observations were obtained too long after the peak of the flares (5–10 yr) to measure the actual shape of the light curves. Follow-up *Hubble Space Telescope* (*HST*) Space Telescope Imaging Spectrograph (STIS) narrow-slit spectroscopy confirmed two of the galaxies as inactive, qualifying them as the most convincing hosts of a tidal disruption event (Gezari et al. 2003). The *ROSAT* flare with the best sampled light curve was successfully modeled as the tidal disruption of a brown dwarf or planet (Li et al. 2002), although its host galaxy was subsequently found to have a low-luminosity Seyfert nucleus (Gezari et al. 2003). Two new candidates from the *XMM-Newton* Slew Survey (XMMSL1) were reported by Esquej et al. (2006) to be optically non-active galaxies that demonstrated a large amplitude (80–90) increase in soft (0.1–2 keV) x-ray luminosity compared to their previous *ROSAT* PSPC All-Sky Survey upper-limits, with XMMSL1 luminosities of 10^{41}–10^{43} ergs s^{-1}. We obtained an optical spectrum of one of these candidates, elliptical galaxy NGC 3599, with the MDM 2.4m telescope on 24 December 2006. The spectrum shows weak emission lines of [O III], [N II], and Hα, indicating that the galaxy may host a low-luminosity AGN, raising doubts about the interpretation its x-ray flare as a tidal disruption event.

All but one of the tidal disruption event detections described above were simply "off"-"on" detections, with no detailed light curve information. We initiated a program to take advantage of the UV sensitivity, large volume, and temporal sampling of the *Galaxy Evolution Explorer* (*GALEX*) Deep Imaging Survey (DIS) to search for stellar disruptions in the nuclei of galaxies over a large range of redshifts, and attempt to measure the detailed properties of a larger sample of tidal disruption events. Here we present two detections by *GALEX* of UV flares from inactive galaxy hosts with a luminosity, UV light curve, and blackbody temperature consistent with the theoretical predictions for a tidal disruption event. Throughout this Conference Proceedings, calculations are made using *Wilkinson Microwave Anisotropy Probe* (*WMAP*) cosmological parameters (Bennett et al. 2003): $H_0 = 71$ km s^{-1} Mpc^{-1}, $\Omega_M = 0.27$, and $\Omega_\Lambda = 0.73$.

2. Selection of Candidates

The *GALEX* DIS covers 80 deg^2 of sky in the far-ultraviolet (FUV; $\lambda = 1344$–1786 Å) and near-ultraviolet (NUV; $\lambda = 1771$–2831 Å) with a total exposure time of 30–150 ks, that is accumulated in visits during $\sim$1.5 ks eclipses (when the satellite's 98.6-minute orbit is in the shadow of the Earth). Due to target visibility and mission planning constraints, some 1.2 deg^2 DIS fields are observed over a baseline of 2 to 4 years to complete the total exposure time. This large range in cadence of the observations allows us to probe variability on timescales from hours to years. We analyzed 4 DIS fields which overlap with the optical Canada-France-Hawaii Telescope Legacy Survey (CFHTLS) Deep Imaging Survey: *XMM*/LSS (CFHTLS D1); COSMOS (CFHTLS D2); GROTH (CFHTLS D3); and CFHTLS D4, resulting in a total overlapping area of 2.88 deg^2. We optimize our sensitivity to UV flares that decay on the timescale of months to years by coadding the visits of each field into yearly epochs. We chose to search for variable sources in the FUV coadds in order to avoid blending issues that are more pronounced in the NUV. We use aperture magnitudes from the SExtractor catalogs and match them to the closest source within a 2″ radius in the other coadds. We then use the standard deviation of matched source magnitudes as a function of magnitude in order to measure the photometric error, and identify sources that are intrinsically variable by selecting sources that vary at the 5σ level.

The 5σ variable UV sources are then matched with the CFHTLS deep survey photometric catalog in u, g, r, i, and z (closest match within a 3″ radius) and the optical colors and morphology of the matched sources are used to identify the hosts of the flares. Sources are flagged as flares if their FUV fluxes are constant within the errors before the start of the flare, decrease monotonically thereafter, and do not fade below the flux before the start of the flare. Optically unresolved sources with $g - r < 0.6$ are classified as quasars, and all resolved sources are classified as galaxies. X-ray catalog data from *XMM-Newton* for the *XMM*/LSS field (Chiappetti et al. 2005) and the COSMOS field (Mainieri et al. 2007), and from *Chandra* for the Groth field (Nandra et al. 2005), are used to classify galaxies with hard x-ray emission as AGNs. We also match the sources to the list of spectroscopically confirmed quasars and active nuclei from Veron-Cetty & Veron (2006) and from the COSMOS Survey (Trump et al. 2007). The CFHTLS fields are observed up to five times a month during the seasonal visibility of each field. Real-time difference imaging is performed on the images to produce a list of optically variable sources from which the CFHT Supernova Legacy Survey (SNLS) selects type Ia supernova candidates for follow-up spectroscopy. We match the variable optical source list (which includes optical variability associated with quasars and AGNs) with our UV variable sources. Contamination by type Ia supernovae and gamma-ray burst after-glows are unlikely since they are intrinsically faint in the UV (Panagia 2003; Brown et al. 2005; Mészáros & Rees 1997). However, we do check for anti-coincidence with the SNLS supernova triggers.

Figure 1 shows the optical color-color diagram of the UV variable source hosts. Many of our UV variability selected sources have quasar hosts, which are also detected in x-rays and as optically variable sources. Our analysis does not detect many RR Lyraes and M-dwarf flare stars since our deep coadds average out their variability signal which occurs on much shorter timescales of days, and minutes, respectively (Welsh et al. 2005, 2006). We define tidal disruption event candidates as UV flares from optically resolved galaxy hosts with no hard x-ray emission or optical emission lines that classify the galaxy as an AGN. We then follow-up these galaxy hosts with optical spectroscopy to search

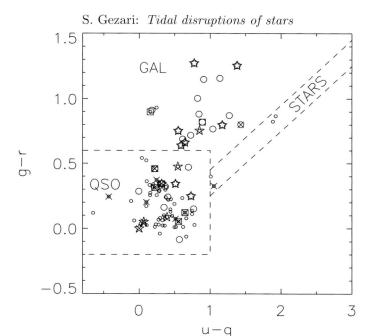

FIGURE 1. Optical color-color diagram of CFHTLS matches to the *GALEX* variable UV sources. Resolved optical hosts are plotted with open circles, and unresolved optical hosts are plotted with dots. Sources detected by CFHT SNLS as variable sources in the optical are plotted with stars. Sources with an x-ray catalog detection are plotted with an "x," and sources that are spectroscopically confirmed AGNs are plotted with squares. Dashed lines show the regions of the diagram dominated by quasars, early-type galaxies, and main sequence stars.

for signs of Seyfert activity in the form of emission lines or a non-stellar continuum, that would cast doubt on their interpretation as tidal disruption events.

3. Tidal Disruption Flare Detections

Two UV flares were found to have galaxy hosts with no signs of a Seyfert nucleus in their optical spectra. The first flare in the GROTH field was reported by Gezari et al. (2006) for having a DEEP2 DEIMOS Keck spectrum that classified its host galaxy as an inactive early-type galaxy at $z = 0.3698$. The second flare, to be presented in Gezari et al. (2008), is a new detection from a galaxy in the *XMM*/LSS field, whose VLT FORS1 spectrum is well fitted by a Bruzual-Charlot early-type galaxy template with an old stellar population and no star formation at $z = 0.326$ (Figure 2). Archival AEGIS (Davis et al. 2007) *Chandra* (0.3–10 keV) x-ray observations of the galaxy in GROTH one year after the peak of the flare detected no hard x-rays, placing an upper limit on the hard x-ray luminosity of a power-law AGN spectrum in the nucleus of $L_X \lesssim 1 \times 10^{42}$ ergs s^{-1}. However, an extremely variable soft x-ray source (<1 keV) was detected during the observations. The *XMM* Medium Deep Survey (XMDS) did not detect a source at the position of the flaring galaxy in the *XMM*/LSS field two years before its UV flare, placing an upper limit on its nuclear x-ray luminosity of L_X (0.5–2 keV) $\lesssim 3 \times 10^{41}$ ergs s^{-1} and L_X (2–10 keV) $\lesssim 2 \times 10^{42}$ ergs s^{-1} for a power-law index ($f_\nu \propto \nu^{-\Gamma}$) of $\Gamma = 1.7$, and a Galactic column density towards the source of $N_H = 2.61 \times 10^{20}$ cm^{-2} (Chiappetti et al. 2005). The lack of hard x-ray emission and the absence of Seyfert-like emission lines in both of these sources securely rules out the presence of a Seyfert nucleus, and leaves a

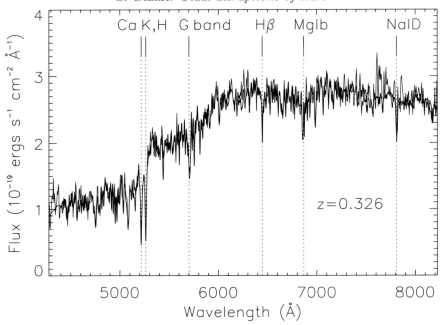

FIGURE 2. VLT FORS1 spectrum of the flaring galaxy in the *XMM*/LSS field obtained on 2006 September 26, smoothed by 4 pixels ($\sim$10 Å). The spectrum shows strong stellar absorption lines typical of an early-type galaxy. The best-fitting Bruzual-Charlot early-type galaxy template (Bruzual & Charlot 2003) with an old stellar population and no star formation is plotted with a thick line, and scaled to the continuum of the spectrum.

tidal disruption event as the most likely scenario for the cause of the UV flare. Thus below, we present an analysis of the properties of the UV flares in the context of a tidal disruption event.

The flare from the galaxy in the *XMM*/LSS field was also detected as an optically variable source in the *g*, *r*, and *i* bands by the CFHTLS. The *r*-band ($\lambda_{\rm eff}$ = 6174 Å) light curve has the best temporal sampling, and is shown along with the UV light curve of the flare in Figure 3. The optical photometry is measured from difference imaging following the method of Alard & Lupton (1998) implemented by Leguillou (2003). The *r*-band light curve catches the steep rise of the flare from a non-detection from 2003 August 1 through 2004 July 24 to its peak on 14 August 2004. We fit the first five months of the *r*-band light curve of the flare with a $(t - t_D)^{-5/3}$ power-law decay, which results in a least-squares fit to the time of disruption, t_D = 2004.04 ± 0.05. If we allow t_D and the power-law index n, from $(t - t_D)^{-n}$, to vary, the fitted values are t_D = 2004.2 ± 0.4 and n = 1.3 ± 0.9, which are both within the errors of the n = 5/3 fit. The *g*- and *r*-band light curves a year after the peak of the flare indicate that the optical emission does not continue to decay with the same power-law, and appears to plateau, or even rise. The UV flux does continue to decay through 2006, which if the optical-UV color is constant, indicates that the optical emission should also decay further. It has been proposed that after the "fallback" phase, the stellar debris will shock and circularize, and spread out to form a thick disk that accretes via viscous processes, resulting in a shallower power-law decay (Cannizzo et al. 1990). The time scale for the debris to shock and circularize, and form a torus with $R = \sqrt{2}R_p$ to conserve angular momentum from a parabolic to circular orbit, should be on the order of the minimum period of the debris, $P_{\rm min} \approx 0.7(M_{\rm BH}/10^7\,M_\odot)^{1/2}$ yr (Cannizzo et al. 1990). The change in exponent of the

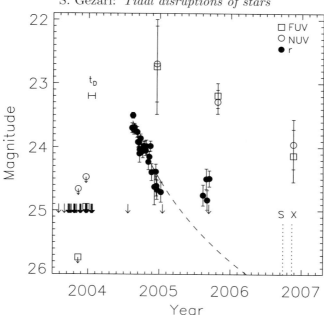

FIGURE 3. The FUV, NUV, and r-band light curve of the flare in the *XMM*/LSS field measured by *GALEX* and CFHTLS. The least-squares fit to the r-band light curve with a $(t - t_D)^{-5/3}$ decay is shown with a thick solid line, yielding a best fit time of disruption $t_D = 2004.04 \pm 0.05$ yr. Dotted line with an "S" indicates the time the VLT spectrum was taken, and dotted line with an "X" indicates the time of our TOO *Chandra* 0.3–10 keV observation which detected an extremely soft x-ray source.

decay occurs after 2004.96, $0.92/(1+z) = 0.70$ yr after the disruption, which is consistent with this interpretation. Numerical simulations from Ayal et al. (2000) also show that the accretion rate can also deviate from a $(t - t_D)^{-5/3}$ power-law during the "fallback" phase due to the expulsion of debris from compression during the second passage through pericenter.

Since the excellent temporal sampling of the CFHTLS observations measure the time of the peak of the flare, $t_0 = 2004.622$, we can determine the rest-frame time delay of the most tightly bound material to return to pericenter to high accuracy, $(t_0 - t_D)/(1 + z) = 0.44 \pm 0.04$ yr, which implies $M_{\mathrm{BH}} = (1.6 \pm 0.3)k^3 \times 10^7 \, M_\odot$, where k depends on the spin-up of the star upon disruption. The UV light curve of the flare in the GROTH field is shown in Figure 4. The NUV light curve is also well-fit with a $(t - t_D)^{-5/3}$ power-law, although the coarse time sampling of the *GALEX* observations puts weaker constraints on the time of the disruption and peak of the flare.

We triggered a 30 ks TOO *Chandra* (0.2–10 keV) x-ray observation of the flare in the *XMM*/LSS field on 2006 November 12, and detected four soft photons with energies between 0.2 and 0.4 keV. The soft x-ray flux strongly constrains the blackbody temperature of the flare, and the spectral energy distribution of the flare is well fitted by a blackbody with a rest-frame $T_{\mathrm{bb}} = 2.3 \times 10^5$ K. The soft x-ray count rate corresponds to an unabsorbed flux density at 0.3 keV assuming a blackbody spectrum with $T_{\mathrm{bb}} = 2.3 \times 10^5$ K of 1.3×10^{-33} ergs s^{-1} cm^{-2} Hz^{-1}. The soft x-ray luminosity of the blackbody at $\nu_e = 0.3$ keV $\times (1 + z)$ from $L_{\nu_e = (1+z)\nu_o} = f_{\nu_o}(4\pi d_L^2)/(1 + z)$, is 3.3×10^{23} erg s^{-1} Hz^{-1}. For a blackbody spectrum with $L_\nu = (2\pi h/c^2)\nu^3/(e^{h\nu/kT} - 1)4\pi R_{\mathrm{bb}}^2$, this implies that $R_{\mathrm{bb}} = 1.8 \times 10^{13}$ cm and $L_{\mathrm{bol}} = \sigma T^4 4\pi R_{\mathrm{bb}}^2 = 6.5 \times 10^{44}$ erg s^{-1}. If

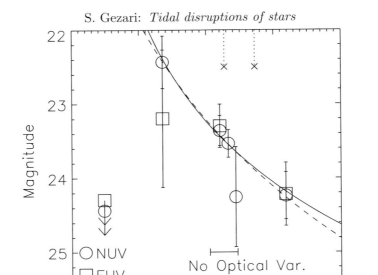

FIGURE 4. *GALEX* FUV and NUV light curve of the flare in GROTH reported by Gezari et al. (2006). The least-squares fit to the NUV light curve with a $(t - t_D)^{-5/3}$ decay is shown with a thick solid line, yielding a best-fit time of disruption $t_D = 2003.3 \pm 0.2$ yr. The least-squares fit to the NUV light curve with the power-law index and t_D allowed to vary is shown with a dashed line. Dotted lines with an "X" indicate times of *Chandra* 0.3–10 keV observations, which detected a variable extremely soft x-ray source. CHFTLS monitoring data from 2005 January to June detected no variable optical source in the nucleus of the galaxy with a sensitivity of $m \sim 25$.

we assume a constant characteristic temperature, then the observed peak g-band flux of $1.63 \pm 0.3 \times 10^{-29}$ ergs s^{-1} cm^{-2} Hz^{-1} corresponds to $L_{bol} = 4.6 \times 10^{45}$ ergs s^{-1}, which is equivalent to the Eddington luminosity of a $\sim 3.5 \times 10^7 \, M_\odot$ black hole.

The radius of the flare emission also places constraints on M_{BH}. The debris disk should form a torus with a pericenter radius within the tidal disruption radius of the central black hole. The blackbody fit to the flare SED has $R_{bb} \sim 1.8 \times 10^{13}$ cm which, if $R_{bb} \lesssim R_T$, places a lower limit on the black hole mass of $M_{BH} \gtrsim 1.7 \times 10^7 \, M_\odot$. The radius of the emission should also be greater than the minimum stable particle orbit for the black hole (R_{ms}), which ranges from $R_{ms} = 6R_g$ for a black hole with no spin, down to $R_{ms} = R_g$ for a maximally spinning black hole, where $R_g = GM_{BH}/c^2$. For $R_{bb} > R_{ms}$, this requires that $M_{BH} < 1.2 \times 10^8 \, M_\odot$, and is consistent with $M_{BH} < M_{crit}$ for a solar-type star. In order for these limits on M_{BH} to be consistent with the light curve, $k \lesssim 2$. If we consider the Eddington luminosity to indicate the minimum black hole mass, then $k \gtrsim 1.3$.

4. Conclusions

The *GALEX* DIS has proven to be a successful survey for discovering stellar disruption flares in the Ultraviolet. With timely multiwavelength follow-up observations, we have been able to measure the broadband spectral energy distribution and early decay of these events for the first time. Understanding the detailed properties of the flares is important for putting tidal disruption theory to the test, and optimizing the next generation of time domain surveys to search for these events. Also for the first time, we have detected a tidal disruption flare in the optical, with the excellent temporal sampling of the CFHTLS. This has important implications for the detection rates of future optical synoptical surveys such as Pan-STARRS and LSST, which will have orders of magnitude larger sky coverage than our study and a regular cadence of days. With a large sample of tidal disruption flares we will be able to probe the evolution of the black hole mass function in normal galaxies, unbiased by the minority of galaxies that are AGNs.

REFERENCES

ALARD, C. & LUPTON, R. H. 1998 *ApJ*, **503**, 325.

AYAL, S., LIVIO, M., & PIRAN, T. 2000 *ApJ*, **545**,772.

BENNETT, C. L., ET AL. 2003 *ApJS*, **148**, 1.

BROWN, P. J., ET AL. 2005 *ApJ*, **635**, 1192.

BRUZUAL, G. & CHARLOT, S. 2003 *MNRAS*, **344**, 1000.

CANNIZZO, J. K., LEE, H. M., & GOODMAN, J. 1990 *ApJ*, **351**, 38.

CHIAPPETTI, L., ET AL. 2005 *A&A*, **439**, 413.

DAVIS, M., ET AL. 2007 *ApJ*, **660**, L1.

DONLEY, J. L., BRANDT, W. N., ERACLEOUS, M. J., & BOLLER, TH. 2002, *AJ*, **124**, 1308.
ESQUEJ, R. D., ET AL. 2006 *A&A*, **462**, L49.

EVANS, C. R. & KOCHANEK, C. S. 1989 *ApJ*, **346**, L13.

FRANK, J. & REES, M. J. 1976 *MNRAS*, **176**, 633.

GEZARI, S., ET AL. 2003 *ApJ*, **592**, 42.

GEZARI, S., ET AL. 2006 *ApJ*, **653**, L25.

GEZARI, S., ET AL. 2008 *ApJ*, **676**, 944.

HALPERN, J. P., GEZARI, S., & KOMOSSA, S. 2004 *ApJ*, **604**, 572.

HILLS, J. G. 1975 *Nature*, **254**, 295.

KOMOSSA, S., ET AL. 2004 *ApJ*, **603**, L17.

LEGUILLOU, L. 2003 PhD thesis at Paris 6 University.

LI, L-X., NARAYAN, R., & MENOU, K. 2002 *ApJ*, **576**, 753.

LIDSKII, V. V. & OZERNOI, L. M. 1979 *Soviet Astron. Lett.*, **5**, 16.

MAGORRIAN, J. & TREMAINE, S. 1999 *MNRAS*, **309**, 447.

MAINIERI, V., ET AL. 2007 *ApJS*, **172**, 368.

MÉSZÁROS, P. & REES, M. J. 1997 *ApJ*, **476**, 232.

NANDRA, K., ET AL. 2005 *MNRAS*, **356**, 568.

PANAGIA, N. 2003. In *Supernovae and Gamma-Ray Bursters* (eds. K. Weiler & A. F. M. Moorwood). Lecture Notes in Physics Vol. 598, p. 113. Springer.

PHINNEY, E. S. 1989. In *The Center of the Galaxy* (ed. M. Morris). IAU Symp. No. 136, p. 543. Kluwer.

REES, M. J. 1988 *Nature*, **333**, 523.

RENZINI, A., ET AL. 1995 *Nature*, **378**, 39.

TRUMP, J. R., ET AL. 2007 *ApJS*, **172**, 383.

ULMER, A. 1999 *ApJ*, **514**, 180.

VERON-CETTY, M.-P. & VERON, P. 2006 *A&A*, **455**, 773.

WANG, J. & MERRITT, D. 2004 *ApJ*, **600**, 149.

WELSH, B. Y., ET AL. 2005 *ApJ*, **130**, 825.

WELSH, B. Y., ET AL. 2006 *A&A*, **458**, 921.

XU, Y., ET AL. 2005 *ApJ*, **631**, 809.

Where to look for radiatively inefficient accretion flows in low-luminosity AGN

By MARCO CHIABERGE[1,2]

[1]Space Telescope Science Institute, 3700 San Martin Drive, Baltimore, MD 21218, USA;
marcoc@stsci.edu

[2]INAF-Istituto di Radioastronomia, Via P. Gobetti 101, 40129 Bologna, Italy

We have studied the nuclear emission detected in *HST* data of carefully selected samples of low-luminosity AGN (LLAGN) in the local universe. We find faint unresolved nuclei in a significant fraction of the objects. FR I radio galaxies' optical nuclei show a tight linear correlation with the radio core emission, which argues for a common synchrotron origin. The nuclear emission in LLAGN is as low as 10^{-8} times the Eddington luminosity, indicating extremely low radiative efficiency for the accretion process and/or an extremely low accretion rate. When the Eddington ratio is plotted against the nuclear "radio-loudness" parameter, sources divide according to their physical properties. It is thus possible to disentangle nuclear jets and accretion disks of different radiative efficiencies. This new diagnostic plane allows us to find objects that are the best candidates for hosting (and showing) radiative inefficient accretion and determine in which ones we cannot see it. The (extremely limited) information available in the *HST* archive to derive the nuclear SEDs strongly supports our results.

1. Introduction

One of the most important results of the last few years has been the realization that most, if not all, galaxies harbor supermassive black holes (BH) in their centers. The presence of a supermassive BH can manifest itself as luminous quasar "activity," powered by accretion of matter onto the BH itself. Such a quasar phase, which peaks somewhere around redshift 2, is likely to play an important role in the build-up of these BHs. However, to avoid over-producing the mass in local BH, accretion has to occur at a very low rate for most of the time, and possibly also with a low radiative efficiency.

Models of radiatively inefficient accretion flows (RIAF) were originally developed to describe the low state of Cyg X-1 (Ichimaru 1977). They were later discussed in relation to supermassive black holes, to account for extremely small levels of activity observed in a large number of nearby galaxies, as well as in the Galactic center (Rees et al. 1982; Fabian & Rees 1995; Quataert & Narayan 1999). The models are based on the possibility that, at low mass-accretion rates, the accreting flow takes the form of a low density, hot, optically thin and geometrically thick gas structure. This radiatively inefficient regime would occur instead of optically thick, geometrically thin ('Shakura-Sunyaev') disk accretion.

Low-luminosity AGN (LLAGN) are the class of extragalactic objects in which RIAFs are most likely to be at work. They show some level of activity (compact radio emission with high brightness temperatures, line emission, nuclear x-ray emission) which, in the majority of the objects, cannot be explained in terms of processes other than a faint active nucleus powered by the central supermassive black hole. However, because of their intrinsic low luminosities, it is extremely difficult to study the emission from the accretion flow. This holds in particular for the wavelengths at which the spectral differences between RIAFs and "standard" accretion disk models are expected to differ the most, i.e., the IR-to-UV spectral region. RIAFs should lack both the "big blue bump" and the IR (reprocessed) bump, which instead characterize optically thick, geometrically thin accretion disk emission and the surrounding heated dust (e.g., Elvis et al. 1994).

Maoz et al. (2005) have shown that among a sample of 17 LLAGN, 15 of them show variability over a timescale of a few months, which demonstrates their non-stellar origin. However, it was still unclear from their work whether the nuclear radiation is from a jet or from the accretion flow. The most effective way to determine the physical nature of these nuclei is to derive their SED. This approach has already been attempted, but only with partial success (see e.g., Ho 1999; Maoz 2007) and with contrasting results. Models of RIAFs (see e.g., Narayan's paper, this proceedings) tell us that the emission mechanisms that are believed to dominate the SED are different from band to band: in the radio-to-far IR spectral region a "peaked" cyclo-synchrotron component; inverse Compton in the IR-to-UV (and possibly extending up to soft x-ray energies); and bremsstrahlung in the x-rays. However, the relative importance of each component in the different bands depends on the physical state of the source. While it is now possible to routinely disentangle the nuclear emission from that of the host—thanks to *HST* (e.g., Ho & Peng 2001; Chiaberge, Capetti, & Celotti 1999; Chiaberge, Capetti, & Macchetto 2005; Capetti & Balmaverde 2005)—emission from the accretion flow cannot be seen when it is swamped by other nuclear radiation processes, or obscured by dust. M 87 (Di Matteo et al. 2003) is a clear example, where the nuclear emission at all wavelengths is dominated by non-thermal synchrotron radiation.

In this paper, we review the current understanding of the nuclear emission in LLAGN, and we describe our method of discriminating between non-thermal jet emission and RIAF. We then derive the SED for some of the nuclei and we show that, despite the incomplete spectral coverage, the SEDs strongly support our picture.

2. The sample of LLAGN

We consider the following samples of low-luminosity AGN:

1. A complete sample of 33 FR I Radio Galaxies (i.e., low-luminosity radio galaxies) from the 3CR catalog (radio selected, Chiaberge et al. 1999);

2. Thirty-two Seyferts from the Palomar Survey of nearby galaxies and from the CfA sample, clearly limiting ourselves to those belonging to the Type 1 class, i.e., unobscured, (optically selected, Ho & Peng 2001);

3. A complete sample of 21 LINERs from the Palomar Survey of nearby galaxies (optically selected, Ho, Filippenko, & Sargent 1997);

4. Fifty-one nearby early-type galaxies (E+S0) with radio emission > 1 mJy at 5 GHz (optical + radio selection, Capetti & Balmaverde 2005, and references therein). The large majority of the galaxies in the sample are spectrally classified as either LINER or Seyfert. A detailed description of this sample is given in Capetti & Balmaverde (2005).

5. The five broad-line radio galaxies with $z < 0.1$ included the 3CR catalog (Chiaberge, Capetti, & Celotti 2002, and references therein).

The sample of nearby ellipticals partially overlaps with samples 1–3. However, there are only 10 objects in common, so the total number of objects considered is 132. Note that being selected according to different criteria, these objects do not constitute a complete sample. However, they perfectly represent the overall properties of low-power active nuclei in the local universe, where RIAFs are most likely to be found.

3. Low-luminosity radio galaxies

In order to understand the origin of the nuclear emission in LLAGN it is useful to start analyzing the information we first achieved on low-luminosity radio galaxies (FR Is). In the large majority of them (~85%), the radial brightness profiles of the nuclear regions

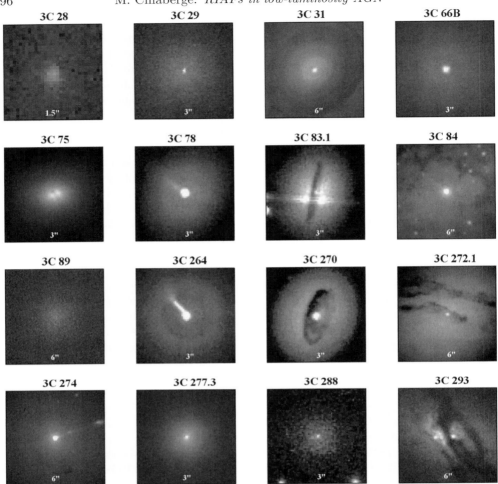

FIGURE 1. *HST*/WFPC2 broad-band images of the central regions of FR I radio galaxies. Note the presence of a central unresolved nucleus in the large majority of the objects.

is clearly indicative of the presence of a central unresolved nucleus (Figure 1). The optical nuclei show a tight correlation with the radio core emission (on the scale provided by the VLA), both in flux and luminosity, which extends over more than four orders of magnitude (Chiaberge et al. 1999). The fact that the correlation is found both in flux and luminosity gives us confidence that it is not induced by either selection effects or a common redshift dependence. As the radio core emission is certainly non-thermal synchrotron radiation, the presence of a tight linear correlation between radio core and optical nuclear emission is a strong clue that the optical nuclei are also synchrotron. The radio-optical spectral indices of the nuclei span the range $\alpha_{\rm ro} = 0.6$–0.9 ($F_\nu \propto \nu^{-\alpha}$), fully consistent with the $\alpha_{\rm ro}$ typical of blazars and optical jets which are indeed dominated by synchrotron radiation.

The interpretation of the optical nuclei of low-luminosity radio galaxies has been recently confirmed by the detection of optical polarized nuclear emission in the nine closest 3C FR Is. The level of polarization has values between 2 and 11%, typical of synchrotron

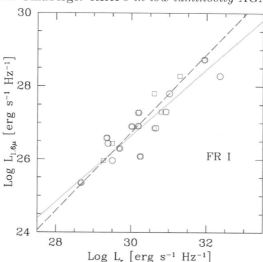

FIGURE 2. IR (1.6 μ) luminosity of FR I nuclei vs. radio core luminosity. Circles refer to VLA cores, while squares are VLBI measurements. Note that most VLBI luminosities perfectly overlap the VLA measurements. The lines represent linear fits to the data. The steepest line (dashed) is obtained using VLBI cores.

jets. The polarization vector shows no clear preferential alignment with the jet axis, thus ruling out scattered radiation for the origin of the optical nuclei (Capetti et al. 2007).

The detection of optical nuclei indicates that we have a direct view of the innermost nuclear regions of FR I. The high detection rate suggests that if obscuring material is present in these objects, it has to be distributed in a geometrically thin structure. Alternatively, the geometrically and optically thick tori expected in the framework of the AGN unification scheme (e.g. Urry & Padovani 1995) are present only in the minority of FR I.

However, it is still possible that compact tori exist in FR I, and the optical radiation we observe is produced on a larger scale. In fact, the upper limits on the size of the optical emission derived from the *HST* resolution is still of the order of a few parsecs (for the closest objects). In order to rule out the presence of strong nuclear emission obscured at optical wavelengths (from e.g., the accretion disk, as in Type 2 AGN) we observed the 3C FR Is with *HST*/NICMOS at 1.6 μ. While most Type 2 AGN show a bright nuclear component in the near-IR (Quillen et al. 2001), FR I do not show any significant IR excess with respect to the optical emission (Chiaberge et al. 2009).

Furthermore, if VLBI radio data are used instead of VLA data, the correlation between IR nuclei and radio cores becomes slightly tighter and more linear (Figure 2). This is in fact a strong clue that the IR-optical radiation is produced very close to the black hole, on a scale similar to that of the VLBI cores (<0.1 pc for the large majority of the galaxies in the sample). This further piece of evidence strongly argues against the presence of pc-scale dusty tori in FR I, which would be expected in analogy with other AGN classes. As a result of that, the absence of broad emission lines in FR Is cannot be attributed to obscuration, therefore these objects intrinsically lack substantial BLR. Clearly, there is still space for some exceptions, such as 3C 120 which is a rather unconventional FR I, showing a typical FR I radio morphology associated to an S0 galaxy with broad emission lines and a strong thermal continuum.

The luminosity of the synchrotron optical nucleus also represents an upper limit to any thermal disk emission. For a $10^9 \, M_\odot$ black hole, this limit translates into a fraction as small as $\lesssim 10^{-5}$ of the Eddington luminosity, suggesting that accretion might take place in a low-efficiency radiative regime. Further evidence for the presence of RIAFs in the giant elliptical hosts of FR Is has been obtained from Bondi accretion rate estimates (Allen et al. 2006). However, even though it appears very likely that FR I radio galaxies host RIAFs, the nuclear emission is dominated by the jet, therefore RIAFs cannot be directly "seen" in these objects. This implies that all FR Is that have always been considered as "RIAF candidates," such as M 87, M 84, NGC 4261, NGC 6251, and so on, are not good targets for obtaining direct evidence for the presence of RIAFs (i.e., by means of deriving their nuclear SED).

4. Seyferts

Following Ho & Peng (2001), we consider a sample of Seyfert 1 which includes objects from the Palomar survey (Ho et al. 1997), that are faint and nearby ($D_{\mathrm{med}} = 20$ Mpc), together with relatively brighter (and slightly more distant) objects from the CfA survey (Huchra et al. 1983; Osterbrock & Martel 1993).

HST/WFPC2 observations are available for the majority (25/32) of the objects. The data analysis and nuclear photometry has been performed by Ho & Peng (2001) by modeling the galaxy brightness profile with multiple components using *GALFIT*. The radio data are also taken from the list in Ho & Peng (2001).

Optical nuclei are detected in all of the observed Seyferts in the sample.

5. Broad-line radio galaxies

BLRG may be similar to Seyferts, as far as the radio-optical properties are concerned. As in Seyferts, the optical emission of these objects is thought to be non-thermal emission from the accretion disk, while the radio core of BLRG is certainly synchrotron from the base of the relativistic jet. However, the relative contribution of jet and disk emission in the optical may vary from one source to another. Recently, Grandi & Palumbo (2007) were able to discriminate between jet and disk emission in a small sample of BLRG using x-ray information. We will show in the following that similar conclusions can be drawn using optical/radio information.

6. LINERs

Of the 21 LINERs imaged with the *HST*, we detect the nucleus in 9 objects, namely NGC 404, NGC 1052, NGC 2681, NGC 3368, NGC 3718, NGC 4143, NGC 4203, NGC 4378, and NGC 4736.

A few details on the nuclear measurements are worth mentioning. For NGC 3368 the nucleus is not detected in the V band, but it is clearly seen in the UV. We argue that this is due to the higher contrast between the nucleus and the host galaxy stellar emission at UV wavelengths compared with the V band. The upper limit in the optical band ($F_\lambda < 9.1 \times 10^{-17}$ erg s^{-1} cm^{-2} Å^{-1}) is, however, in substantial agreement with the observed UV nuclear flux (within less than 1σ). Conversely, NGC 3718 has no detected nucleus in the optical, while it is seen in the IR 1.6 μm NICMOS image. Here the absence of the optical nucleus is most likely due to the presence of an extended nuclear dust lane, clearly seen in the optical images.

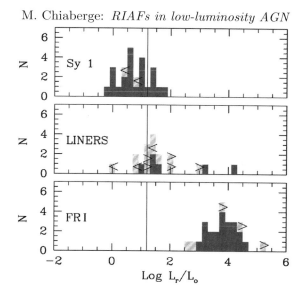

FIGURE 3. Radio core to optical core luminosity ratio. Shaded areas refer to object with upper limits to the nuclear emission upper. The solid lines correspond to the "historical" dividing line between radio-loud and radio-quiet AGN. Because of the different observing band (R band instead of B band), our definition of that dividing line corresponds to $R = L_r/L_o \sim 16$. The sample of early-type galaxies of Capetti & Balmaverde (2005) is not included in this figure.

Radio core data for LINERs are taken from VLA observations at 2 cm (Nagar et al. 2000, 2002).

7. The nuclear radio-loudness parameter

We can infer important information by studying the connection between radio and optical properties considering a nuclear "radio-loudness" parameter, defined as $R = L_R/L_o$, i.e., the ratio between radio and optical nuclear luminosity. Although the presence of a clear dichotomy in the radio-loudness distribution of powerful AGN is still a matter of debate, Seyfert 1 and LINERs are historically considered as radio quiet with respect to their global properties. Intriguingly, Ho & Peng (2001) have shown that when the nuclear emission of Seyfert (type 1 through 1.9) is disentangled from the host galaxy stellar component, the majority of their nuclei have $R' = F_{5\mathrm{GHz}}/F_B > 10$, i.e., they fall into the standard definition of radio-loud objects (Kellermann et al. 1989). But how does this compare with the nuclei of standard radio-loud objects? The radio-optical correlation found for FR I is approximately linear and it translates into a "radio-loudness" distribution clustered around $R \sim 10^4$. The usual value of R' adopted in the literature as the separation between radio-quiet and radio-loud objects does not correspond to any significant distinction between Seyferts and radio galaxies nuclei (see Figure 3). Although some Seyfert 1s are indeed above this *threshold*, the median value for Seyferts is more than a factor of ~ 1000 lower than that measured in FR I, and thus they appear to be fundamentally different from LLRG. While "radio-loud" nuclei like those of FR Is are jet-dominated both in the radio and in the optical, the nuclear emission of Seyferts is different in the two bands. The radio core is most likely non-thermal synchrotron radiation from a small jet (e.g., Nagar et al. 2000), and the optical nucleus is most likely thermal emission from the accretion disk. In this sense, Seyfert 1 nuclei are not "radio-loud."

Thus, the nuclear radio-loudness parameter for LLAGNs has now physical meaning: radio-loud objects are jet dominated. This is true in the radio and in the optical, but also possibly at other wavelengths, such as the x-rays (e.g., Fabbiano et al. 1984; Hardcastle & Worrall 1999; Balmaverde, Capetti, & Grandi 2006). Radio-quiet objects, instead, show emission from the accretion process in the optical. Of course, this is of great interest if our aim is to look for different accretion regimes at work.

LINERs span a large range in R. Some of them are similar to Seyferts, while at least three of the objects of the sample have R in the region spanned by *bona fide* radio-loud objects. This analysis suggests that there may be two distinct populations of LINERs. A first group extends the behavior of FR I radio galaxies to even lower powers. The second group of LINERs (which possibly represents the majority of the objects of the complete sample) behave like Seyfert 1 at the lower end of their luminosity function. We will thus refer to them as to "radio-quiet" LINERs. However, the number statistics are not sufficient to test whether there is a continuous transition between the two subclasses, or the population is bi-modal. Note that the radio-quiet LINERs are associated with a mixed population of host galaxy morphologies and optical spectral classifications. However, x-ray studies show that LINERs nuclei are basically unobscured (Terashima et al. 2002). On the other hand, the three radio-loud LINERs are hosted by early type galaxies as it is for FR Is.

The sample of early-type galaxies behaves similarly to LINERs and low-luminosity Seyferts. The sample is not included in Figure 3, but the objects are reported in Figure 5. Capetti & Balmaverde (2005) have shown that there is a striking correspondence between the isophotal shape of the hosts and the radio loudness of their nuclei. Radio-loud nuclei are exclusively associated to core galaxies (see e.g., D. Merrit's paper, this proceedings) while radio-quiet nuclei are present in power-law galaxies.

8. Black hole masses in LLAGN

If we want to find low radiatively efficient accretion flows we also need an estimate of the mass of the central supermassive black hole. When more direct measurements (Ho 2002, and ref. therein) are not available, the black hole masses are estimated through the correlation with the central velocity dispersion determined by Tremaine et al. (2002). It is worth noting a trend between radio-loudness and black hole mass. We stress that, since the three samples are selected according to completely different criteria and observational properties, it is dangerous to carry out any statistical test on the total sample of local AGN. Thus, no correlation can be derived between these two quantities. Nonetheless, it is clear that "radio-loud" nuclei exclusively appear in galaxies with the most massive black holes. On the other hand, radio-quiet nuclei are in general associated to low-mass black hole objects, although a few exceptions are present.

It is tempting to speculate that the enhanced efficiency in producing powerful jets is somehow related to the black hole mass. An alternative, possibly more realistic, scenario is that a combination of the BH mass and (following e.g., Blandford et al. 1990; Sikora, Stawarz, & Lasota 2007) the BH spin determines the radio loudness of each nucleus.

9. How to find RIAFs: nuclear radio loudness and Eddington ratio

In Chiaberge et al. (2005) we showed that the nuclear properties of LLAGN are best understood when the ratio between optical luminosity and Eddington luminosity $L_o/L_{\rm Edd}$ is plotted against the "nuclear radio-loudness parameter" R, defined as the ratio between the *nuclear* radio (core) luminosity at 5 GHz ($L_{\rm 5GHz}$) as measured from high-resolution

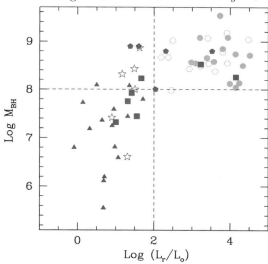

FIGURE 4. Black hole mass estimate vs. radio loudness of the nuclei for the samples of local AGN. LINERs are marked as squares, Seyferts as triangles, FR I radio galaxies as filled circles and BLRG are pentagons. The sample of early-type galaxies are marked as empty symbols: radio-loud nuclei are hexagons, radio-quiet nuclei are stars. Only detected nuclei are plotted, but upper limits behave consistently (see Chiaberge et al. 2005).

(VLA or higher resolution) data and the *nuclear* optical luminosity L_o as measured from *HST* images (Figure 5). Although the "Eddington ratio" is formally defined using the bolometric luminosity instead of the optical luminosity, here we do not perform any bolometric correction. This is because i) we do not know the SED of the nuclei "*a priori*"; and ii) a bolometric correction of a factor of ~ 10, or even slightly larger, would not significantly change the location of the points in the logarithmic planes described in this paper.

The power of this diagnostic plane resides in the fact that we only need two measured quantities, plus an estimate of the black hole mass, to clearly separate sources of different physical origin. This allows us to discriminate between jets and accretion disks of different radiative efficiency (and/or different accretion rates). The drawback is that we need *HST* images to detect the optical nuclei.

Since we are looking for objects harboring RIAFs, in Figure 5 we plot galaxies for which the "active nucleus" has been detected both in the radio and the optical (or near-IR) bands. The lack of detection of compact radio emission is considered as lack of evidence for AGN activity. On the other hand, the lack of detection of the optical nucleus would set a double upper limit in the plot and, most importantly, the location of the data point along the radio-loudness axis would remain undetermined. Therefore, we discard such objects. We also only plot the objects for which a black hole mass estimate is available, either through gas kinematics or by using the relation with the stellar velocity dispersion.

Sources separate into four quadrants: Seyfert 1s occupy the top-left quadrant, FR I radio galaxies are in the bottom-right quadrant. These two classes define the regions characterized by "radio-quiet" high-radiative efficiency accretion and "radio-loud" jet emission, respectively. LINERs split into two sub-samples: both are in the region of low $L_o/L_{\rm Edd}$, some of them being radio quiet and some radio loud. Nuclei of ellipticals separate according to the properties of the radial brightness profile of the host (Capetti & Balmaverde 2006), and behave similarly to LINERs: core galaxies are radio loud, power-

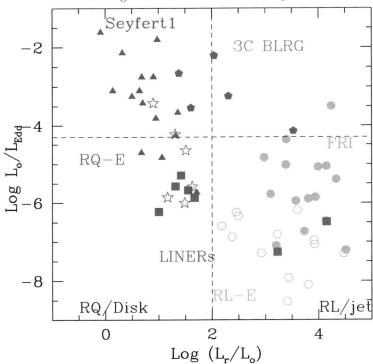

FIGURE 5. The diagnostic plane for the nuclear emission of LLAGN. The ratio between the optical nuclear luminosity and the Eddington luminosity is plotted against the radio-loudness parameters. Symbols are as follows: Seyfert 1 are marked as triangles, pentagons are BLRG and QSO from the 3CR catalog, circles are FR I radio galaxies, squares are LINERs, hexagons are early-type "core-galaxies" and stars are early-type "power-law" galaxies. The dashed lines are only plotted to guide the eye in discriminating between different "species" of nuclei.

law galaxies are radio quiet (for a definition of core and power-law galaxies, see e.g., Faber et al. 1997). Broad-line radio galaxies, instead, are found at the top of the plot, at high values of $L_o/L_{\rm Edd}$, while their location on the radio-loudness axis most likely depends on the relative importance of the jet and disk emission in the optical band (the radio core being dominated by radiation from the jet). Recently, Grandi & Palumbo (2007) were able to study the relative contribution of jet and disk emission in a small sample of BLRG, based on x-ray data. They found that, among others, 3C 390.3 and 3C 382 are dominated by disk emission, the jet emission contributing only by less than 45%. Intriguingly, of the five BLRG in our sample, 3C 382 is well inside the top-left quadrant, while 3C 390.3 is very close to the dividing line between jet-dominated and disk-dominated nuclei. 3C 382 is also the most "disk-dominated" BLRG in the sample of Grandi & Palumbo (2007). This result might not only imply that the plane of Figure 5 has diagnostic power for BLRG, but also that objects that are disk dominated in the optical are also disk-dominated in the x-rays. On the other hand, the most "radio-loud" BLRG is 3C 111, a source in which the jet contribution is known to be strong.

Although it is not easy to identify any "bi-modality" when plotting all the samples together as done in Figure 5, we draw dashed lines just to guide the eye in discriminating between physical processes that are clearly different. Clearly, there is a bit of ambiguity for the nuclei that fall close to those "dividing lines."

Let us now focus on the right-hand side of the panel. Here we find jet-dominated nuclei (LLRG and a subsample of LINERs). In these objects, both the radio and the optical radiation is most likely synchrotron emission from the base of the jet, as it has been established for low-luminosity radio galaxies (see Section 3). The radiation from the accretion flow is swamped by the jet emission, and cannot be studied directly (at least in the observing bands considered here). We should point out that in this case we have an "upper limit" to the Eddington ratio. Nevertheless, for several objects, this upper limit is as low as $L_o/L_{\rm Edd} \sim 10^{-8}$. On the other hand, none of the detected objects on the left side of the plot reaches such low values. Interestingly, these extremely low-efficiency accreting black holes are still capable of producing a "jet." This confirms that although RIAFs (or whatever these objects are!) are inefficient from the point of view of producing radiation, under certain circumstances (i.e., for the "radio-loud" nuclei) they can be very efficient in producing outflows or even (relativistic) jets, as in the case of LLRG (e.g., Rees et al. 1982). Note that the FR I with the highest value of $L_o/L_{\rm Edd}$ is 3C 84 (NGC 1275). This is a peculiar source, often classified as a blazar. The nucleus in this object is highly variable and the relativistic jet is most likely seen at a very small angle to the line of sight. The nuclear flux is thus probably enhanced because of relativistic beaming, thus leading to a higher L_o.

Let us now focus on the left side of the plot. In that region we find nuclei that show a low value of L_r/L_o, i.e., an optical excess with respect to the optical counterpart of the radio (synchrotron) radiation. For Seyfert 1s, such an optical excess is readily interpreted as emission from the accretion disk. A few bright Seyferts for which the nuclear SED has been derived support this interpretation (e.g., Alonso-Herrero et al. 2003; Chiaberge et al. 2006). For the brightest nuclei, the limits on the Eddington ratio are still compatible with radiatively efficient accretion. Furthermore, we should stress that for those objects, a bolometric correction of a factor ~ 10 (Elvis et al. 1994; Marconi et al. 2004) should be performed, because we know their SED and we know that the optical R or V bands do not correspond to the peak of radiatively efficient accretion-energy output. On the other hand, for objects of the lowest Eddington ratios, we don't know what the bolometric correction should be, but a factor of 10 (or even slightly higher) would not change our conclusions. Objects with Eddington ratios as low as 10^{-6} cannot be reconciled with "standard" accretion-disk models. Therefore, they are the best candidates to host radiatively inefficient accretion disks.

We stress that since we want to identify the objects in which radiation from a RIAF can be detected and studied in the IR-to-UV spectral region, in order to derive the SED and constrain the models, our best RIAF candidates are objects in which a nuclear source has been detected in archival *HST* images. This reduces the number of objects to nine candidates. We thus predict that RIAFs can be detected and studied in the following objects: M 81, NGC 3245, NGC 3414, NGC 3718, NGC 3998, NGC 4143, NGC 4203, NGC 4565 and NGC 4736.

10. The SEDs of LLAGN

In order to confirm that the plane of Figure 5 has diagnostic power, we should derive the SED of objects that lie in different regions of the plane and see if they look different. First of all, it is easy to predict that objects in the top-left quadrant (relatively high Eddington ratio and disk dominated in the optical) should show the signatures of "standard accretion." In Figure 6a we show the SED of NGC 4151, a Seyfert 1 galaxy located in that side of the diagram. The comparison with the average QSO SED clearly shows the

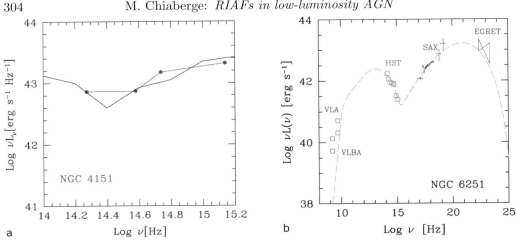

FIGURE 6. a) Nuclear SED of the Seyfert galaxy NGC 4151. The solid line is the average SED of radio-quiet QSOs (Elvis et al. 1994) rescaled to the luminosity of the object. b) Nuclear SED of the FR I radio galaxy NGC 6251. The dashed line corresponds to a synchrotron self-Compton model (Chiaberge et al. 2003).

signature of radiatively efficient accretion surrounded by hot dust, i.e., both the big blue bump and the IR bump in this object, as expected.

On the other hand, NGC 6251 shows a completely different SED (Figure 6b), that can be modeled with synchrotron self-Compton radiation from a mildly relativistic component of a jet (Chiaberge et al. 2003). This is again expected from the location of this object in the diagnostic plane, i.e., on the bottom-right quadrant, among jet-dominated nuclei.

In the following sections we will focus on the bottom-left quadrant of the plane, where disk-dominated objects of extremely low Eddington ratios are located. If RIAFs are present in that region of the plane, we expect the SED of those objects to lack the signatures of optically thick accretion. More details are given in (Chiaberge et al. 2009).

10.1. *NGC 4565*

In Chiaberge et al. (2006) we have studied the case of NGC 4565, a low-luminosity Seyfert that resides in the bottom-left region of the diagnostic plane. We derived the SED and we showed that it is unusual, lacking the typical signatures of radiatively-efficient accretion disks (i.e., lacks both the UV and the IR bumps).

This is one of the very few objects for which *HST* data are available in the archive from the IR to the UV band, so that the SED can be effectively derived. Note that the observations were not taken simultaneously in all bands, therefore variability may affect the shape of the SED. However, the typical variability observed in LLAGN is small, of the order of 20–30% (Maoz et al. 2005), and comparable or only slightly larger than the uncertainty of the measurements. Furthermore, two of the optical filters used (F450W and F555W) include relatively strong emission lines (mainly [O III]5007 and Hβ). However, since the pass bands are ~1000 Å wide, the observed flux is likely to be dominated by continuum emission.

The absorption-corrected nuclear spectral energy distribution is shown in Figure 7. The *HST* data are de-reddened using our measurement of $N_H = 2.5 \times 10^{21}$ cm^{-2} obtained from *Chandra* data, which converts to $A_V = 1.25$, assuming Galactic gas-to-dust ratio. Although in AGN the gas-to-dust ratio may differ from the local value, we believe that this choice is justified in the case of NGC 4565. In fact, this object is a spiral seen almost

FIGURE 7. Top: *HST* WFPC2 "true-color" RGB image. The R channel corresponds to the F814W filter, the B channel is the F450W, and the G channel is the average of the two filters. Direction to the North is indicated by the arrow. Bottom: Absorption-corrected nuclear spectral energy distribution of NGC 4565 in the IT-to-UV region. The solid line is the average SED of radio-quiet QSO from Elvis et al. (1994), normalized to the flux of NGC 4565 at 1.6 μm.

edge on. Therefore, it is reasonable to assume that a significant amount of dust and gas in the disk of the galaxy projects on our line of sight to the nucleus. An A_V of ∼2 mag can be easily explained with material in the disk of the galaxy, without the need for any nuclear-absorbing structure. This implies that NGC 4565 is a nice example of the objects called "true Seyfert 2s," i.e., unobscured AGN intrinsically lacking the BLR.

The nuclear SED appears basically flat ($\alpha \sim 1$, $F_\nu \propto \nu^{-\alpha}$) from the 1.6 μm to 4500 Å, with possibly a small peak between 5000 and 4000 Å and a small drop-off in the U band. This peak may be real, or due to a possible contamination from emission lines (mainly [O III] and Hβ) that fall in the F555W and F450W filters pass bands. Whatever the nature of such a small peak, it is clear that neither a significant UV bump nor IR thermal emission from hot dust, which are characteristic of AGNs, are visible in NGC 4565. For comparison, in Figure 7 we also show the average SED of radio-quiet QSO (solid line) as taken from Elvis et al. (1994).

In the diagnostic diagram of Figure 5, NGC 4565 lies in the lower-right quadrant, among "radio-quiet" LLAGN with low Eddington ratio. In order to move NGC 4565 to the top-left quadrant, where most Seyferts are found, the central black-hole mass would have to be at least a factor of ∼100 lower. This would substantially violate the $M_{\rm BH}$–σ relation. We conclude that the nucleus of NGC 4565 is a very low-efficiency accretion object and that we are observing the accretion process directly in the optical.

On the other hand, the "unusual" shape of the SED of NGC 4565 strongly supports our prediction that objects in that part of the plot are not "standard" disks.

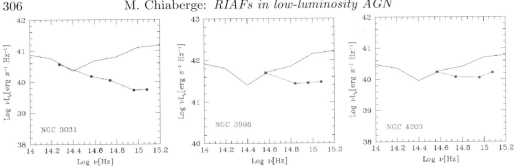

FIGURE 8. De-reddened nuclear SEDs of three LLAGN. All of the objects lie in the bottom-left region of the diagnostic plane of Figure 5, where RIAFs are expected to be found. The solid line is the average SED of radio-quiet QSOs (Elvis et al. 1994) rescaled to the luminosity of each object.

10.2. *SEDs of RIAFs?*

Unfortunately, the data currently available in the *HST* archive are not suitable to derive the nuclear SED of a substantial number of objects that are located in the bottom-left quadrant, where RIAFs are expected to be found. This is mainly because either the spectral coverage is very limited or the central region of the galaxy is saturated in some images. Furthermore, the observations at the different wavelengths were not taken simultaneously. However, for a few objects, we were able to derive the nuclear SED in the IR-to-UV spectral region (Chiaberge et al. 2009). In Figure 8 we show the results, again compared with the average QSO SED of Elvis et al. (1994), rescaled to the luminosity of each object. Typical errors are between 10 and 20%. The fluxes were de-reddened assuming the N_H column density derived from x-ray data and converted to A_V using the standard Galactic gas-to-dust ratio.

NGC 3031 (M 81) is the object with the best spectral coverage. Data are available in the *HST* archive from the IR (1.6 μ) to the UV (∼2500 Å). The data were taken between 1995 and 2003 (only the UV and *U*-band data are quasi-simultaneous). Note that the UV luminosity is a factor of ∼20 lower than expected in the case of a standard QSO spectrum, for the same IR luminosity. The SED we obtain using broad-band photometry is substantially in agreement with the results of Ho, Filippenko, & Sargent (1996) which, based on *HST*/FOS spectra, ruled out significant contribution from stellar light to the nuclear flux at all UV-optical wavelengths.

NGC 3998 and NGC 4203 show similar SEDs. Both objects lack IR observations, therefore the spectral coverage is not appropriate to draw conclusions. However, it is clear that even for these nuclei a "big blue bump" is not present. Lawrence (2005) pointed out that for different parameters such as temperature and accretion rate, the SED of a standard accretion disk should change significantly, as the peak of both the blue and IR bumps shifts to lower frequencies. This scenario cannot be ruled out with the present observations—in particular for M 81—in which we might just be seeing the high energy cutoff of a low-temperature standard disk, where the blue bump has shifted to the IR. However, we point out that for both NGC 3998 and NGC 4203 the SED seems to be flat, or even slightly concave. This cannot be easily explained in terms of a simple shift of the bumps toward longer wavelengths.

Summarizing, although the information in the *HST* archive is very limited, we showed that the SED of objects located in the region of the $L_o/L_{Edd} - L_r/L_o$ plane where RIAFs are expected to be observed are indeed very different both from those of higher luminosity Seyfert and FR I radio galaxies. With the observations available now it is dangerous to model the SED and draw conclusions on their nature, because variability can substantially affect the spectral shape. However, it is clear that these objects do not show the SED typical of standard accretion disks.

11. Conclusions

We have shown that the nature of the nuclei of LLAGN in the local universe can be understood when the radio-optical properties are considered. By combining the Eddington ratio as measured from detected unresolved nuclei in *HST* images and the nuclear "radio-loudness" parameter, we find a straightforward way of distinguishing jet-dominated from disk-dominated nuclei. The ultimate scope of this research is to find objects in which the emission from a radiatively efficient accretion flow can be directly detected. We have shown that FR I radio galaxies are jet-dominated (at all wavelengths), therefore such objects should be excluded from samples of RIAF candidates. This is not because a RIAF may not be present at their center, but just because its emission cannot be detected, being swamped by the jet at most, if not all, wavelengths.

From a sample of 132 LLAGN, only in 9 objects we can hope to detect RIAF emission in the IR to UV. We have derived the SED for 4 of these objects and we have shown that, despite the limited information available to date in the *HST* archive, their SEDs are extremely unusual, and they are not consistent with radiation from "standard accretion." These objects are therefore the best targets for follow-up studies, not only in the IR-to-UV spectral region with *HST*, but also in other bands such as far-IR, mm, and x-rays.

Of course, the results presented in this paper were achieved thanks to the effort of several people. In particular, I wish to thank Alessandro Capetti, Duccio Macchetto, Annalisa Celotti and Roberto Gilli for their valuable contribution to the work presented here, and for providing continuous unlimited support. I also thank the participants of the Symposium, in particular Andy Fabian, Matt Malkan, and Ari Laor for interesting discussion after my talk. I cannot forget to acknowledge Mario Livio, who organized this extremely stimulating conference and gave me the opportunity to present this work.

REFERENCES

ALLEN, S. W., DUNN, R. J. H., FABIAN, A. C., TAYLOR, G. B., & REYNOLDS, C. S. 2006 *MNRAS* **372**, 21.

ALONSO-HERRERO, A., QUILLEN, A. C., RIEKE, G. H., IVANOV, V. D., & EFSTATHIOU, A. 2003 *AJ* **126**, 81.

BALMAVERDE, B., CAPETTI, A., & GRANDI, P. 2006 *A&A* **451**, 35.

BLANDFORD, R. D., NETZER, H., WOLTJER, L., COURVOISIER, T. J.-L., & MAYOR, M., EDS. 1990 *Active Galactic Nuclei.* Saas-Fee Advanced Course 20. Lecture Notes 1990. Swiss Society for Astrophysics and Astronomy, XII, 280 pp. Springer-Verlag.

CAPETTI, A., AXON, D. J., CHIABERGE, M., SPARKS, W. B., MACCHETTO, F. D., CRACRAFT, M., & CELOTTI, A. 2007 *A&A* **471**, 137.

CAPETTI, A. & BALMAVERDE, B. 2005 *A&A* **440**, 73.

CAPETTI, A. & BALMAVERDE, B. 2006 *A&A* **453**, 27.

CHIABERGE, M., CAPETTI, A., & CELOTTI, A. 1999 *A&A* **349**, 77.

CHIABERGE, M., CAPETTI, A., & CELOTTI, A. 2002 *A&A* **394**, 791.

CHIABERGE, M., CAPETTI, A., & MACCHETTO, F. D. 2005 *ApJ* **625**, 716.

CHIABERGE, M., GILLI, R., CAPETTI, A., & MACCHETTO, F. D. 2003 *ApJ* **597**, 166.

CHIABERGE, M., GILLI, R., MACCHETTO, F. D., & SPARKS, W. B. 2006 *ApJ* **651**, 728.

CHIABERGE, M., TREMBLAY, G., CAPETTI, A., MACCHETTO, F. D., TOZZI, P., & SPARKS, W. B. 2009 *ApJ* **696**, 1103.

DI MATTEO, T., ALLEN, S. W., FABIAN, A. C., WILSON, A. S., & YOUNG, A. J. 2003 *ApJ* **582**, 133.

ELVIS, M., WILKES, B. J., MCDOWELL, J. C., GREEN, R. F., BECHTOLD, J., WILLNER, S. P., OEY, M. S., POLOMSKI, E., & CUTRI, R. 1994 *ApJS* **95**, 1.

FABBIANO, G., TRINCHIERI, G., ELVIS, M., MILLER, L., & LONGAIR, M. 1984 *ApJ* **277**, 115.

FABER, S. M., TREMAINE, S., AJHAR, E. A., BYUN, Y.-I., DRESSLER, A., GEBHARDT, K., GRILLMAIR, C., KORMENDY, J., LAUER, T. R., & RICHSTONE, D. 1997 *AJ* **114**, 1771.

FABIAN, A. C. & REES, M. J. 1995 *MNRAS* **277**, L55.

GRANDI, P. & PALUMBO, G. G. C. 2007 *ApJ* **659**, 235.

HARDCASTLE, M. J. & WORRALL, D. M. 1999 *MNRAS* **309**, 969.

HO, L. C. 1999 *ApJ* **516**, 672.

HO, L. C. 2002 *ApJ* **564**, 120.

HO, L. C., FILIPPENKO, A. V., & SARGENT, W. L. W. 1996 *ApJ* **462**, 183.

HO, L. C., FILIPPENKO, A. V., & SARGENT, W. L. W. 1997 *ApJS* **112**, 315.

HO, L. C. & PENG, C. Y. 2001 *ApJ* **555**, 650.

HUCHRA, J., DAVIS, M., LATHAM, D., & TONRY, J. 1983 *ApJS* **52**, 89.

ICHIMARU, S. 1977 *ApJ* **214**, 840.

KELLERMANN, K. I., SRAMEK, R., SCHMIDT, M., SHAFFER, D. B., & GREEN, R. 1989 *AJ* **98**, 1195.

LAWRENCE, A. 2005 *MNRAS* **363**, 57.

MAOZ, D. 2007 *MNRAS* **377**, 1696.

MAOZ, D., NAGAR, N. M., FALCKE, H., & WILSON, A. S. 2005 *ApJ* **625**, 699.

MARCONI, A., RISALITI, G., GILLI, R., HUNT, L. K., MAIOLINO, R., & SALVATI, M. 2004 *MNRAS* **351**, 169.

NAGAR, N. M., FALCKE, H., WILSON, A. S., & HO, L. C. 2000 *ApJ* **542**, 186.

NAGAR, N. M., FALCKE, H., WILSON, A. S., & ULVESTAD, J. S. 2002 *A&A* **392**, 53.

OSTERBROCK, D. E. & MARTEL, A. 1993 *ApJ* **414**, 552.

QUATAERT, E. & NARAYAN, R. 1999 *ApJ* **520**, 298.

QUILLEN, A. C., MCDONALD, C., ALONSO-HERRERO, A., LEE, A., SHAKED, S., RIEKE, M. J., & RIEKE, G. H. 2001 *ApJ* **547**, 129.

REES, M. J., PHINNEY, E. S., BEGELMAN, M. C., & BLANDFORD, R. D. 1982 *Nature* **295**, 17.

SIKORA, M., STAWARZ, L., & LASOTA, J.-P. 2007 *ApJ* **658**, 815.

TERASHIMA, Y., IYOMOTO, N., HO, L. C., & PTAK, A. F. 2002 *ApJS* **139**, 1.

TREMAINE, S., GEBHARDT, K., BENDER, R., BOWER, G., DRESSLER, A., FABER, S. M., FILIPPENKO, A. V., GREEN, R., GRILLMAIR, C., HO, L. C., KORMENDY, J., LAUER, T. R., MAGORRIAN, J., PINKNEY, J., & RICHSTONE, D. 2002 *ApJ* **574**, 740.

URRY, C. M. & PADOVANI, P. 1995 *PASP* **107**, 803.

Making black holes visible: Accretion, radiation, and jets

By JULIAN H. KROLIK

Department of Physics and Astronomy, Johns Hopkins University, Baltimore, MD 21218, USA

With the fundamental stress mechanism of accretion disks identified—correlated MHD turbulence driven by the magneto-rotational instability—it has become possible to make numerical simulations of accretion disk dynamics based on well-understood physics. A sampling of results from both Newtonian 3-d shearing box and general relativistic global disk MHD simulations is reported. Among other things, these simulations have shown that: contrary to long-held assumptions, stress is continuous through the marginally stable and plunging regions around black holes, so that rotating black holes can electromagnetically give substantial amounts of angular momentum to surrounding matter; the upper layers of accretion disks are primarily supported by magnetic pressure, potentially leading to interesting departures from local black-body emitted spectra; and initially local magnetic fields in accretion flows can, in some cases, spontaneously generate large-scale fields that connect rotating black holes to infinity and mediate strong relativistic jets.

1. Prolog: The classical view of accretion disks

It has been understood for decades that accretion through disks can be an extremely powerful source of energy for the generation of both photons and material outflows. When the central object is a black hole, the gravitational potential at the center of the disk is relativistically deep, so that the amount of energy that might be released per unit rest-mass accreted can be a substantial fraction of unity. If the central black hole spins, an additional store of tappable energy resides in its rotation.

At the same time, however, the physical processes by which matter inflow is transmuted into observable outputs has long remained extremely murky. For matter to move inward, it must somehow lose its orbital angular momentum. That there might be some sort of inter-ring friction seems plausible, given the orbital shear, but the way this friction is generally envisaged is in terms of an imaginary "viscosity coefficient" famously parameterized by Shakura & Sunyaev (1973) as $\alpha c_s h$, for local sound speed c_s and vertical scale-height h. This *ansatz* is based entirely on dimensional analysis: the (unknown) local stress is set equal to a dimensionless number α times the local pressure solely because the local pressure has the same units as stress. Although there is no particular reason why the stress, measured in local pressure units, should always have the same value, α is very frequently assumed to be a constant at all places and at all times. Moreover, despite the fact that ordinary molecular viscosity fails miserably to explain the friction, it is often assumed that the stress is some sort of intrinsically dissipative kinetic process whose operation is at least analogous to that of conventional viscosity.

Inter-ring torques do work, moving energy outward. Simultaneously, matter moves inward, carrying its orbital energy. In a steady-state disk, where the mass inflow rate is the same at all radii, these two energy fluxes do not cancel. Instead, there is a net amount of energy that must be deposited in each ring. If one thinks of the torques as due to some sort of kinetic process like viscosity, it is natural to suppose that this energy imbalance is deposited as heat. The radial profile of this heating in a steady-state disk can be easily written down when the disk is in steady state, provided one is able to guess a boundary condition at the inner edge of the disk (we will return to this issue in greater detail later). Integrating over the radial heating profile then immediately yields the total amount of energy per accreted rest-mass that could, in principle, be used for radiation, i.e., the radiative efficiency. Unfortunately, its value depends strongly on the guessed boundary condition.

When the matter density is large enough (as it often should be), atomic collisional processes can efficiently transform the heat into photons, which can then escape after diffusing vertically through the disk. In conditions of high density and large optical depth, the emergent spectrum should be nearly thermal, an argument that has led many to assume that the spectrum is locally Planckian. Note the use of "it is natural to suppose" and "can" and "should be" and generic terminology such as "collisional processes" here; the specific mechanisms for all these steps are as little understood as the actual torque is when analyzed through the α-model.

How exactly to make use of the black hole's rotational energy has been in a similarly unsatisfactory state. Although Blandford & Znajek (1977) pointed out 30 years ago that magnetic fields can, in principle, efficiently convey black hole rotational energy from deep within the black hole's ergosphere all the way to infinity, knowledge about this mechanism's details has been almost as scanty as for accretion. The particular solution found by Blandford and Znajek and extended by Phinney (1984) assumes that the magnetic field is essentially force-free, with (almost) negligible matter inertia everywhere (including in the plane of the accretion disk) and is valid only for small spin parameter a/M. In addition, nothing in that theory specifies the strength of the magnetic field, or explains how it came to extend to infinity.

2. Genuine disk physics

I would not have painted such a gloomy picture if I did not quickly intend to adopt a very different attitude and report on some very significant recent progress. As a result of this progress, many of the mysteries bemoaned in the preceding section can now be substantially solved by the application of well-understood physical processes. In some cases, the main stumbling block is not lack of physics knowledge, but lack of computing power. With this recently-gained understanding, it has become possible to outline a program by which the entire process, from mechanics of mass inflow to photon generation to jet launching, might reasonably be followed as a sequence of connected events.

The story behind these advances begins 15 years ago when Steven Balbus and John Hawley pointed out that weak magnetic fields destabilize an orbiting disk and lead to the rapid growth of MHD turbulence (Balbus & Hawley 1991; Hawley & Balbus 1991). Orbital shear is what drives this "magneto-rotational instability"; orbital shear also enforces a correlation in the resulting MHD turbulence such that the magnetic stress $-B_r B_\phi/4\pi$ is always on average positive. That is, the shear itself ensures that the magnetic forces transmit angular momentum outward.

It is important to note that magnetic stress, unlike viscosity, is *not* intrinsically dissipative. The local rate of heat generation is *not* always proportional to the local stress. On the other hand, it is intimately connected to dissipation. Nonlinear mode-mode interactions can convey turbulent energy from the relatively large lengthscales on which the turbulence is stirred to much shorter lengthscales on which a variety of genuinely dissipative processes can act efficiently. When integrated over a large enough volume (vertically integrated through the disk and wide enough to make averaging well defined), the heating rate must agree with the one predicted by the time-steady disk picture described above, but there is no requirement for it to match up with the stress rate locally, either in time or space.

The radiative output can then be described by taking a position- and time-dependent heating rate from turbulence calculations and solving the radiation transfer equation using physical opacities. If the vertical structure of the disk is known well enough, non-LTE effects in the disk atmosphere can be incorporated, and departures from locally Planckian spectra can be predicted.

To find the global radial profile of heating and radiation, it is still necessary to understand that inner boundary condition. Thirty years ago, heuristic arguments based on hydrodynamic intuition pointed toward a boundary condition requiring the stress to be zero at and inside the marginally stable orbit, but even then it was recognized that if magnetic fields were important, a different choice might be necessary (Page & Thorne 1974). Now that we know magnetic fields are essential to the entire accretion process, what Page and Thorne saw as a back-of-the-mind worry is now front-and-center. Fortunately, as we shall shortly see, it is now possible to *calculate*, rather than guess, what happens to the magnetic stress in the marginally stable region.

If we can compute the magnetic field in the marginally stable region, then we can also compute the magnetic field even closer to the black hole's event horizon. Part of its nature is determined by dynamics within the accretion flow itself; part may be constrained by boundary conditions at infinity, where it is possible that some fieldlines are anchored.

3. Numerical simulations

The only fly in the ointment is that analytic methods for calculating the properties of any turbulent system are extremely limited in their power. Numerical simulation is really the only tool we have for examining fully developed turbulence, and it has its own limitations. Nonetheless, some 15 years after the first attempt to numerically study the properties of MHD turbulence in accretion disks, a great deal has been learned.

The simulations that have been done to date can all be divided into two classes: shearing boxes and global disks. To make a shearing box, imagine cutting out from a complete disk a narrow radial annulus of limited azimuthal extent. When its azimuthal length is small compared to a radian, it can be well approximated as straight along the tangential direction. Rather than describing the orbital motion by a rotational frequency $\Omega(r) = \Omega_0(r/r_0)^{-3/2}$ for radius r and annular central radius r_0, we can instead approximate it by moving into a frame rotating with the orbital frequency Ω_0 of the center of the box. In this frame of motion, the fluid velocity can be described as a sum of a local velocity and the underlying orbital motion $v_y = -(3/2)\Omega_0(r - r_0)$. The equations of motion can then be written with appropriate centrifugal and Coriolis terms. At a similar level of approximation, the vertical gravity is $g_z = z\Omega_0^2$.

Shearing boxes are best for wide dynamic range studies of the turbulent cascade, well-resolved exploration of internal vertical structure within the disk, and tracing disk thermodynamics. The advantage for the latter subject is that the relatively large dynamic range in lengthscale for turbulent dynamics allows one to localize comparatively well where dissipation occurs and then follow the diffusion of radiation away from its source regions.

In a global disk simulation, one places a large amount of mass on the grid in an initial state of hydrodynamic equilibrium. To avoid noise propagation across the boundary of the problem area, the outer boundary is placed well outside the outer edge of the initial mass distribution. When angular momentum begins to flow outward through the disk as a result of the MHD turbulent torques, matter from the inner part moves inward while a small amount of mass on the outside moves outward, soaking up the angular momentum that has been carried outward to it. Thus, these global disks are truly "accretion disks" only in their inner portions. For this reason, it is necessary to locate their initial centers far enough outside the innermost stable circular orbit (the ISCO, i.e., the radius of marginal stability) that there can be a reasonable radial dynamic range for the accretion flow proper.

In the current state-of-the-art, shearing box simulations employ 3-d Newtonian dynamics in the MHD approximation including radiation forces. By means of integrating both an internal energy and a total energy equation, they can track the numerical dissipation rate as a function of time in each cell of the simulation; this numerical dissipation rate is assumed to mimic the physical dissipation and is used to increase the heat content of the cells where and when it occurs. Radiation transfer is computed in the approximation of flux-limited diffusion, using thermally-averaged opacities (Hirose et al. 2006).

Global disk simulations are best for following the inflow dynamics, the radial profile of magnetic stress, the surface density profile that the stress produces, and non-local magnetic field effects. They also permit identification of typical global structures (the main disk body, disk "coronae," etc.) and the study of jets.

The most physically complete global disk simulations now available use 3-d fully general relativistic dynamics in the MHD approximation, but they take no account of radiation. In one version (Gammie et al. 2003), the total energy equation is integrated, so energy is rigorously conserved; in another (De Villiers & Hawley 2003), an internal energy equation is solved, permitting numerical energy losses. The advantage of the former method is that numerical dissipation does not lead to energy loss; its disadvantage is that physical radiation losses don't occur either. The advantages and disadvantages of the latter method are more or less reversed, if one is willing to accept numerical energy losses as approximating genuine radiative losses.

In all cases, both shearing box and global simulations, the magnetic field is nearly always assumed to have zero net flux. This choice is made largely because it's the simplest and involves the smallest number of arbitrary choices: the initial field on the boundary of the simulation is always zero. On the other hand, it is possible that there can be large-scale fields running through real accretion flows, and they may have substantial effects on the character of those flows; this question remains to be investigated.

4. Selected results

The body of this talk will be devoted to a brief summary of some of the principal achievements of these simulations. Consistent with my title, I will focus on three topics: what we have learned from the global simulations about the radial profile of stress (and possibly of dissipation); what shearing box simulations have shown us about the vertical structure of disks, and how that can influence the character of the emitted spectrum; and how magnetically-driven accretion can (or perhaps can not) launch relativistic jets.

4.1. *Radial stress profiles*

The De Villiers-Hawley simulation code is designed to do an excellent job of reliably conserving angular momentum and propagating magnetic fields, but is less good at conserving energy. Consequently, we believe its description of the electromagnetic stress and its relation to mass inflow should be reliable, but it is much more difficult to use the data from these simulations to predict dissipation and the radiation that may follow from it. For the time being, then, we can discuss the stress with some confidence, while using it as an indirect and approximate indicator of dissipation.

Figure 1 shows the instantaneous shell-integrated electromagnetic stress (i.e., $-b^r b_\phi + ||b||^2 u^r u_\phi$, for magnetic four-vector b_μ and four-velocity u_μ) evaluated in the local fluid

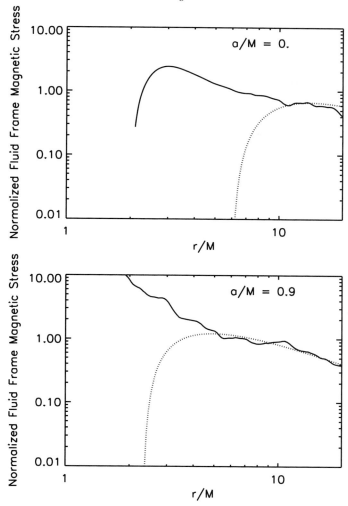

FIGURE 1. Fluid-frame stress, integrated over spherical shells. Upper panel shows a simulation with a non-rotating black hole, lower panel one with a black hole having a spin parameter $a/M = 0.9$. Solid curves are the electromagnetic stress as found in the simulations at a particular time; dotted curves are the prediction of the Novikov-Thorne model for an accretion rate matching the time-average of that simulation.

frame at late times in two simulations, one with a non-rotating black hole, the other with a black hole having spin parameter $a/M = 0.9$. For comparison, the figure also shows the stress predicted by the Novikov-Thorne model (i.e., a time-steady disk with a zero-stress inner boundary condition) when the accretion rate is the same as the time-averaged value in the corresponding simulation. As can readily be seen, the zero-stress boundary condition drastically fails to describe what actually happens. In both cases, the electromagnetic stress continues quite smoothly inward through the marginally stable region. In the Schwarzschild case, the stress rises slowly inward until just outside the event horizon, and then plummets as the event horizon is approached. In the Kerr case, the stress rises sharply upward and does not diminish even very near the horizon.

It is easy to understand qualitatively both stress profiles. Because there is no reason for the magnetic field to disappear suddenly at the ISCO, while orbital shear continues to stretch any radial components in the azimuthal direction, there is no physical mechanism to eliminate magnetic stress there (Krolik 1999; Gammie 1999); rather, the stress continues and, if anything, strengthens. The contrast in behavior between the spinning and non-spinning cases can just as easily be understood when one thinks of stress as momentum flux. The electromagnetic stress is nothing else than an outward flux of angular momentum, carried in the electromagnetic field. A non-rotating black hole has no angular momentum to give up, so it cannot act as a source for outgoing electromagnetic angular momentum flux; on the other hand, the angular momentum of a rotating black hole can be tapped, and we see this process in action here. Those concerned by an outflow of anything from an event horizon should have their qualms removed by the recognition that there is nothing to prevent a rotating black hole from swallowing negative angular momentum, i.e., angular momentum corresponding to rotation in the opposite sense. This is, of course, completely equivalent to releasing positive angular momentum. Indeed, deep in the ergosphere, the electromagnetic energy-at-infinity is frequently negative (Krolik et al. 2005).

As previously discussed, although the work done by magnetic stress does not necessarily correspond to any particular local rate of heating, there is a relationship on a more globally-averaged level. Thus, the curves of integrated fluid-frame stress shown in Figure 1 hint that dissipation may also continue smoothly across the marginally stable orbit, also contrary to the guessed boundary condition of the Novikov-Thorne model.

A further suggestion that this is so comes from a different argument. Many of the specific physical mechanisms of dissipation in this context are associated with regions of high current density. If, for example, there is a (small: we make the MHD approximation, after all) uniform resistivity η in the fluid, the local heating rate is $\eta ||J||^2$, where J^μ is the electric current four-vector and $||J||$ is its scalar magnitude. In fact, the dissipation may be even more closely associated with $||J||^2$ than this simple guess would suggest, because there are a number of plasma instabilities that create anomalous resistivity precisely where the current is strong. Maps of the current density show that it is strongly concentrated toward the center of the accretion flow, rising rapidly into the plunging region inside the ISCO (Hirose et al. 2004). To the degree that current density indicates candidate regions for rapid magnetic energy losses, this signal, too, suggests that there may be a great deal of dissipation in and within the marginally stable region.

The continuation of stress through the plunging region can also be looked at from a different point of view: in the language of the Shakura-Sunyaev α model. Their argument from dimensional analysis was that the time-averaged vertically-integrated stress should be comparable to the time-averaged vertically-integrated pressure. Our data confirm that this is so, provided one interprets "comparable" loosely. In the disk body, that is, at radii well outside the ISCO and well inside the initial pressure maximum (beyond which there is no accretion), the time-average ratio at a single radius of vertically-integrated stress to vertically-integrated pressure is generally in the range ~ 0.01–0.1. However, the instantaneous value of this ratio can easily change by factors of several over an orbital period.

Moreover, if one tracks this ratio from somewhat outside the ISCO to well inside it, there is a consistent trend for the ratio between stress and pressure to increase. As shown in Figure 1, the stress generally increases inward in this region; because the radial speed of the accretion flow also increases inward, the vertically-integrated density and pressure of the matter tend to decrease. The result is that the ratio of stress to pressure increases sharply, often rising by factors of 10–100 from the disk body to deep inside the plunging region. Thus, a description of inflow dynamics near the ISCO in terms of a constant α parameter is strongly in conflict with the results of these simulations. Claims (as are

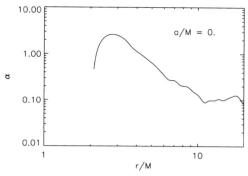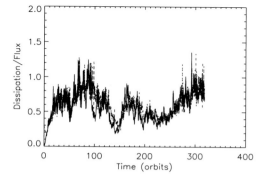

FIGURE 2. Heating and radiative output from two shearing box simulations. The left panel shows a case in which $p_r/p_g \simeq 0.2$ (taken from Hirose et al. 2006), the right panel a case in which $0.5 \leqslant p_r/p_g \leqslant 2$, depending on the time within the simulation (Krolik et al. 2007). In both figures, the solid curve shows the volume-integrated dissipation rate, while the dashed curve shows the radiative output. After initial transients, heating and radiative flux are nearly identical.

often made) based on assuming a constant value of α within this region are therefore on very shaky ground.

4.2. *Internal vertical structure*

Having thus emphasized the consequences of magnetically-driven accretion in the inner part of the accretion flow, it is now time to turn to its implications for the internal structure of accretion disks at larger radii. The best tool for studying this problem is shearing box simulations that both accurately conserve energy and follow radiation transfer (as described in (Hirose et al. 2006). In this review, we will briefly discuss two of the principal results of these simulations: their implications for thermal stability in disks and the fact that disk upper layers are generically supported primarily by magnetic fields.

In their classic paper on the α-model, Shakura and Sunyaev (1973) also predicted that the inner regions of all disks surrounding black holes in which the accretion rate is more than a small fraction of the Eddington rate should be dominated by radiation pressure. Three years later Shakura & Sunyaev (1976), the same two authors, demonstrated that, within the approximation scheme of the vertically-integrated α-model, radiation pressure dominance leads directly to thermal instability. If so, the standard equilibrium solution for the region from which most of the light is generated in the brightest accreting black hole systems is unstable. To this day, no satisfactory resolution to the question, "What actually happens in these circumstances?" has emerged.

One of the principal motivations for our program of simulating shearing boxes with radiation generation and transport is to answer this question. At this stage, there is progress to report, but not yet any firm answer to the big question. When the gas pressure is dominant, it is clear that a truly stable steady-state can be found. Figure 2a illustrates this point by showing the "light-curve" of a shearing box disk segment in which the radiation pressure p_r is only about 20% of the gas pressure p_g. The fluctuations in radiative output are only at the tens of percent level.

On the other hand, increasing radiation pressure does tend to drive fluctuations. When the radiation and gas pressures are comparable (Figure 2b), the output flux varies over a range of a factor of 3–4. Intriguingly, although p_r/p_g at its greatest is above the threshold for instability suggested by Shakura and Sunyaev, and stays there for as long as five cooling times, the disk exhibits large limit-cycle oscillations, but no unstable runaway.

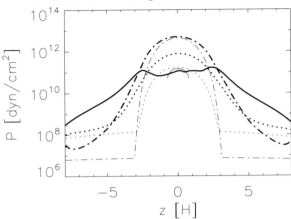

FIGURE 3. Vertical profiles of several kinds of pressure in a simulation whose volume- and time-averaged ratio of radiation pressure to gas pressure was $\simeq 0.2$ (Hirose et al. 2006). The thick curves are time-averaged over fifty orbital periods starting from the end of initial transients; the thin curves are the initial conditions. Magnetic pressure is shown by the solid curves, radiation by the dotted curves, and gas pressure by the dash-dot curves.

We are actively investigating whether the instability remains under control at still higher values of p_r/p_g (Hirose et al. 2009).

A consistent result of all vertically-stratified shearing box studies is that their upper layers are magnetically-dominated (Miller & Stone 2000, and as shown in Figure 3, taken from Hirose et al. 2006). Although the data shown here are from a particular gas-dominated simulation, more recent work studying shearing boxes with radiation pressure comparable to gas pressure (Krolik et al. 2007) and radiation pressure considerably greater than gas pressure (Hirose et al. 2009) shows very much the same pattern: independent of whether gas or radiation pressure dominates near the midplane, by a few scale-heights from the center, magnetic pressure is larger than either one.

A somewhat surprising corollary of magnetic dominance in the upper layers is that "coronal" heating is rather limited. Strong hard x-ray emission is so commonly seen from accreting black holes, no matter whether the central black hole has a mass $\sim 1\ M_\odot$ or $\sim 10^9\ M_\odot$, that somewhere in the system there must be a region of intense heating with only small matter density and optical depth. Otherwise, there would be no way to heat electrons to the ~ 100 keV temperatures required to produce the x-rays. This region is generally called the accretion disk "corona," and it is often thought of in conceptual terms derived from experience with the Solar corona: it is imagined that somehow magnetic field loops emerge buoyantly from the nearby disk, twist and cross, and release energy at reconnection points.

Unfortunately for this popular scenario, these simulations, the first to treat the dynamics of the upper layers of disks in a consistent fashion, show no sign of anything resembling it. The very fact that magnetic fields dominate the energy density of the upper strata of disks means that these regions are comparatively quiet, and the strong magnetic fields themselves prevent any twisting or field line crossing that might lead to reconnection. Rather, orbital shear stretches radial field components into azimuthal field with great regularity and smoothness (Hirose et al. 2004). At the same time, the nonlinearly saturated Parker instability (i.e., magnetic buoyancy counter-balanced by magnetic tension) supports smooth "plateaus" of field interrupted by occasional narrow cusps (Blaes et al.

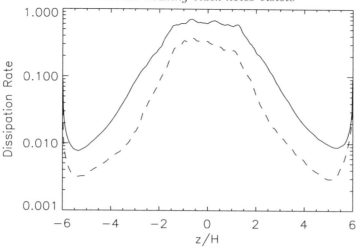

FIGURE 4. Time-averaged vertical profiles of the dissipation rate in a shearing box simulation. In this simulation, the mean gas and radiation pressures were comparable, but there were order unity fluctuations. The solid curve pertains to those times when the radiation pressure was particularly high, the dashed curve to those times in which it was particularly low.

2007). The smoothness and steadiness of the magnetic field in this kind of corona tends to suppress dissipation.

Indeed, the dissipation in shearing box segments of accretion disks is typically confined to the central regions of the disk. Figure 4 (Krolik et al. 2007) illustrates this fact in a shearing box whose radiation and gas pressures were, on a time-averaged basis comparable, but in which the ratio of radiation to gas pressure fluctuated over the range 0.5–2. Whether the energy content of the disk was high (and the radiation pressure dominated the gas pressure) or low (and the ratio went the other way), the dissipation was still confined to the inner few scale-heights of the disk.

Although magnetic dominance in the upper layers of disks likely does not lead to strong coronal heating, it does have other potentially important observational consequences. Chief among them is the fact that when magnetic pressure gradients replace gas pressure gradients as the matter's principal support against gravity, the density of the gas must fall. Because the photosphere of the disk is located where magnetic support is so important, we immediately infer that previously-estimated photospheric densities were too high, and that LTE may not be enforced as thoroughly as previously thought (Blaes et al. 2006). The locally emitted spectrum may therefore have larger deviations from Planckian, and if these features are sufficiently strong, may be visible in the disk-integrated spectrum. An example is shown in Figure 5; when magnetic support is properly included in the atmosphere structure, a prominent C VI edge appears. Still further departures from conventional spectral predictions may arise from the fact that at any given moment, the atmosphere can depart substantially from the usual picture of plane-parallel symmetry.

The surprising quietness of the magnetically dominated regions of shearing box atmospheres motivates a search for other places to supply the intense heating required to explain the observed hard x-ray emission. Better places to look might include the plunging region, which is both strongly magnetized and highly dynamical, and the region just above it from which jets are launched.

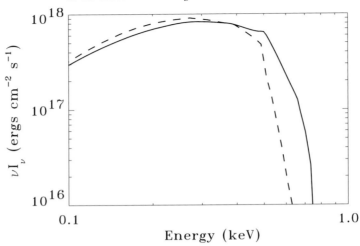

FIGURE 5. Output spectrum from a gas-dominated shearing box at 55° from the local vertical direction. The solid curve shows the spectrum predicted when the atmosphere is magnetically supported, the dashed curve shows the predicted spectrum when magnetic support is neglected (Blaes et al. 2006).

4.3. *Jet launching*

For many years, two models have dominated thinking about the launching of jets: the Blandford-Znajek mechanism (Blandford & Znajek 1977) and the Blandford-Payne scheme (Blandford & Payne 1983). The two models share two central elements: large-scale poloidal magnetic fields that extend from infinity and pierce the midplane of the accretion flow; and dynamically enforced rotation. They are distinguished by whether the rotation is enforced by space-time frame-dragging (Blandford-Znajek) or the orbiting matter of the accretion disk (Blandford-Payne). Because the latter depends in an essential way on accretion, whereas the other (at least in principle) does not, they are also distinguished by whether the inertia of matter is significant. Lastly, they differ in the source of power for the jets: orbital energy of accreting matter for Blandford-Payne, the rotational kinetic energy of the black hole itself for Blandford-Znajek.

Both models' dependence on externally imposed large-scale fields is problematic, because the most natural way to bring magnetic fields into either the inner parts of accretion disks or all the way to the black hole event horizon is by advection along with accreted fluid. However, we have no way of knowing whether such large-scale connections might survive the many orders of magnitude in compression suffered by the matter; reconnection might destroy large-scale connections far from the black hole. On the other hand, if even a small fraction of the flux survives, over time it might still build up to be significant. This remains an open question.

The simulations we have done so far have all assumed *zero* large-scale field, primarily as a result of our effort to minimize the number of arbitrarily chosen free parameters. Interestingly, we have found that even when the magnetic field in the accreting matter has no net flux at all, and therefore no externally imposed connections to infinity, under the right circumstances it can spontaneously *create* such connections within a limited volume.

To be specific, when the accretion flow contains closed dipolar field loops large enough that the outer ends are accreted long ($\gg 10^3 GM/c^3$) after the inner ones, a jet is automatically created from the flux provided by the inner half of the field loop (McKinney &

a/M	η_{EM}	η_{NT}
−0.9	0.023	0.039
0.0	0.0003	0.057
0.5	0.0063	0.081
0.9	0.046	0.16
0.93	0.038	0.17
0.95	0.072	0.18
0.99	0.21	0.26

TABLE 1. Jet Poynting flux efficiency in rest-mass units η_{EM}, as a function of black hole spin parameter a/M (Hawley & Krolik 2006). These numbers can be compared with the radiative efficiency predicted by the Novikov-Thorne model, i.e., the specific binding energy of a particle in the innermost stable circular orbit.

Gammie 2004; Hawley & Krolik 2006). When flux is brought toward the black hole along with the accretion flow, as soon as field lines begin to thread the event horizon, matter drains off them, and the field lines, freed of the matter's inertia, float upward. Because a centrifugal barrier prevents any matter with non-zero angular momentum from penetrating into a cone surrounding the rotation axis, there is little inertia above these field lines. The field lines then rapidly expand upward. This process of filling the region around the rotation axis with magnetic field ceases only when there is enough field intensity that the magnetic pressure distribution reaches equilibrium. If the central black hole spins, the portions of the field lines within the ergosphere are forced to rotate along with the black hole, imposing a twist on the field lines. The result is Poynting flux traveling outward through the evacuated cone around the rotation axis.

When the black hole rotates rapidly, the electromagnetic luminosity can be quite sizable. Table 1 presents that luminosity, normalized by the rest-mass accretion rate, as a function of black hole spin. The radiative efficiency predicted by the Novikov-Thorne model is also given in that table in order to provide a standard of comparison. As can be seen, when the black hole rotates rapidly, the jet efficiency becomes comparable to the putative radiative efficiency. Thus, consistent with what many have long speculated, black hole rotation does seem to enhance jet luminosity.

On the other hand, black hole spin may not be the only relevant parameter. For example, large dipolar loops are not the only imaginable field structure for an accretion flow. One could just as easily imagine narrower dipolar loops, or quadrupolar loops (loops that don't cross the equatorial plane) or toroidal loops. These other geometries are, in general, less favorable to jet support than the initial form explored, the large dipolar loops (Beckwith et al. 2008). Real systems are likely to exhibit some mixture of these sorts of field topologies, and that mixture could easily vary from one object to another, or from one time to another in a single object. Some of the observed variability in jets may conceivably reflect varying field structures in the matter fed to the central black hole.

5. Conclusions

Thanks to the fundamental discovery that stresses in accretion disks come from correlated MHD turbulence—driven by the magneto-rotational instability—we can now begin to speak with confidence about a number of aspects of their operation. With the aid of ever-more-detailed and realistic numerical simulations, we have taken the first steps toward connecting their internal dynamics with observable properties.

In this talk, advances in this direction in three areas have been reported:

- We now see that angular momentum transport is accomplished quasi-coherently, by magnetic stress. Because this mechanism is not a kinetic process like viscosity, it is *not* intrinsically dissipative, although the associated MHD turbulence does eventually dissipate.

Moreover, far from ceasing at the innermost stable circular orbit, as has been generally assumed for more than 30 years, magnetic stress continues through the marginally stable region and deep into the plunging region. When the black hole spins, the stress can be continuous all the way to the event horizon. At the very least, the ability of electromagnetic stresses to carry angular momentum away from the black hole and into the accretion flow means that the spin-up rate of black holes can be rather less than would have been estimated on the basis of accreting matter with the specific angular momentum of the last stable orbit. It is also possible, although quantitative determination of this effect remains a job for the future, that these extended stresses lead to extended dissipation as well, augmenting the radiative efficiency of black hole accretion beyond the traditional values.

- Detailed study of the vertical structure of disks subject to the MRI has shown that their upper layers are supported primarily by magnetic pressure gradients. In addition, these upper layers can be far from the smooth time-steady plane-parallel condition in which they are commonly imagined. As a result, the density at the photosphere is likely to be rather smaller than previously estimated, and the locally-emitted spectrum may have significant departures from black-body form.

Ongoing work promises to clear up the long-standing mystery of whether radiation-dominated disk regions are thermally unstable, and if they are, what happens in the nonlinear stage of this instability.

- As had been initially pointed out in the mid-1970s, when large-scale magnetic fields pass close by rotating black holes, it is possible for very energetic relativistic jets to be driven, deriving their energy from the rotational kinetic energy of the black hole itself. We can now begin to compute the detailed structure of these jets, as functions of both space and time. In addition, we now see that it is possible to create large-scale magnetic field threading the ergosphere of the black hole and stretching out to infinity from much smaller-scale field embedded in the accretion flow—but not all small-scale field structures are capable of doing this.

I am indebted to my many collaborators on the work reported here. In alphabetical order, they are: Kris Beckwith, Omer Blaes, Shane Davis, Jean-Pierre De Villiers, John Hawley, Shigenobu Hirose, Ivan Hubeny, Yawei Hui, Jeremy Schnittman, and Jim Stone. This work was partially supported by NSF Grants AST-0313031 and AST-0507455 and NASA ATP Grant NAG5-13228.

REFERENCES

BALBUS, S. A. & HAWLEY, J. F. 1991 *ApJ* **376**, 214.

BECKWITH, K., HAWLEY, J. F., & KROLIK, J. H. 2008 *ApJ* **678**, 1180.

BLAES, O. M., DAVIS, S. W., HIROSE, S., KROLIK, J. H., & STONE, J. M. 2006 *ApJ* **645**, 1402.

BLAES, O. M., HIROSE, S., & KROLIK, J. H. 2007 *ApJ* **664**, 1057.

BLANDFORD, R. D. & PAYNE, D. G. 1983 *MNRAS* **199**, 883.

BLANDFORD, R. D. & ZNAJEK, R. L. 1977 *MNRAS* **179**, 433.

DE VILLIERS, J.-P. & HAWLEY, J. F. 2003 *ApJ* **589**, 458.

GAMMIE, C. F. 1999 *ApJ* **522**, L57.

GAMMIE, C. F., MCKINNEY, J. C., & TÓTH, G. 2003 *ApJ* **589**, 444.

HAWLEY, J. F. & BALBUS, S. A. 1991 *ApJ* **376**, 223.

HAWLEY, J. F. & KROLIK, J. H. 2006 *ApJ* **641**, 103.

HIROSE, S., KROLIK, J. H., & BLAES, O. 2009 *ApJ* **691**, 16.

HIROSE, S., KROLIK, J. H., DE VILLIERS, J.-P., & HAWLEY, J. F. 2004 *ApJ* **606**, 1083.

HIROSE, S., KROLIK, J. H., & STONE, J. M. 2006 *ApJ* **640**, 901.

KROLIK, J. H. 1999 *ApJ* **515**, L73.

KROLIK, J. H., HAWLEY, J .F., & HIROSE, S. 2005 *ApJ* **622**, 1008.

KROLIK, J. H., HIROSE, S., & BLAES, O. M. 2007 *ApJ* **664**, 1045.

MCKINNEY, J. C. & GAMMIE, C. F. 2004 *ApJ* **611**, 977.

MILLER, K. A. & STONE, J. M. 2000 *ApJ* **534**, 398.

PAGE, D. N. & THORNE, K. S. 1974 *ApJ* **191**, 499.

PHINNEY, E. S. 1984 unpublished Cambridge University Ph.D. thesis.

SHAKURA, N. I. & SUNYAEV, R. A. 1973 *A&A* **24**, 337.

SHAKURA, N. I. & SUNYAEV, R. A. 1976 *MNRAS* **175**, 613.